DNA and Protein Sequence Analysis

FROM BENCHTOP TO BOOKSHELF

This book belongs to

DR. BASAK

725 Parkdale Ave, Ottawa, ON

K1Y 4E9

(613) 798-5555 (6128)

The Practical Approach Series

SERIES EDITOR

D. RICKWOOD
Department of Biology, University of Essex
Wivenhoe Park, Colchester, Essex CO4 3SQ, UK

B. D. HAMES
Department of Biochemistry and Molecular Biology
University of Leeds, Leeds LS2 9JT, UK

★ **indicates new and forthcoming titles**

Affinity Chromatography
Anaerobic Microbiology
Animal Cell Culture (2nd Edition)
Animal Virus Pathogenesis
Antibodies I and II
★ Antibody Engineering
Basic Cell Culture
Behavioural Neuroscience
Biochemical Toxicology
Bioenergetics
Biological Data Analysis
Biological Membranes
Biomechanics–Materials
Biomechanics–Structures and
  Systems
Biosensors
Carbohydrate Analysis
  (2nd Edition)
Cell–Cell Interactions
The Cell Cycle
Cell Growth and Apoptosis
Cellular Calcium
Cellular Interactions in
  Development
Cellular Neurobiology
Clinical Immunology
Crystallization of Nucleic
  Acids and Proteins
★ Cytokines (2nd Edition)
The Cytoskeleton

Diagnostic Molecular Pathology
  I and II
Directed Mutagenesis
★ DNA and Protein Sequence
  Analysis
★ DNA Cloning 1: Core
  Techniques (2nd Edition)
★ DNA Cloning 2: Expression
  Systems (2nd Edition)
★ DNA Cloning 3: Complex
  Genomes (2nd Edition)
★ DNA Cloning 4: Mammalian
  Systems (2nd Edition)
Electron Microscopy in Biology
Electron Microscopy in
  Molecular Biology
Electrophysiology
Enzyme Assays
★ Epithelial Cell Culture
Essential Developmental Biology
Essential Molecular Biology I and II
Experimental Neuroanatomy
★ Extracellular Matrix
Flow Cytometry (2nd Edition)
★ Free Radicals
Gas Chromatography
Gel Electrophoresis of Nucleic
  Acids (2nd Edition)
Gel Electrophoresis of Proteins
  (2nd Edition)
★ Gene Probes 1 and 2

DNA and Protein Sequence Analysis

A Practical Approach

Edited by

M. J. BISHOP

UK HGMP Resource Centre,
Hinxton, Cambridge CB10 1SB, UK

and

C. J. RAWLINGS

SmithKline Beecham Pharmaceuticals
New Frontiers Science Park
Third Avenue, Harlow, Essex CM19 5AW, UK

OXFORD UNIVERSITY PRESS
Oxford New York Tokyo

Oxford University Press, Great Clarendon Street, Oxford OX2 6DP

Oxford   New York

Athens   Auckland   Bangkok   Bogota   Bombay   Buenos Aires
Calcutta   Cape Town   Dar es Salaam   Delhi   Florence   Hong Kong
Istanbul   Karachi   Kuala Lumpur   Madras   Madrid   Melbourne
Mexico City   Nairobi   Paris   Singapore   Taipei   Tokyo   Toronto

and associated companies in
Berlin   Ibadan

Oxford is a trade mark of Oxford University Press

Published in the United States
by Oxford University Press Inc., New York

A catalogue record for this book is available from the British Library

Library of Congress Cataloging in Publication Data

DNA and protein sequence analysis: a practical approach/edited by
M. J. Bishop and C. J. Rawlings.—1st ed.
(The practical approach series; 171)
Includes bibliographical references and index.
1. Nucleotide sequence—Data processing.   2. Amino acid sequence—
Data processing.   I. Bishop, M. J. (Martin J.)   II. Rawlings,
Christopher J., 1954–   .   III. Series: Practical approach series; 171.
[DNLM: 1. Sequence Analysis, DNA—methods.   2. Proteins—analysis.
3. Sequence Analysis—methods.   4. Software.   QH 441 D628 1997]
QP620.D58 1997
574.87'328'0285—dc20
DNLM/DLC
for Library of Congress   96–18782   CIP

ISBN 0 19 963464 5 (Hbk)
ISBN 0 19 963463 7 (Pbk)

Typeset by Footnote Graphics, Warminster, Wilts
Printed in Great Britain by Information Press, Ltd, Eynsham, Oxon.

Preface

It is ten years since our previous title in the Practical Approach Series was published. The present volume is not a revision, having only a few chapter subjects and three authors in common. We are seeing a progressive evolution from the use of the computer to store and analyse nucleic acid and protein sequences and to suggest some worthwhile laboratory experiments, towards the use of the computer systematically to model biological knowledge about a whole organism. Complete genome sequences of unicellular organisms including a bacterium and a yeast (*Haemophilus influenzae, Saccharomyces cerevisiae*) are available now, the nematode (*Caenorhabditis elegans*) sequence will be available in two years, and the human genome will be largely characterized in five years.

This book is a practical aid to biologists wishing to have access to the wealth of (relatively ill-organized) molecular biology data, to have an understanding of some elegant results and conclusions from sequence comparison, approach the questions of sequence/structure/function relations, and to put these in the all important evolutionary context. Databases are described, as well as access to on-line information on the World Wide Web, which has come to dominate the scene in the last few years. There are contributions on simple sequences, repetitive sequences, and isochores. The subject of prediction of sequence function for DNA, mRNA, and proteins is elaborated, as is DNA and RNA structure. A chapter on phylogenetic estimation helps clarify fact from fantasy and the book concludes with synthetic thoughts on the evolution of protein families. We have attempted to emphasize topics which we believe will be of lasting value in this fast moving field.

Cambridge   M. J. B.
Harlow   C. J. R.
April 1996

Contents

Contents

3. EBI databases and tools

Rainer Fuchs and Graham N. Cameron

4. Networked services

G. Williams

Contents

5. DNA sequencing methodology and software 75

William D. Rawlinson and Barclay G. Barrell

Contents

9. Repetitive sequences in DNA 185

Jörg T. Epplen and Olaf Riess

10. Isochores and synonymous substitutions in mammalian genes 197

Giorgio Bernardi, Dominique Mouchiroud, and Christian Gautier

11. Identifying genes in genomic DNA sequences
209

Eric E. Snyder and Gary D. Stormo

12. Prediction of mRNA sequence function
225

Keith Vass

Contributors

STEPHEN F. ALTSCHUL
National Center for Biotechnology Information, National Library of Medicine, Building 38A, 8600 Rockville Pike, Bethesda, MD 20894, USA.

PASCAL AUFFINGER
Institut de Biologie Moléculaire et Cellulaire, CNRS, 15 Rue Descartes, 67084 Strasbourg Cedex, France.

BARCLAY G. BARRELL
Sanger Centre, Hinxton Hall, Hinxton, Cambridge CB10 1RQ, UK.

GIORGIO BERNARDI
Laboratoire de Genetique Moleculaire, Institut Jacques Monod, Université Paris VII-Tour 43, 2 Place Jussieu, F-75005 Paris Cedex 05, France.

CHRISTIAN BURKS
Theoretical Biology and Biophysics Group, Los Alamos National Laboratory, T-10, M-K710, Los Alamos, New Mexico, NM 87545, USA.

GRAHAM N. CAMERON
European Bioinformatics Institute, EMBL Outstation, Wellcome Trust Genome Campus, Hinxton, Cambridge CB10 1SD, UK.

JÖRG T. EPPLEN
Molecular Human Genetics, Faculty of Medicine, MA 5, Ruhr University, Universitaetsstrasse 150, 44780 Bochum, Germany.

RAINER FUCHS
Glaxo–Wellcome Research Institute, Five Moore Drive, Research Triangle Park, NC 22709, USA.

CHRISTINE GASPIN
SBIA/INRA, Chemin de Borde Rouge, BP27, 31326, Castanet Tobsan, Cedex, France.

CHRISTIAN GAUTIER
Laboratoire de Biometrie, Genetique et Biologie des Populations, URA 243, Université Claude Bernard, 69600 Villeurbanne, France.

M. GINSBURG
Laboratory Research IT Unit, Imperial Cancer Research Fund, PO Box 123, Lincoln's Inn Fields, London WC2A 3PX, UK.

NICK GOLDMAN
Department of Genetics, University of Cambridge, Downing Street, Cambridge CB2 3EH, UK.

Contributors

T. C. HODGMAN
Glaxo–Wellcome Medicines Research Centre, Advanced Technologies and Informatics Unit, Gunnels Wood Road, Stevenage, Herts SG1 2NY, UK.

M. P. MITCHELL
Laboratory Research IT Unit, Imperial Cancer Research Fund, PO Box 123, Lincoln's Inn Fields, London WC2A 3PX, UK.

DOMINIQUE MOUCHIROUD
Laboratoire de Biometrie, Genetique et Biologie des Populations, URA 243, Université Claude Bernard, 6900 Villeurbanne, France.

J. M. OSTELL
National Center for Biotechnology Information, National Library of Medicine, Building 38A, 8600 Rockville Pike, Bethesda, MD 20894, USA.

WILLIAM D. RAWLINSON
Department of Virology, ICPMR, Westmead, NSW 2145, Australia.

OLAF RIESS
Molecular Human Genetics, Faculty of Medicine, MA5, Ruhr University, Universitaetsstrasse 150, 44780 Bochum, Germany.

ERIC E. SNYDER
Biocomputational Scientist, Sequana Therapeutics Inc., 11099 North Torrey Pines Road, Suite 160, La Jolla, CA 92037, USA.

GARY D. STORMO
Department of Molecular, Cellular and Developmental Biology, University of Colorado, Campus Box 347, Boulder, CO 80309-0347, USA.

WILLIAM R. TAYLOR
Division of Mathematical Biology, National Institute of Medical Research, The Ridgeway, Mill Hill, London NW7 1AA, UK.

KEITH VASS
Beatson Institute for Cancer Research, Garscube Estate, Switchback Road, Bearsden, Glasgow G61 1BD, UK; Glasgow University Computing Service, Glasgow University, Glasgow G12 8QQ, UK.

ERIC WESTHOFF
Institut de Biologie Moléculaire et Cellulaire, CNRS, 15 Rue Descartes, 67084 Strasbourg Cedex, France.

G. WILLIAMS
Computing Services, UK HGMP Resource Centre, Hinxton, Cambridge CB10 1SB, UK.

JOHN C. WOOTTON
National Center for Biotechnology Information, National Library of Medicine, Building 38A, 8600 Rockville Pike, Bethesda, MD 20894, USA.

Abbreviations

ASN.1	Abstract Syntax Notation 1
CDS	Coding Sequence
DDBJ	DNA Databank of Japan
EBI	European Bioinformatics Institute
EMBL	European Molecular Biology Laboratory
EMBnet	European Molecular Biology network
FTP	file transfer protocol
GDB	genome database
HCMV	human cytomegalovirus
HGMP	Human Genome Mapping Project
i.i.d.	independent identically distributed
LINE	long interspersed nucleotide element
MCMV	murine cytomegalovirus
MD	molecular dynamics
MM	molecular mechanics
MSP	maximum segment pair
MSS	maximum segment scoring
NCBI	National Center for Biotechnology Information
NDB	nucleic database
OODBMS	object-oriented database management system
ORF	open reading frame
OSI	open systems interconnection
PAM	point accepted mutation
PDB	protein database
RDBMS	relational database management system
RFLP	restriction fragment length polymorphism
RISC	reduced instruction set computer
SCF	standard chromatogram file
SDF	synonymous difference frequency
SINE	short interspersed nucleotide element
SRS	sequence retrieval system
URF	unidentified reading frame
URL	uniform resource locator
UWGCG	University of Wisconsin Genetics Computer Group
VEP	vector excision program
VNTR	variable number of tandem repeat
WAIS	wide area information server
WWW	World Wide Web

1

Molecular biology databases

CHRISTIAN BURKS

1. Overview

1.1 Summary

Over the past decade, it has been widely recognized that computers—in their application to informatics, analysis, and modelling of biological data and systems—have had and will continue to have a profound effect on the practice of molecular biology (1–4). One aspect of this revolution has been the storage and analysis of molecular biological data sets on electronic media, and in particular the appearance and use of molecular biological computer databases established as community-wide resources maintained by centralized databank operations.[1]

There are now hundreds of molecular biology databases that are available in electronic form for retrieval or analysis, and that are dedicated to collecting relevant data sets from the laboratories generating them. This chapter provides an overview of the data domains covered by these databases, a description of the pipeline (and the assumptions underlying it) that carries data from research laboratories into databank resources and back out to research laboratories, and guidelines for the use of the data found in databases.

Our approach, rather than providing detailed protocols (e.g. of how to retrieve specific information from a specific database using a specific electronic platform), is to provide a general overview of the available databases, and interactions with them, along with information on where to retrieve the most current specific protocols. This approach is driven in part by the sheer volume of alternative protocols that are available, and in part by the dynamism of databases and their underlying technology, which render most specific instructions out-of-date within a matter of months.

The databases and topics associated with them are often cited here in the context of the Human Genome Project, in part because that is the author's

---

[1] In this chapter, 'data set' denotes information, 'database' denotes a data set embedded in and accessed by software tools on a computer, and 'databank' denotes a project (staff, funding, etc.) maintaining and distributing one or more databases as a community resource.

immediate frame of reference, but also because the Human Genome Project has, since the beginning, prompted database discussions because the end product (DNA sequences) of the project is a large data set intended to be delivered to and accessible by the scientific community. This goal will rely on mechanisms and solutions that had not at the beginning of the project been agreed upon (or in some cases even identified), and which are still being worked out. It should be noted, however, that other areas of biology (e.g. the neurosciences and, in particular, human brain mapping) (5, 6) are contemplating the directed generation and collection of very large data sets that precipitate very similar discussions of problems and solutions.

1.2 Molecular biology databases

The amount of available molecular biological data, initially in the literature and then through databases, has been growing rapidly for a number of years. One such example is nucleic acid (DNA and RNA) sequences, which began appearing in the literature in the 1960s: the number of bases reported has been doubling every one to two years over the past decade, and is predicted to continue at this rate into the next millennium (7). The number of sequences (and the number of contiguous bases they contained) exploded in the mid-1970s as the efficiency of sequencing methodologies improved, and has continued to grow as a result of both:

(a) the widespread application of molecular biological techniques (in particular, the sequencing of DNA and mRNA to determine encoded protein sequences) and

(b) the ever-increasing methodological efficiency arising from improved chemistry and automation.

In addition to the large amounts of a given kind of data, the breadth of categories of biological information which are now available in database form is rapidly expanding. The range of available information includes: bibliographic sources; macromolecular sequences and structures; biomolecular chemical properties; genetic and physical genomic maps and polymorphisms; taxonomic and phylogenetic schema; experimental resource listings (e.g. organism strains, clone libraries, etc.); and meta-directories to the above resources. *Table 1* gives an impression of the wide range of information available.

1.3 Sequence databases

Biopolymer sequence databases exist for nucleic acid sequences (polynucleotides composed of nucleotide monomeric units), proteins (polypeptides composed of amino acid monomeric units), and complex carbohydrates (polysaccharides composed of carbohydrate monomeric units). In the case of nucleic acid and protein sequences, the chemical structure is represented as a single sequence, or string, of characters, with one character corresponding to

Table 1. Subject areas in or related to molecular biology that are covered, at least in part, in databases; drawn from the keyword index of release 3.0 of the LiMB database (16, 17)

abstracts
  [magazines]
  [newsletters]
  [newspapers]
  [press releases]
  [scientific journals]
accession [germplasm samples]
allele frequency for interspecies or intraspecies study
allele number
amino acid
  [loci]
  [properties in protein]
amino acid sequences
  [HIV]
  [artificial variants]
  [class covering pattern]
  [cluster trees]
  [diagnostic sequence patterns]
  [immunoglobulin]
  [pattern information content]
  [patterns]
  [protein kinases]
  [signal peptide]
  [wild-type]
amino acid substitutions
  [SV40 large T antigen mutant]
  [unsequenced SV40 large T antigen mutant]
annotation [protein]
antibodies [monoclonal]
antibody specificities
atomic coordinates
  [biomacromolecule]
  [small molecule]
authors/editors [books]
base composition
  [host gene]
  [transposable element ORF]
bibliographic information
biological activity and physical properties [variant]
breakpoints
carbohydrate sequences
cell lines
  [characteristics]
  [human]
chemical modification [amino acid]
chemicals [toxic]
chromosome number [*Drosophila*]
chromosome numbers [plants]
chromosome rearrangements [*D. melanogaster*]

3

Table 1. *continued*

clones
  [chromosome specific libraries]
  [descriptions]
  [distribution lists]
  [human cDNA]
  [human genomic]
  [immuno-]
  [library characterizations]
  [mouse]
codon usage
codon usage [*Drosophila*]
conferences
cross-references
  [EMBL]
  [GENBANK]
  [database]
  [microbial strain database]
cultures
  [curator]
  [distribution]
database access information
database characteristics
database contribution information
database maintenance
  [hardware]
  [software]
databases
  [biotechnology]
  [life sciences]
  [molecular biology]
diseases [human inherited]
Drosophila [genotype]
enzyme
  [catalytic activity]
  [substrate modifications]
enzymes
  [EC numbers]
  [cofactor]
functional features
  [16S rRNA nucleic acid sequence]
  [amino acid sequence]
  [biomacromolecule]
  [carbohydrate sequence]
  [nucleic acid sequence]
  [restriction enzyme]
  [amino acid sequence, immunoglobulin]
  [nucleic acid sequence, immunoglobulin]
gel electrophoretic data
  [protein, spot coordinates]
  [protein, spot quality]
  [protein, spot representation]

4

gene [location]
gene frequencies [domestic cat]
gene name
genes
genetic maps
  [*C. elegans*]
  [human]
  [man on mouse homologous loci]
  [man on mouse homology maps]
  [mouse clone and probe]
genetic mutations [*D. melanogaster*]
genotypes [CEPH panel families]
germplasm [evaluation information]
germplasms
hazardous properties [chemicals]
hybridomas
index terms
  [biological]
  [chemical]
  [medical]
  [microbial strain]
  [organizations]
inventory information
laboratories
  [computer capabilities]
  [focus]
  [location]
legume
  [conservation status]
  [economic importance]
  [geographical distribution]
  [life form]
  [name]
  [notes]
  [synonyms]
  [tribe membership]
  [vernacular names]
lipids
  [aqueous phase composition]
  [enthalpy change]
  [mesomorphic phase transition behavior]
  [miscibility properties]
  [transition temperature]
  [transition type]
literature abstracts
  [amino acid sequence]
  [biological]
  [chemical]
  [life sciences]
  [medical]
literature citations
  [AIDS]

5

Table 1. *continued*

   [agriculture]
   [amino acid sequence]
   [biological]
   [biomacromolecule]
   [chemical compound]
   [chemical]
   [crystallization]
   [disease, human inherited]
   [eukaryotic POL II promoter sequence]
   [legume]
   [life science]
   [life sciences]
   [medical]
   [molecular biology]
   [names of organizations]
   [nucleic acid sequence, cloned]
   [pharmacology]
   [plants]
   [protein substrate cleavage site]
   [sequence analysis]
   [sequences]
   [taxonomy]
   [transcriptional element]
methods
   [amino acid mutagenizing]
   [restriction enzyme isolation]
microbial strain collection information
micro-organisms
   [cultured]
   [history]
molecular biological activity [protein]
molecular formulae
molecular properties
   [amino acid]
   [biomacromolecule]
   [protein]
molecular structure [chemical representations]
morphology [plants]
mutants
   [SV40 large T antigen deletion & insertion]
   [SV40 large T antigen mutant parents]
   [expression system]
   [protein]
   [truncated SV40 large T antigen mutants]
   [unsequenced SV40 large T antigen deletion & insertion]
   [wild-type activity or structure difference]
NMR
   [NOEs]
   [coupling constants]
   [experimental conditions]
   [shift assignments]

nucleic acid sequences
  [features]
  [16S rRNA]
  [5S rRNA]
  [HIV]
  [chemically synthesized oligonucleotide]
  [eukaryotic POL II promoter]
  [host gene]
  [immunoglobulin]
  [small RNA]
  [small subunit rRNA]
  [tRNA gene]
  [tRNA]
  [transcriptional element]
  [transposable element ORF]
  [variants]
oligonucleotide sequences [features]
organisms
  [*C. elegans* strains]
  [*D. melanogaster*, wild-type strains]
  [bacterial, synonyms]
  [bacterial]
  [conferences]
people [AI researchers]
peptides [synthetic substrates]
phenotypes [*D. melanogaster* non-chromosomal]
plasmid prefixes
  [locale of registrants]
  [name of registrants]
polymorphic information
polymorphic markers [RLFPs]
probes [mouse]
profiles
  [from company literature]
  [from questionnaires]
program names [software]
protein
  [coding regions]
  [crystal composition]
  [crystallization conditions]
  [function]
  [optimal PH]
  [organization]
  [sources]
  [transcription factor, names]
proteinase inhibitors
proteinase sources
proteinase specificity
publishers
restriction enzymes
RFLPs
scientific names of plants

Table 1. *continued*

secondary structure features [nucleic acid sequence, small subunit rRNA]
secondary structure
  [nucleic acid sequence, rRNA]
  [protein]
secondary structures [tRNA]
sequence alignments
sequence location
sequences
software [acquisition information]
software [functions]
software [system requirements]
somatic cell hybrids [human & rodent]
sources
  [carbohydrate sequence]
  [micro-organism]
  [protein variant]
  [sequence]
  [tRNA]
statistical analysis [cat gene frequency distribution]
stock number [*Drosophila*]
substances [chemical]
synonomy [plants]
taxonomic classification
taxonomy
3D coordinates
toxicity [chemical]
toxin properties
transposons
tray number [*Drosophila*]
viruses [AIDS-related animal]

one monomeric unit. However, complex carbohydrate sequences, with the complication arising of representing a larger number of monomeric units, several alternatives for linking chemistry, and branched polymeric structures, have required a more complicated representation.

In general, these databases contain much more information than just the sequence data. This additional information, or *annotation*, provides bibliographic, biological, and administrative contexts for the sequence data. Where and when were the sequence and/or annotation published in the scientific literature? What positions along the sequence correspond to sites or regions that have a specific biological function, mediated by the cellular machinery with which the corresponding polymer comes in contact? When was this database entry last updated, and to what extent? *Figures 1–3* contain sample entries from nucleic acid, protein, and complex carbohydrate databases, and illustrates the kinds of annotation provided with these databases.

```
ID   HSGONA       standard; RNA; PRI; 621 BP.
XX
AC   V00518;
XX
DT   09-JUN-1982 (Rel. 01, Created)
DT   07-JAN-1995 (Rel. 42, Last updated, Version 5)
XX
DE   Human messenger RNA for chorionic gonadotropin.
XX
KW   complementary DNA; gonadotropin.
XX
OS   Homo sapiens (human)
OC   Eukaryota; Animalia; Metazoa; Chordata; Vertebrata; Mammalia;
OC   Theria; Eutheria; Primates; Haplorhini; Catarrhini; Hominidae.
XX
RN   [1]
RP   1-621
RL   Fiddes J.C., Goodman H.M.;
RL   "Isolation, cloning and sequence analysis of the cDNA for the
RL   alpha-subunit of human chorionic gonadotropin";
RL   Nature 281:351-356(1979).
XX
DR   SWISS-PROT; P01215; GLHA_HUMAN.
XX
CC   KST HSA.GONADOTROPIN [621]
XX
FH   Key              Location/Qualifiers
FH
FT   source           1..621
FT                    /organism="Homo sapiens"
FT   mRNA             <1..>398
FT                    /note="messenger RNA"
FT   CDS              51..401
FT                    /product="chorionic gonadotropin"
XX
SQ   Sequence 621 BP;  165 A;  152 C;  124 G;  180 T;  0 other;
     cagtaaccgc cctgaacaca tcctgcaaaa agcccagaga aaggagcgcc atggattact        60
     acagaaaata tgcagctatc tttctggtca cattgtcggt gtttctgcat gttctccatt       120
     ccgctcctga tgtgcaggat tgcccagaat gcacgctaca ggaaaaccca ttcttctccc       180
     agccgggtgc cccaatactt cagtgcatgg gctgctgctt ctctagagca tatcccactc       240
     cactaaggtc caagaagacg atgttggtcc aaaagaacgt cacctcagag tccacttgct       300
     gtgtagctaa atcatataac agggtcacag taatggggg tttcaaagtg gagaaccaca       360
     cggcgtgcca ctgcagtact tgttattatc acaaatctta aatgtttac caagtgctgt       420
     cttgatgact gctgattttc tggaatggaa aattaagttg tttagtgttt atggctttgt       480
     gagataaaac tctccttttc cttaccatac cactttgaca cgcttcaagg atatactgca       540
     gctttactgc cttcctcctt atcctacagt acaatcagca gtctagttct tttcatttgg       600
     aatgaataca gcattaagct t                                                 621
```

Figure 1. Nucleic acid sequence database entry. This entry, for the mRNA sequence encoding the human chorionic gonadotropin alpha subunit, is from the EMBL database, and was retrieved from the e-mail server at the National Center for Biotechnology Information (NCBI).

```
[GLHA_HUMAN]      GLYCOPROTEIN HORMONES ALPHA CHAIN PRECURSOR.
ID [LOC]
ACCESSION [ACC]      GLHA_HUMAN      STANDARD;      PRT;      116 AA.
   P01215;
DATES [DAT]
   21-JUL-1986 (REL. 01, CREATED)
   21-JUL-1986 (REL. 01, LAST SEQUENCE UPDATE)
   01-OCT-1994 (REL. 30, LAST ANNOTATION UPDATE)
KEYWORDS [KEY]
   HORMONE; GLYCOPROTEIN; SIGNAL.
GENE NAME [GEN]
   CGA.
SOURCE [SRC]
   HOMO SAPIENS (HUMAN).
ORGANISM CLASSIFICATION [CLS]
   EUKARYOTA; METAZOA; CHORDATA; VERTEBRATA; TETRAPODA; MAMMALIA;
   EUTHERIA; PRIMATES.
CROSS REFERENCE [DCR]
   EMBL; V00518; HSGONA.
   EMBL; V00485; HSAGC2.
   EMBL; V00486; HSAGC3.
   EMBL; V00487; HSAGC4.
   PIR; A01481; TTHUAP.
   MIM; 118850; 11TH EDITION.
   MIM; 188530; 11TH EDITION.
   PROSITE; PS00779; GLYCO_HORMONE_ALPHA_1.
   PROSITE; PS00780; GLYCO_HORMONE_ALPHA_2.
REFERENCE [REF]
   [1]
   80011660
   FIDDES J.C., GOODMAN H.M.;
   NATURE 281:351-356(1979).
   [2]
   SEQUENCE OF 1-98 FROM N.A.
   82267643
   FIDDES J.C., GOODMAN H.M.;
   J. MOL. APPL. GENET. 1:3-18(1981).
   [3]
```

```
SEQUENCE OF 1-24.
   81117268
   BIRKEN S., FETHERSTON J., CANFIELD R.E., BOIME I.;
   J. BIOL. CHEM. 256:1816-1823(1981).
   .
   .
   .
   [13]
   STRUCTURE OF CARBOHYDRATES.
   91122088
   WEISSHAAR G., HIYAMA J., RENWICK A.G.C., NIMTZ M.;
   EUR. J. BIOCHEM. 195:257-268(1991).
   [14]
   X-RAY CRYSTALLOGRAPHY (3.0 ANGSTROMS).
   94261179
   LAPTHORN A.J., HARRIS D.C., LITTLEJOHN A., LUSTBADER J.W.,
   CANFIELD R.E., MACHIN K.J., MORGAN F.J., ISAACS N.W.,
   NATURE 369:455-461(1994).
COMMENT [COM]
   -!- SUBUNIT: HETERODIMER OF A COMMON ALPHA CHAIN AND A UNIQUE BETA
       CHAIN WHICH CONFERS BIOLOGICAL SPECIFICITY TO THYROTROPIN,
       LUTROPIN, FOLLITROPIN AND GONADOTROPIN.
FEATURES [FEA]
   SIGNAL       1     24
   CHAIN       25    116          GLYCOPROTEIN HORMONES ALPHA CHAIN
   DISULFID    31     55
   DISULFID    34     84
   DISULFID    52    106
   DISULFID    56    108
   DISULFID    83    111
   CARBOHYD    76     76
   CARBOHYD   102    102
   CONFLICT    29     29          Q -> E (IN REF. 7).
   CONFLICT   108    109          CS -> SC (IN REF. 4 AND 5).
SEQUENCE DATA [BAS]
   SEQUENCE    116 AA;   13075 MW;   77052 CN;
SEQUENCE
   MDYYRKYAAI FLVTLSVFLH VLHSAPDVQD CPECTLQENP FFSQPGAPIL QCMGCCFSRA
   YPTPLRSKKT MLVQKNVTSE STCCVAKSYN RVTVMGGFKV ENHTACHCST CYYHKS
```

Figure 2. Protein sequence database entry. This entry, for the human chorionic gonadotropin alpha subunit protein sequence (coded for by the nucleic acid sequence in *Figure 1*), is from the SWISS-PROT database, and was retrieved from the e-mail server at NCBI. To compress the entry vertically, blank lines were removed, as were several citations (as indicated by the ellipsis).

```
; start of record
; db=CCSD12
CCSD accession number:
    CCSD:20872
Author:
    Weisshaar G; Hiyama J; Renwick AGC
Title:
    Site-specific N-glycosylation of human chorionic gonadotropin.
    Structural analysis of glycopeptides by one- and two-
    dimensional proton NMR spectroscopy
Citation:
    Glycobiology (1991) 1: 393-404
Structure Code (used in citation):
    N1-A
Biological Source:
    (CN) human
Parent Molecule:
    human chorionic gonadotropin
Molecular Type
    N-linked glycoprotein
Aglycon:
    Asparagine
Protein attachment site:
    hCGalpha, Asn-52
Verified (by):
    Doubet S
Date of entry:
    02-22-1995
-----------------
structure:

                                  a-D-Manp-(1-3)-a-D-Manp-(1-6)+
                                                              |
                                      b-D-Manp-(1-4)-b-D-GlcpNAc-(1-4)-b-D-GlcpNAc-(1-4)-Asn
                                                              |
a-D-Neup5Ac-(2-3)-b-D-Galp-(1-4)-b-D-GlcpNAc-(1-2)-a-D-Manp-(1-3)+
===========end of record
```

Figure 3. Complex carbohydrate sequence database entry. This entry, for a polysaccharide sequence covalently bound to the human chorionic gonadotropin alpha subunit protein shown in *Figure 2*, is from the Complex Carbohydrate Structure Database, and was provided by CarbBank staff in a 'transliterated' ASCII text format for use here.

1.3.1 Nucleic acid sequence databases

Because of the interest in highly specialized annotation and/or organization of nucleic acid sequence information, there are probably more databanks in this domain operating in a full-scale or partial public resource mode than with any of the other categories discussed in this chapter. *Table 2* provides contact information for four major nucleic acid sequence databanks, and *Figure 1* presents a sample database entry. These four databanks originated and continue to be administered independently, but work together collaboratively to provide the same, comprehensive set of nucleic acid sequences and associated annotation. This effort is successful in that they do all contain most of the sequence data that have been published to date, and relatively soon (days to weeks) after publication if not prior to publication. Additions or corrections to previously entered information, however, have not always migrated into all of the databases and so if it is extremely important in a given context to retrieve *everything* that is known about a particular gene's sequence, one is forced to scan all four databases to be safe.

The annotation in these databases focuses particularly on the genes (proteins and structural RNAs) encoded by the nucleic acid sequences (DNA, mRNA, and structural RNAs), and on the processing sites associated with transcription and translation. There are a number of other nucleic acid sequence databases providing annotation (and, occasionally, sequences) outside the charter of the databases above, focusing, for example, on more in-depth information on a particular organism's genome, on population variants, on signals for gene expression, etc.

The data set covered by these databases corresponds to over 7.4×10^5 nucleic acid sequences, containing 5.0×10^8 bases (April 1996, GenBank).

1.3.2 Protein sequence databases

Protein sequence data became available as a databank resource well before nucleic acid sequences because amino acid sequences were—at that time—easier to determine directly than were nucleic acid sequences. However, in recent years, the majority of new amino acid sequences have been determined by isolation and sequencing (followed by symbolic translation) of the corresponding genes. Though the distinction between the primary data items has therefore become less significant, the kind and extent of annotation and contextual analysis associated with amino acid sequences remains distinct from nucleotide sequences. In particular, the major protein sequence databanks have focused much of their annotation effort on categorization of the sequences into protein superfamilies.

Table 3 provides contact information for four comprehensive protein sequence databanks, and a sample entry is shown in *Figure 2*. Three of these databanks form a partnership as the single Protein Identification Resource, and among their efforts there is—unlike the case with the international data-

Table 2. Contact information for nucleic acid sequence databanks

	DNA databank of Japan	**EMBL database**
Address:	DDBJ, National Institute of Genetics, Mishima,Shizuoka 411, Japan	European Bioinformatics Institute, Wellcome Trust Genome Campus, Hinxton, Cambridge CB10 1SD, UK
Telephone:	+81-559-81-6853	+44-1223-494-400
Telefax	+81-559-81-6849	+44-1223-494-468
e-mail,		
queries:	ddbj@ddbj.nig.ac.jp	datalib@ebi.ac.uk
data submissions:	ddbjsub@ddbj.nig.ac.jp	datasubs@ebi.ac.uk
data updates:	ddbjupdt@ddbj.nig.ac.jp	update@ebi.ac.uk
World Wide Web:	http://www.ddbj.nig.ac.jp	http://www.ebi.ac.uk
	GenBank	**Genome Sequence Database**
Address:	National Center for Biotechnology Information, National Library of Medicine, 8600 Rockville Pike, Bldg 38A, Room 8N-803, Bethesda, MD 20894, USA	GSDB, National Center for Genome Resources, 1800 Old Pecos Trail, Santa Fe, NM 87505, USA
Telephone:	+1-301-496-2475	+1-505-982-7840
Telefax:	+1-301-480-9241	+1-505-982-7690
e-mail,		
queries:	info@ncbi.nlm.nih.gov	gsdb@ncgr.org
data submissions:	gb-sub@ncbi.nlm.nih.gov	datasubs@ncgr.org
data updates:	update@ncbi.nlm .nih.gov	update@ncgr.org
World Wide Web:	http://www.ncbi.nlm.nih.gov //	http://www.ncgr.org/gsdb

Table 3. Contact information for protein sequence databanks

	PIR-Japan	**PIR-Europe**
Address:	JIPID, Science University of Tokyo, 2641 Yamazaki, Noda 278, Japan	MIPS, Max-Planck-Institut für Biochemie, 82152 Martinsried bei München, Germany
Telephone:	+81-559-75-0771	+49-89-8578-2656
Telefax:	+81-559-75-6040	+49-89-8578-2655
e-mail,		
queries:	tsugita@jpnsut31.bitnet	mips@ehpmic.mips.biochem.dbp.de
data submissions:	tsugita@jpnsut31.bitnet	mips@ehpmic.mips.biochem.dbp.de
data updates:	tsugita@jpnsut31.bitnet	mips@ehpmic.mips.biochem.dbp.de
World Wide Web:	—	http://www.mips.biochem.mpg.de/mips/pir_48_0.html

	PIR-USA	**SWISS-PROT**
Address:	PIR Technical Services Coordinator, National Biomedical Research Foundation, 3900 Reservoir Road, NW, Washington, DC 20007, USA	SWISS-PROT, Medical Biochemistry Department, Centre Medical Universitaire, 1211 Geneva 4, Switzerland
Telephone:	+1-202-687-2121	+41-22-784-40-82
Telefax:	+1-202-687-1662	+41-22-347-33-34
e-mail		
queries:	pirmail@nbrf.georgetown.edu	bairoch@cmu.unige.ch
data submissions:	pirmail@nbrf.georgetown.edu	datasubs@ebi.ac.uk
data updates:	pirmail@nbrf.georgetown.edu	update@ebi.ac.uk
World Wide Web:	—	http://expasy.hcuge.ch/sprot/sprot-top.html

banks listed above for nucleic acid sequences—identity with respect to form and content.

As the possible sources of protein sequences multiplied, including direct sequencing and sequences derived from nucleic acid sequences, so did the appearance of databases focusing more on comprehensive, non-redundant collections of protein sequences than on any particular biological application or criteria for inclusion. The SWISS-PROT database (*Table 3*) is one such example that is widely used and cited. Similarly to the case with nucleic acid sequences, though in many if not most applications it may suffice to search one of the comprehensive databases, it is unfortunately the case that if it is extremely important in a given context to retrieve everything that is known about a particular protein sequence, one should consider scanning multiple databases, including computational 'translations' of the nucleic acid databases.

The annotation in these databases focuses particularly on post-translational processing, either in converting a precursor into an active polypeptide, or identifying intramolecular modification sites (e.g. disulphide bonds), or identifying intermolecular modification sites (e.g. carbohydrate binding sites on glycoproteins).

The data sets covered by these databases corresponds to about 1×10^5 protein sequences, containing about 3×10^7 amino acid residues (April 1996, SWISS-PROT and TREMBL).

1.3.3 Complex carbohydrate sequence databases

Though they have not attracted nearly the attention as have the nucleic acid and protein sequence databases, the complex carbohydrate sequencing community has quietly been accumulating thousands of sequences and reached the same divide—the need for computationally-based archiving and retrieval resources—that the other communities arrived at earlier. Contact information for the primary resource for these data, the Complex Carbohydrate Structure Database, is given in *Table 4*, and a sample entry in *Figure 3*.

In this case, thus far, the research community has stood together and helped create a single, comprehensive community resource.

Annotation focuses primarily on the kind and source of protein to which

---

Table 4. Contact information for a complex carbohydrate sequence databank

	CarbBank
Address:	CarbBank Director, 114 W. Magnolia, Suite 305, Bellingham, WA 98225, USA
Telephone:	+1-360-733-7183
Telefax:	+1-360-733-7283
e-mail:	`CarbBank@PacificRim.net`
	`CarbBank@UGA.bitnet`
World Wide Web:	`ftp://ncbi.nlm.nih.gov/repository/carbbank/`

the polysaccharide is attached, and the point of attachment along the polypeptide chain. This database contains about 4.9×10^4 entries, with about 2×10^4 unique sequences (July 1996).

1.4 What are the uses of databases?

The original and simplest reason for establishing databanks was for archival purposes (compact and durable because electronic, and standardized because centralized); however, as the scientific community has adjusted to the presence of the databases and computer tools for accessing and analysing them, a wide range of uses has become prevalent (these tools, the algorithms underlying them, and the interpretation of their output are the topics of other chapters in this book).

First, retrieval of sequences and associated annotation is an important use. As sequences have become longer, and journal editors less eager to fill pages with them, it is increasingly the case that the full sequence being discussed is not published in the paper, but is submitted in its entirety to one of the databases.

Second, the databases provide a context for interpreting one's own sequence (e.g. assigning one or more functions to the protein encoded by a gene being sequenced), eliminating artefacts (e.g. noticing that a region of the eukaryotic gene being sequenced is uncannily similar to a bacterial vector sequence), and/or determining whether or not it is being worked on by others in the community. If one has sequenced gene A in organism B, a hit in a database on gene A from organism B, or gene A' from organism B, or gene A from organism C, all are of interest and provide insights into one's own work. Increasingly, this kind of analysis is occurring earlier in, and as part of, the continuum of the experimental process, so that the analysis can have an effect—for the sake of directional efficiency, if nothing else—on subsequent experiments (see, for example, the discussion of scanning databases with individual sequencing runs during shotgun sequencing by Claverie) (8).

Finally, databases can provide an 'experimental' platform by allowing for global views of the data that are linked by common threads, but which are not available in individual articles in the literature. A popular example is the examination of large sets of sequences surrounding a given common functional site (e.g. the transcript splice donor site) across a large number of genes and/or a large number of organisms.

2. Contributing data to the databases

2.1 Community pipelines

The molecular biology community is rapidly moving toward the ideal of establishing a continuum of computer-based storage and computer-mediated transfer of data along the path from raw data collection, to project data interpretation and resolution, to public databank (9, 10). Several developments have contributed to this movement, including the availability of cheap, decentralized computing; the widespread implementation of reliable, high-bandwith networking; the need to develop paradigms for databank operations that allowed the amount of data being managed to be independent of the amount of resources available (10, 11); and the conscious effort to separate the medium used for presenting nucleic acid sequences and the medium used for presenting scientific discussion of the sequences (11).

This has increased the speed, efficiency, and accuracy of the effort to collect data into databanks. However, it also impinges on a laboratory's strategy for harvesting and developing project data locally: the expectation that the data should migrate into the data banks electronically and—wherever possible—automatically puts constraints on the tools used, data items collected, and data structures used at the project level. Thus, the likelihood of meeting this expectation efficiently should be included in the planning process when setting up the computational support for local data management.

2.2 Direct, electronic submission

Data that are going to end up in a database should be entered into a primary electronic form as early in the progress of their being determined, and every effort made to avoid using hardcopy and/or manual re-entry as an intermediate step for further processing or transmission thereafter. Furthermore, each investigator should consider it their responsibility to contribute data through direct submission to the database. As noted above, this is the only model by which a databank operating with relatively fixed resources can be expected to keep up with ever-increasing amounts of data.

Almost all databanks prefer receiving data in electronic form. Each databank has its own preference for electronic medium for submitting data, file structure, etc. Several databanks provide software that can be used locally for creating a submission, with assistance in choice of correct vocabulary in the appropriate fields, etc. Inquire with the databank staff that is being targeted for submission at the time that a manuscript describing the data is being drafted.

This, in addition to the altruistic benefits to the community at large, brings considerable benefit to the individual investigator. It is a long-established observation that the error rate associated with publication (when relying on hardcopy and/or re-entry for transmitting, composing, or formatting figures) is far greater than the orginal, experimental error. In addition, the routine consistency checks that databanks apply to submitted data often turn up errors that one would like to correct prior to publication.

2.3 Timeliness of release of data to databanks

There is a historical and natural tension between, on the one hand, the eagerness of interested groups in the scientific community and funding sources to see new data deposited in public databanks and, on the other hand, the reluctance of those generating the data to release it until they have convinced themselves both:

- that the data are unlikely to change
- that they have extracted maximum scientific value from them.

The balance between these opposing pressures can be difficult to delineate, but it is extremely disappointing to all parties involved when a decided upon release schedule is confounded by technical or bottleneck problems at the stage of transferring the data from the project to the databank. This can best be avoided when those generating the data work co-operatively with the relevant databank(s), well ahead of the planned release date, to ensure that the quantity and quality of the data being transferred are as near optimal as possible.

The molecular biology community has swung considerably in the direction of earlier submission and release of data, led by the nucleic acid sequencing community and, most recently, particularly by those doing genomic sequencing, where it is clear that data are being released before the single laboratory or project generating the data has done all possible analyses and follow-ups. This has been achieved through co-operation among the authors in the community, the editors at journals publishing data that belonged in the databases, and the databank staffs. Other subsets of the commmunity (e.g. those generating macromolecular crystal structures) have been slower to follow, but are clearly moving in the direction of making primary data associated with a publication available no later than the date of publication, and often before that, so that the databank can arrange for its release no later than the publication date.

2.4 Promulgating data revisions and extensions

With the increased emphasis in the Human Genome Project on large-scale generation of data sets using what are often less-well established protocols, and the increased emphasis on timely deposition of data, the already high rate of revision of 'published' data will be further increased. The process of getting these revisions into the public version of the data set in a database can be greatly facilitated if the mechanisms supporting the transfer of data from projects to databanks are discussed and established early on in project planning.

At the least, it is extremely desirable that as the members of a laboratory, based on feedback from the community or their own further work, correct or augment primary data and/or annotation associated with them, they directly submit these revisions to the relevant databank(s).

The general problem of updating and revising individual reports or general annotation schemes (e.g. protein nomenclature) is considerable for the databanks. The model for how best to accomplish this has evolved considerably over the past few years. Originally, it was left to the databank staffs to scan the literature and to search for and implement corrections or additions to earlier reports. As this became too time-consuming, individual authors were encouraged to contact the databanks (e.g. by letter or telephone) and inform the staff of changes that needed to be made. Again, as evaluating, standardizing, and implementing individually suggested changes became impractical for large and broad collections, several databanks experimented with bringing in curators from the community to periodically sweep through a subsection of the database and update all associated information (much as one would do in writing a review article). Finally, this, too, has proven to be inefficient as a general approach that can be applied to large and rapidly growing databases, and several databanks are experimenting with the possibility—consistent with the notion of a community pipeline described above—of allowing for and encouraging individual groups that generated original data to have ongoing and continuous 'ownership' of the data in the database that they contributed, and providing them with technical tools that allow them to modify 'their' data directly.

3. Retrieving data from the databases

The options for retrieving data from molecular biology databases are manifold; about the only mechanism that has fallen by the wayside is hardcopy (printed) form for any but the small, specialized collections. Thus, if one would rather look at the data on paper, it will usually be necessary to retrieve the data electronically from the databank, and then print it out locally.

3.1 Finding databases of interest

There are a number of options, once one has had the inspiration either to track down a particular database or to take a topic into 'database-space', for finding databases of interest. We describe several options below.

3.1.1 *Nucleic Acids Research*

This journal publishes an annual issue—the most recent in 1996 (volume 24, issue no. 1)—that includes brief database summaries and contact information for a number of databanks. This includes small and/or new databases that might not yet be apparent through the mechanisms below.

3.1.2 Net surfing

The widespread availability of electronic networks (collectively referred to as the 'Internet') as well as resources and tools for navigating the networks have been taken full advantage of by the molecular biology community and in particular those interested in databases. Krol (12) has provided a useful introduction to this medium and the various specific tools that are available—Wide Area

Information Services (WAIS), Gopher, or the World Wide Web (WWW)—
for identifying, locating, and retrieving information of interest (for briefer
but more up-to-date overviews, see ref. 13 and Chapter 4). It is worth noting
that the end section of Krol's book, which provides a partial compendium of
Internet resources, includes more entries for biology and biology-related topics
(including molecular biology) than for any other scientific discipline.

The essence of what is now possible with this medium is the ability, sitting
in one's office and using familiar software tools on one's own computer, to
look for and find resources of interest (using, for example, keyword
searches), electronically migrate to them, and electronically retrieve data sets
of interest from them. One does have to spend a little time learning how the
relevant tools operate, but the subsequent ease of both open-ended exploring
and directed retrieval makes the investment of time worthwhile.

3.1.3 The LiMB database

The LiMB database, which was created to help facilitate and simplify tracking,
use, and development of molecular biology databases, is a database of
databases, or a database directory (14–16). Each LiMB entry (see the sample
entry in *Figure 4*) contains a number of fields covering: a description of the
database content, size, and scope; contact information for general information,
data submission, data access, and LiMB questionnaires; citations describing the
database; and information about the computational platform (hardware and
software) used to maintain and query the database. This broad scope of information was designed into LiMB to support those in the community who are
interested in providing tools for database maintenance and use, and especially
for those working on the problem of providing access to distributed, heterogeneous databases. It provides a systematic and co-ordinated approach to identifying, linking, and accessing heterogeneous, distributed databases relevant to
molecular biology by providing an overview of these databases and the data
sets they cover. Release 3.0 of LiMB contained 113 entries (17).

As of this writing, it has been several years since the last LiMB update, and
some of the information in the database is out of date; therefore, this
database is no longer being distributed by its primary source.

However, the last version of LiMB as well as alternative sources of similar
information can be found on a number of World Wide Web and anonymous
FTP sites. Examples of a good starting point, with repositories and/or links to
repositories, include:

```
http://expasy.hcuge.ch/cgi-bin/listdoc
http://ifcsun2.ifisiol.unam.mx/html/easy_access.html
http://molbio.info.nih.gov/molbio/
http://www.ai.sri.com/~pkarp/mimbd/rsmith.html
http://www.pasteur.fr/other/biology/english/
  bio-databases-browsers-uk.html
```

```
entry        LIMB
number       10012
history      fm 04/07/87 initial entry
             jl 12/10/87 update
             jl 02/08/88 update using memo from cb
             jl 12/06/89 updated for release 1.2
             gk 06/21/90 updated entry from returned questionnaire
             cb 10/01/94 updated entry in preparation for next release
status       questionnaire
.
.
.

gen.nam      LiMB Database
gen.add      T-10, MS K710
             Los Alamos National Laboratory
             Los Alamos, NM 87545
             U.S.A.
gen.tel      -
gen.fax      505-665-2598
gen.net      limb@t10.lanl.gov
.
.
.

name.now     LiMB
nam.alt      Listing of Molecular Biology Databases
nam.obs      -
source       standard questionnaires completed by the staffs of other
             databases
funding      LANL
citation     [1] Burks, C., Lawton, J., Bell, G.  (1988) The LiMB database.
             <Science> 241, 888.
             [2] Lawton, J., Burks, C., Martinez, F.  (1989) Overview of the
             LiMB database.  <Nucleic Acids Research> 17, 5885-5899.
             [3] Keen, G., Redgrave, G., Lawton, J., Mishra, S., Fickett, J.,
             Burks, C.  (1992) Access to molecular biology databases.  <Mathl.
             Comput.  Modelling> 16, 93-101.
charter      The goal of LiMB is to provide the scientific community with a
             comprehensive overview of databases relevant to molecular biology
             and related data sets.
crossref     -
data.pri     databases [molecular biology]
data.sec     database access information; database contribution information;
             database characteristics; literature citations [molecular
             biology]; database maintenance [hardware]; database maintenance
             [software]
hardware     Sun SPARCstation 1
op.sys       UNIX
dbms         SYBASE
language     -
software     -
format       flat text file: line type record
access       no limitations
.
.
.
///
```

Figure 4. Sample entry from the LiMB database. This is drawn from a pre-release version (4.0) of LiMB. To compress the entry vertically, several fields were removed (as indicated by the ellipses).

3.2 Media

With a specific database in mind, how does one access and retrieve the information of interest? What software and hardware does one need to have on hand to take advantage of the mechanisms for distributing data from a database? Unfortunately, there is no simple answer to the latter question because of the variety of media and mechanisms that are used for different purposes within a single resource, and that vary considerably from resource to resource.

In general, a laboratory (or group of laboratories) planning to use databases provided by external databanks will need a computer (PC, Macintosh, or UNIX workstation), peripheral hardware for downloading information from shipped media (e.g. diskettes, high density tape cartridges, and/or CD-ROMs), an Internet connection, an electronic mail (e-mail) tool, and a tool for interfacing to the World Wide Web (the Mosaic tool is useful and freely available (13), and other tools, both free and vendor-supplied, have also recently become available). The choice among computers is unimportant for access to the comprehensive sequence databanks (they all cater to all three platforms), but some of the smaller molecular biology databanks only support one of the three platforms.

For more specialized or more ambitious tasks, a laboratory may consider acquiring a vendor-supplied software environment for accessing the data sets of interest. In addition, either a relational database management system (RDBMS) or an object-oriented database management system (OODBMS) may be desirable; see the discussion of these systems in refs 18–20. Finally, various stand-alone tools can be acquired for analysing specific kinds of data (e.g. scanning a database for sequence similarity) that may augment or improve upon what is available in a particular packaged environment. The choice among strategies and products for these specialized tools is driven by the particular application in mind, degree of regional support from vendors, hardware and operating system platforms already in use, etc.

3.3 Mechanisms

What mechanisms are available for accessing databank databases? There is a range of answers to this question, and a detailed description is in many cases specific to a particular database.

In general, the most direct mechanism is on-line access to the database residing in a centralized database management system. This was a standard approach when local computing hardware was not prevalent, but local terminals and modems were available and could be used to dial up the central database. As local hardware (and the graphical interfaces they provided) became more common, and for reasons of autonomy and convenience, laboratories began importing databases (and packages for managing them) onto their own machines. Recently, this trend is again reversing, at least in part, because of the ease of accessing other computers through network connections and because of the move—based on client-server software architectures—towards use on local (distributed) computers of graphical interfaces that are connected to and driven by a centralized database package on a remote computer. These local access tools may be embedded in World Wide Web interfaces or custom software interfaces developed by the databanks or third parties.

A second option is to import copies of databases of interest onto a local computer, with access software supplied by either the databank, or by an independent, third party (usually, a vendor), or by the computationally ambitious in one's own group. The lattermost option is not recommended as a basis for supporting a local resource unless one is in the business of developing software systems. Third party packages often include copies of databases as part of the package, and updates of the databases as part of the price of the package. This approach has the advantage of immediacy and autonomy, but importing, loading, and troubleshooting several databases and/or packages can rapidly become overbearing unless the local economy of scale allows for supporting system administrators/analysts.

A final option is to use the growing number of analysis servers that databank and more general, related resources are providing that allow one to query or probe a database remotely using network and e-mail based protocols. One composes, on one's local computer system, an e-mail message containing a simple, structured query (a request for an entry containing a specific keyword, or a request to scan an appended sequence against the database), sends the e-mail to a remote address, and (usually within minutes) gets back a response with results (or an explanation of why no results are forthcoming) in an e-mail message from the server resource machine. This approach has the marked advantage of requiring mastery of only a local e-mail tool and a few syntactical rules for composing the e-mail message; it has the obvious disadvantage of being limited to a relatively fixed set of queries or analyses.

3.4 Which databases should I get?

In terms of sequence databases, the drive behind answering this question for the individual laboratory comes from determining: Which databases provide data in a medium that matches one's local computer platform(s)? Which databases offer geographical convenience (though this is less important now than in the past because of networking)? What software tools are provided by the databank or a third party for accessing, querying, and analysing data from a particular database? Once these questions have been resolved, one should identify a few 'typical queries' that one can imagine (e.g. 'Tell me what you have on protein X', 'Send me entries that match the following sequence ...', or 'Send me recent entries deposited by my competitor, scoundrel Y') representing a good sampling of the anticipated daily use of a database, and run this query through as many of the databases being considered as possible. Sampling this way will give one an impression of the completeness and timeliness of the given database, the depth of annotation, and style of presentation in text file format.

This is most conveniently done through one of the several resources supplying a smorgasbord of databases that can be simply queried or retrieved with a common e-mail based format. One such example is the service provided by the National Center for Biotechnology Information. Instructions for retrieving entries from a number of databases through this server acquired by sending an e-mail message with the single word 'help' in the text of the message to `retrieve@ncbi.nlm.nih.gov`. Similarly, instructions for scanning the same databases with a query sequence are acquired by sending an e-mail message with the single word 'help' in the text of the message to `blast@ncbi.nlm.nih.gov` (see Chapter 4 for more detailed and alternative protocols).

4. Using the data

The other chapters of this book are devoted to the algorithms, tools, and protocols for analysing and interpreting sequence data. We note here a few points for contemplation, rooted in the databases, that should be considered (and re-considered) as one begins exploring and using the databases.

4.1 Do I have a current version of the database?

How often is the version updated on which your analysis or retrieval is based, and how far behind databank's current version is the updated copy? If you get different results than another's query, compare the release numbers for the underlying databases. Finally, keep track of which version you used for a particular set of analyses (and cite it when you report your results), and consider whether or not it is going to be important to be able to re-analyse the same version of the database at some later date (in which case, you might consider preserving that version of the database).

4.2 How often should I repeat routine queries?

In view of the fact that the body of sequence data has been doubling roughly every year or two over the past decade, exploratory queries that are important should be repeated every few months. For the sake of computational efficiency, try to set up these repeated queries to probe only new entries (and, depending on the query, revised entries) in the target databases.

4.3 How redundant is the database?

Does the database being queried contain many copies of the same sequence (perhaps because it has been sequenced independently by several groups) or closely related sequences (as in the case of multi-gene families)? For some queries, probing a database that has eliminated or at least decreased redundancy is better, merely from the point of view of efficiency, both for doing the original query and for examining the output results. However, for other queries, the presence of redundancy can, more importantly, skew the results in a way that can be misleading unless the redundancy is taken into account in post-processing. If one wanted to know how much of the *Escherichia coli* genome had been sequenced, or what the consensus donor splice signal is for all exon/intron boundaries, simply counting the number and length of *E. coli* entries in one of the large nucleic acid sequence databases (for the former query), or extracting all sequences surrounding annotated splice donor sites (for the latter query), would lead to misleading results unless redundancy was taken into account.

Note that 'redundancy' cannot be simply defined in the context of sequence searches in a way that can be generally applied to all possible queries. Most genomes are naturally redundant at some level, especially higher eukaryotic genomes, some of which have a high percentage of redundancy (involving highly similar though seldom identical tracts of sequence). Removing all sequence redundancy, based simply on sequence similarity, can also create an 'unnatural' representation. If a database is being used that has had redundancy reduced, understand how this was done and determine whether or not it effects the query at hand either positively or negatively.

4.4 Are there errors in the database?

The answer is 'yes', regardless of what database you are using. Some of these errors originate with the laboratories submitting the information, others resulted during transcription of the data from one publication format to another, and others originate with data entry or interpretation by the data-bank staffs. The latter two categories have decreased significantly as the rate of direct, electronic submission has gone up. The importance of a potential error in a given result will determine to what extent database information should be double-checked against the original publications and/or other databases.

For the purposes of building up data sets for testing query algorithms, especially those involving pattern recognition in database sequences, researchers have often post-processed data from the databases, based on careful examination, reference to the original literature, etc. so that the search algorithms will have 'clean' data on which to train and test. However, one should, in the course of such development, test the algorithm (and the process of interpreting results) on error-containing data sets, as the typical user of such algorithms will be using it in such an environment.

4.5 How did I get that result?

For a given set of software tools for scanning a database (e.g. for sequence similarities), one should understand the underlying algorithms well enough to evaluate whether or not the various parameter settings, underlying assumptions about statistical models for significance, and published rates for false-positive and false-negative results are a good match to the database being studied.

5. Queries across multiple databases

Currently, there are hundreds of databases both of interest to and available to the molecular biology community. These databases are generally relatively up-to-date, relatively complete, and relatively easy to retrieve or examine electronically.

However, there are a number of difficulties with this cornucopia. For example, many—if not most—of the databases currently available are not structured and stored in systems that are easily accessible to the user community and that would allow them to execute more sophisticated queries on the wealth of information within a given database.

An even more severe problem is the extreme inefficiency (if not near impossibility) of querying across databases, grounded in the lack of standardization among them (15, 16, 21–23). Deciding what entries to include in *Figures 1–3* is a very simple example: the goal was to identify an entry from each of the three databases corresponding to the same gene product in the same species. Thus, the conceptual query was 'Tell me about entries in the nucleic acid, protein, and complex carbohydrate sequence databases that correspond to the same protein from the same organism'. Unfortunately, there is not currently any publically accessible server that provides a query language capable of such logical construct and that accesses all the databases of interest for that query. Putting together the examples in *Figures 1–3* involved consultation with the staff of one databank, reformatting of one entry for text presentation, probing several databases with keyword searches (and experimenting with synonyms for protein names for use as keywords), and sifting through the large number of retrieved entries to filter out entries other than those specifically sought. Waterman *et al.* (23) have developed a list of benchmark queries that are representative of those which an investigator might typically want to answer, and which are currently very difficult to answer automatically by probing across databases.

This problem is the result of the conflict between two important facts about building databases:

(a) Autonomy allows for speed and flexibility in individual databank operations.

(b) Global standardization is necessary across databanks to enable and increase the efficiency of interoperability/interconnectivity.

One solution to the problem of database interconnectivity is to centralize all the databases of potential interest, and organize them under a single query system. Though several groups have experimented with this, and it has the advantage of not requiring any co-ordination among the targeted databanks, it is an impractical solution, involving a tremendous amount of resources to keep up with the task of importing and reformulating a large number of very dynamic databases. A related approach, virtual centralization, stores information about the structures and query procedures for a number of databases in a centralized tool (though not the component databases themselves), and uses this information to probe various databases remotely, as necessary, in response to a user query formulated in a single interface to the collection of access information.

An alternative intermediate solution is for new databases—especially those that are derivative of or augmenting another, primary database—to be constructed with the same structure and controlled vocabulary as pre-existing database to which the new database can be related. This model (the 'backbone' model) (24) increases the likelihood of keyword-based inter-database queries, and reduces the redundancy—and accompanying potential for error and asynchrony—among the backbone and derivative databases (for example, where a given nucleic acid sequence database will itself contain no nucleic acid sequences, but only pointers to sequences contained in a separate database, accompanied by value-added annotation) (24, 25). This can still be a relatively resource-intensive approach for the derivative databases if the primary database is changing rapidly, and this strategy also limits flexibility for the databank staff responsible for the derivative database.

In the face of the prohibitive overhead associated with reconfiguring existing databases in a centralized system, and the unlikelihood that all databanks will adopt (or even enunciate) common standards for vocabulary, information schemas, and database management systems, an alternative approach, database 'federation', is being explored by the database research community at large (e.g. see refs 26, 27). Federation is a philosophy that recognizes the conflicting importance of both standardization and autonomy, and which promotes the consensual agreement among participating databanks to conform to minimal standards for internal structure, semantics and self-description of their structure that will support an agreed-upon level of cross-database queries. The biology database community, including particularly the Human Genome Project databanks, are discussing and exploring initial implementations of this philosophy (5, 23).

6. Keeping up and going further

The following are a number of locations where databases as well as database problems, solutions, and tools are discussed regularly.

Several periodicals regularly publish articles related to molecular biological database description, development, or use:

- *Computer Applications in the Biosciences*
- *Journal of Computational Biology*
- *Nucleic Acids Research*
- *Genomics*
- *Genome News* (published by Oakridge National Laboratory, Oakridge, TN)

A number of electronic bulletin boards (e.g. those collected under the BIOSCI rubric) provide forums for questions and answers about biological databases. For information about how to subscribe to the BIOSCI bulletin boards, send an e-mail message to `biosci-server@net.bio.net`, containing in the text of the message the single phrase '`info ukinfo`' (if you are located in Europe, Africa, or Central Asia) or '`info usinfo`' (if located in the Americas or Pacific Rim) (see also Chapter 4).

There are several research groups and deliberative organizations that are focused at least in part on database and related issues relevant to molecular biology in general and the Human Genome Project in particular, including (for example):

- CODATA Task Group on Macromolecular Sequence Databases (Paris, France)
- European BioInformatics Institute, an outstation of the European Molecular Biology Laboratory (Cambridge, UK)
- National Center for Biotechnology Information, part of the National Library of Medicine (Bethesda, MD, USA)
- National Center for Genome Resources (Santa Fe, NM, USA)
- Theoretical Biology and Biophysics Group, part of Los Alamos National Laboratory (Los Alamos, NM, USA)

These groups sponsor workshops, policy forums, long- and short-term visitors, and/or scientific collaborations.

Finally, several annual scientific meetings and workshops have emphasized molecular biology databases as a single or one of several primary topics, including:

- *Intelligent Systems for Molecular Biology* (28, 29, 34, 35). Most recently in St. Louis, USA, in June 1996.

- *Pacific Symposium on Biocomputing* (30, 33). Next in Hawaii, USA, in January 1997.
- *Genome Informatics Workshop* (31, 36). Most recently in Yokohama, Japan, in December 1995.
- *Meeting on the Interconnection of Molecular Biology Databases* (32). Most recently in St. Louis, USA, in June 1996.

Acknowledgements

This work depended heavily on information and insights gained in the context of the LiMB Database project, and specifically on the staff that have contributed to that effort, including M. Engle, G. Keen, J. Lawton, F. Martinez, S. Mishra, and G. Redgrave. I have benefited immensely from conversations on database topics with G. Cameron, M. Cinkosky, J. Fickett, P. Gilna, and R. Robbins. I am grateful to the various databank staff and meeting organizers for providing information, and particularly to S. Doubet for providing the customized version of a CCSD entry in *Figure 3*. This work was done under the auspices of the US Dept. of Energy.

References

1. DeLisi, C. (1988). *Science*, 24, 47–52.
2. Kingsbury, D. T. (1989). *Trends Biotechnol.*, 7, 82–7.
3. Wooley, J. C. (1989). *Trends Biotechnol.*, 7, 126–32.
4. Lander, E. S., Langridge, R., and Saccocio, D. M. (1991). Commun. ACM, 34, 33–9.
5. Fox, P. T. and Lancaster, J. L. (1994). *Science*, 266, 994–6.
6. Fox, P. T., Mikiten, S., Davis, G. and Lancaster, J. L. (1994). In *Functional neuroimaging: technical foundations* (ed. R. W. Thatcher, M. Hallett, T. Zeffiro, E. R. John, and M. Huerta), pp. 95–105. Academic Press, San Diego, CA.
7. Burks, C., Cinkosky, M. J. and Gilna, P. (1994). In *The Human Genome Project: deciphering the blueprint of heredity* (ed. N. G. Cooper), pp. 254–5. University Science Books, Mill Valley, CA.
8. Claverie, J.-M. (1994). *Genomics*, 23, 575–81.
9. Burks, C. (1989). In *Computers and DNA* (ed. G. I. Bell and T. Marr), pp. 35–45. Addison-Wesley, Reading, MA.
10. Fickett, J. W. (1989). *Biomolecular data: a resource in transition* (ed. R. Colwell), pp. 295–302. Oxford University Press, Oxford, UK.
11. Cinkosky, M. J., Fickett, J. W., Gilna, P., and Burks, C. (1991). *Science*, 252, 1273–7.
12. Krol, E. (1992). *The whole Internet user's guide and catalogue*. O'Reilly & Associates, Sebastopol, CA.
13. Schatz, B. R. and Hardin, J. B. (1994). *Science*, 265, 895–901.
14. Burks, C., Lawton, J. R. and Bell, G. I. (1988). *Science*, 241, 888.

15. Lawton, J. R., Martinez, F., and Burks, C. (1989). *Nucleic Acids Res.*, 17, 5885–99.
16. Keen, G. M., Redgrave, G. W., Lawton, J. R., Cinkosky, M. J., Fickett, J. W., Mishra, S. K., *et al.* (1992). *Math. Comput. Model.*, 16, 93–102.
17. Redgrave, G. W. and Burks, C. (1992). *LiMB Release 3.0. Technical report.* Los Alamos National Laboratory, Los Alamos, NM.
18. Frenkel, K. A. (1991). *Commun. ACM*, 34, 41–51.
19. Lewis, S. (1994). In *Automated DNA sequencing and analysis* (ed. M. D. Adams, C. Fields, and J. C. Venter), pp. 329–38. Academic Press, London, UK.
20. Cuticchia, A. J. (1994). In *Automated DNA sequencing and analysis* (ed. M. D. Adams, C. Fields, and J. C. Venter), pp. 339–46. Academic Press, London, UK.
21. Kamel, N. N. (1992). *Comput. Appl. Biosci.*, 8, 311–21.
22. Fields, C. (1992). *Trends Biotechnol.*, 10, 58–61.
23. Waterman, M., Uberbacher, E., Spengler, S., Smith, F. R., Slezak, T., Robbins, R. J., *et al.* (1994). *J. Computat. Biol.*, 1, 173–90.
24. Cameron, G. and Kahn, P. (1990). *BOSPHORE Actualites*, 2, 3–6.
25. Benson, D., Boguski, M., Lipman, D. J., and Ostell, J. (1990). *Genomics*, 6, 389–91.
26. Heimbigner D. and McLeod, D. (1985). *A federated architecture for information management. ACM Trans. Office Inf. Syst.*, 3, 253–78.
27. Litwin, W., Mark, L., and Roussopoulos, N. (1990). *ACM Comput. Surveys*, 22, 267–93.
28. Clark, D. A. and Rawlings, C. J. (1994). *Comput. Appl. Biosci.*, 10, 199–205.
29. Altman, R., Brutlag, D., Karp, P., Lathrop, R., and Searls, D. (ed.) (1994). *Proceedings: second international conference on intelligent systems for molecular biology.* AAAI Press, Menlo Park, CA.
30. Hunter, L. (ed.) (1995). *Proceedings of the twenty-eighth annual Hawaii international conference on system sciences.* Volume V: *Biotechnology Computing.* IEEE Computer Society Press, Los Alamitos, CA.
31. Miyano, S., Akutsu, T., Imai, H., Gotoh, O., and Takagi, T. (ed.) (1994). *Proceedings: genome informatics workshop V.* Universal Academy Press, Tokyo, Japan.
32. Karp, P. (1994). *Technical Report 549, SRI International AI Center.*
33. Hunter, L., and Klein, T. (ed.) (1996). *Biocomputing: proceedings of the 1996 Pacific symposium.* World Scientific Publishing Co., Singapore.
34. Rawlings, C., Clark, D., Altman, R., Hunter, L., Lengaver, T., and Wodak, S. (ed.) (1995). *Proceedings: third international conference on intelligent systems for molecular biology.* AAAI Press, Menlo Park, CA.
35. States, D. J., Agarwal, P., Gaasterland, T., Hunter, L., and Smith, R. (ed.) (1996). *Proceedings: fourth international conference on intelligent systems for molecular biology.* AAAI Press, Menlo Park, CA.
36. Hagiya, M., Suyama, A., Takagi, T., Nakai, K., Miyano, S., and Yokomori, T. (ed.) (1995). *Proceedings of genome informatics workshop 1995.* Universal Academy Press, Tokyo.

2

The NCBI software tools

1. Introduction

The National Center for Biotechnology Information (NCBI), a part of the National Institutes of Health (NIH), is responsible for the production of the GenBank DNA sequence database. NCBI has also made a number of other tools and databases available to molecular biologists, the best known being *Entrez*, *BLAST*, and dbEST. NCBI software processes thousands of sequence and bibliographic records on a regular basis in house and is also used to build software products such as *Entrez* and *Sequin*, a new sequence data direct submission tool, for distribution. The software toolkit used for this supports data access from local CD-ROM and over Internet. It provides a portable graphical interface and a rich set of data manipulation tools. It can run on a large number of hardware and software platforms. The NCBI data model underlying the toolkit defines a variety of nucleic acid and protein sequence types, physical and genetic maps, a number of classes of sequence annotations, alignments between sequences and/or maps, bibliographic information, and more recently, macromolecular structure and taxonomic trees. All software and databases from NCBI are freely available to the public.

2. The software toolkit

2.1 Portable core library

The NCBI software tools have a small library of 'C' language routines (*Core-Lib*) which allow us to write software which is source code identical on a large number of hardware/software platforms, including DOS, Microsoft Windows, Windows NT, Apple Macintosh, various UNIX platforms, and Digital VAX/VMS, among others. Given the variety of platforms used by molecular biologists and the rate of change of scientific software needs, the ability to write an application once which will run without change on many platforms has obvious advantages. The *CoreLib* covers routine programming functions such as memory allocation, string handling, error reporting, file handling, and so on. The philosophy has been to adhere to ANSI C whenever it was adequate for portability, and to write replacement functions only when it was not. When we were forced to write a function we made its name and argument conventions as similar to the ANSI cognate as possible. This library has been successfully used on all platforms for many years.

2.2 Data encoding in ASN.1

NCBI uses Abstract Syntax Notation 1 (ASN.1) to define its data model and to exchange data between databases and software tools. ASN.1 is an International Standards Organization standard (ISO 8824, ISO 8825). It is the Open Systems Interconnection (OSI) standard for data exchange between the top two layers of the OSI, between the application layer (defined as any application, on any computer, in any language) and the presentation layer (which passes any data/messages between applications as a formally encoded linear stream of bytes). As such, ASN.1 is designed to allow formal specification of structured data in a way which is domain, machine, programming language, and storage technology independent. Thus formal specifications for scientific data can be developed and validated, software tools for reading, writing, and validating such data can be automatically generated by software from the specification, yet the data can still be independent of a particular database management system or software tool. This means that such data can be easily exchanged among many systems using different architectures and can migrate as technologies evolve. It is used in other systems such as the X.400 email protocol, the Z39.50 bibliographic data exchange standard, and WAIS which has seen wide use in the scientific community.

An ASN.1 specification is very similar in appearance and syntax to a Backus-Naur form used to define programming languages (*Figure 1*). Data encoded in print form ASN.1 looks very similar to a C program. The binary encoding for ASN.1 is more compact, machine independent, and faster to transmit electronically, but is not human readable. So we typically develop applications using the print form of ASN.1 which is easy to read and enables use of text based tools such as *grep* and *awk* to monitor behaviour and debug. Once the application is complete, it is simply a matter of setting a flag in the *AsnLib* routines to have the final program read and write the binary form for performance.

3. The NCBI data model

3.1 Introduction

NCBI has developed a formal data model for bibliographic and sequence data. It has a formal specification, in ASN.1, which is independent of programming language or storage technology. The definition is subdivided into several modules, just as good software design breaks a program into well

```
Gene-ref ::= SEQUENCE {
    locus VisibleString OPTIONAL ,      -- Official gene symbol
    allele VisibleString OPTIONAL ,     -- Official allele designation
    desc VisibleString OPTIONAL ,       -- descriptive name
    maploc VisibleString OPTIONAL ,     -- descriptive map location
    pseudo BOOLEAN DEFAULT FALSE ,      -- pseudogene
    db SET OF Dbtag OPTIONAL ,          -- ids in other dbases
    syn SET OF VisibleString OPTIONAL }     -- synonyms for locus
Gene-ref ::= {
    locus "ADH" ,
    desc "Alcohol Dehydrogenase" ,
    maploc "1q21.2",
    db {
        { db "GDB" , tag id 11214 } } }
/**********************************************************************
 *
 *   GeneRef
 *
 **********************************************************************/
typedef struct generef {
    CharPtr locus,
        allele,
        desc,
        maploc;
    Boolean pseudo;
    ValNodePtr db;              /* ids in other databases */
    ValNodePtr syn;             /* synonyms for locus */
} GeneRef, PNTR GeneRefPtr;

GeneRefPtr LIBCALL GeneRefNew PROTO((void));
Boolean    LIBCALL GeneRefAsnWrite PROTO((GeneRefPtr grp, AsnIoPtr aip, AsnTypePtr
    atp));
GeneRefPtr LIBCALL GeneRefAsnRead PROTO((AsnIoPtr aip, AsnTypePtr atp));
GeneRefPtr LIBCALL GeneRefFree PROTO((GeneRefPtr grp));
GeneRefPtr LIBCALL GeneRefDup PROTO((GeneRefPtr grp));
```

Figure 1. A *Gene-ref* (reference to a gene) specified in ASN.1 (top), an example of a *Gene-ref* datum encoded in print form ASN.1 (middle), and the corresponding C structure in the Object Loader layer of the toolkit (bottom).

defined modules. Because of its modularity, it is possible to add new modules to the model without affecting the rest of the specification, or to modify modules with defined consequences. The model is designed to be as stable as possible in a rapidly evolving scientific discipline. The most stable concepts are placed at the core of the model and less stable concepts are built around them. Primary core concepts are a publication (*Pub*) and a biological sequence (*Bioseq*). They are described in more detail below to illustrate the advantages of this design.

3.2 *Pub*

Publication citations are stable in that they are well understood and unlikely to change in form as rapidly as scientific concepts. Yet all important scientific results are usually published. Other data elements in virtually any scientific database must ultimately be connected to scientific literature. A database is very commonly queried based on a paper someone has read. Conversely, exploration in a scientific database routinely requires access to the papers cited by the results. An additional use which is not commonly realized is that citations are a potential 'shared key' to link together scientific databases that have no formally specified connection with each other but whose data elements can be implicitly connected via their use of the same bibliographic source. NCBI has made extensive use of this property in integrating the various sequence databases and linking them to MEDLINE to build *Entrez*.

The NCBI bibliographic specification defines a '*Pub*' as a choice of virtually any form of publication. This includes articles in journals, books, and proceedings, the journals, books, or proceedings themselves (as opposed to an article in them), theses, manuscripts, letters, unpublished works, submissions to databases, and patents. The relevant portions follow as closely as possible the guidelines in the American National Standard for Bibliographic References, ANSI Z39.29-1977. Provision is made for journal titles from a number of sources including MEDLINE, the International Standards Organization (ISO), International Standard Serial Number (ISSN), codens, and full titles. NCBI maintains mapping tables to convert between these types for mapping citations from many sources to standard forms. Once in standard form, citations can then truly be used as shared keys.

3.3 *Bioseq*

The most basic interpretation of a biological sequence, as a single contiguous (at least conceptually) string of amino acids or nucleic acids, is also a core concept which is unlikely to change in a substantial way, at least for a large majority of such molecules. As such it can serve as another important and stable key linking other data together. The *Bioseq* plays this central role in the NCBI data model (*Figure 2*). Its minimal required elements are just an identifier (*Seq-id*), such as an accession number, and the 'instance' (*Seq-inst*) which is defined as such physical characteristics as the type of molecule (DNA, RNA, protein), length, topology, and the sequence itself. The sequence identifier is critical for using the *Bioseq* as a shared key and is dis-

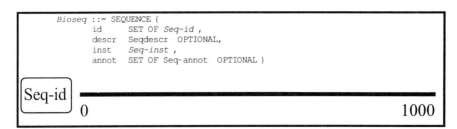

```
Bioseq ::= SEQUENCE {
        id      SET OF Seq-id ,
        descr   Seqdescr OPTIONAL,
        inst    Seq-inst ,
        annot   SET OF Seq-annot  OPTIONAL }
```

Figure 2. A minimal *Bioseq* is at least one identifier and the sequence data itself. This defines a simple, linear coordinate system which is citable. Additional OPTIONAL elements supply information about the minimal element, such as references, features, or alignments.

cussed further below. The instance is confined to the physical facts regarding the sequence data itself. As such one would not expect it to be subject to interpretation beyond that inherent in the sequencing process itself. Thus a *Bioseq* meets the requirements for a citable, stable, core data element.

The optional fields contain information about the *Bioseq*. Descriptors apply generally to the whole *Bioseq*, and may include title, comments, references, etc. Annotations are information about the *Bioseq* which cite specific locations on its co-ordinate system, such as features, alignments, or graphs of sequence properties. These elements are expected to be more diverse and to change much more rapidly than the core *Bioseq* elements. They are extremely important, since they set the context for the *Bioseq* and record scientific assertions about it.

Since the *Bioseq* is a stable, citable entity, it is possible for data acquired and stored outside the public databases, such as alignments or features developed by independent investigators, to reliably refer to sequence data in the public database even as the sequences are revised or extended. Such third party databases can be integrated and compared since they refer to a common, stable co-ordinate system.

3.3.1 Classes of *Bioseq*

A *Bioseq* can represent any molecule class (DNA, RNA, or protein). It can also represent a wide range of sequence-related objects as subclasses of the general *Bioseq* concept (*Figures 3* and *4*). A virtual *Bioseq* is one in which we know the molecule and its length, but do not yet have sequence data (e.g. a band on a gel). A raw *Bioseq* is the familiar sequence entry, in which we also have the sequence data. A segmented sequence is really a set of instructions on how to build a (conceptually) contiguous molecule from other *Bioseqs* (e.g. we know a sequence is a certain length overall, but have only sequenced parts of the length, say around the exons). A physical or genetic map is just a virtual *Bioseq* (we know roughly how long the chromosome is) which has landmarks, represented as features on the *Bioseq*. This design means that tools to display *Bioseqs* can easily show sequences and maps together using the same software. Sequence editing utilities, such as those in the toolkit, operate on a map or a piece of sequence with the same single function. An alignment computed between two proteins and an alignment of a genetic to a physical map provided by a human are the same data structure.

3.3.2 *Seq-id*

A *Bioseq* supports a SET OF *Seq-id*, not just one. This is because sequence identifiers can have different forms and different semantics depending on the source of the *Bioseq*. In addition, NCBI adds a *Seq-id* to all *Bioseqs* it processes, called a GenInfo (*GI*) number. The NCBI *GI* is an arbitrary integer id which:

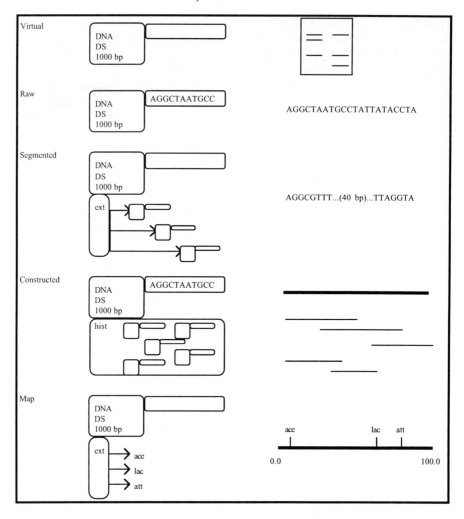

Figure 3. There are many subclasses of *Bioseq*.

- explicitly identifies a specific sequence
- is stable and retrievable over time
- has the same form over all sequence databases
- is used to provide a history of changes to the sequence.

If a sequence changes but its source id does not (usually due to corrections or updates), then a new *GI* number is assigned to it. The old entry receives an annotation in its history which points to the new *GI*. The new entry receives an annotation which indicates the old *GI*. Therefore, an annotation database based on *GI* numbers will remain consistent with the original sequence data

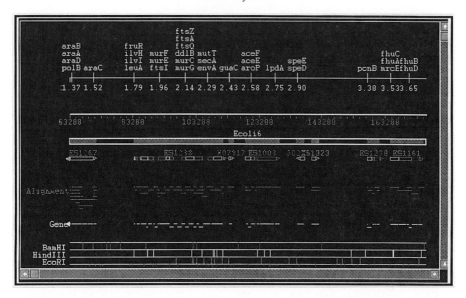

Figure 4. A screen from the *Chromoscope* application using selectable graphical objects implemented with the *Vibrant* portable interface. This view shows a section of the complete *E. coli* chromosome as a segmented *Bioseq* (centre). The dark regions are filled with sequence and the empty regions refer to the physical map (also shown completely at the bottom). The top line is a *Bioseq* of type genetic map using genetic minutes as a numbering scheme. Below that is a *Bioseq* of type physical map, showing base pair coordinates. Then comes the segmented sequence described above. Below each filled area of the segmented sequence are the contigs which fill that region, themselves implemented as segmented sequences to show how they are constructed from parts of GenBank entries (smaller boxes within the contig). The line labelled 'Alignment' shows the alignments (*Seq-align* objects) connecting the assembled and corrected contig with all the GenBank entries which overlap it.

and so it will be possible to retrieve the old sequence (through the NCBI network servers), determine the changes, and update the annotations to conform with the new version of the sequence. This critical element for allowing scientists to personally analyse and annotate the public databases is now in place and available.

3.3.3 *Seq-loc*

A *Seq-loc* is a location on a *Bioseq*. As such it is essentially a *Seq-id* and an offset on the co-ordinate system defined by that *Seq-id*. There are subclasses of *Seq-loc* for points, intervals, collections of intervals, and so on. Since *Bioseq*s may be of many types, one can use a *Seq-loc* to refer to a coding region on a DNA sequence, an active site on a protein, or a gene on a genetic map with the same data structure. If one uses the NCBI *GI* number as the *Seq-id*, then the location will also be stable and traceable over time.

3.3.4 *Seq-align*

A sequence alignment, *Seq-align*, is just a correlated set of *Seq-loc*, with optional scores or comments. A *Seq-align* can be computed between proteins or used to define the relationship of a genetic to a physical map, without change. It is also a valid annotation on a *Bioseq*, just as one may have a feature table as an annotation in the more traditional sequence formats. The value of showing the relationship of one *Bioseq* to another in a readily accessible way is obvious. *Seq-align* can also be used within a *Bioseq* as part of its history to show, for example, how a contig was constructed from other entries, or how an updated sequence relates to an earlier version.

3.3.5 *Seq-feat*

A feature on a sequence, *Seq-feat*, is constructed from a number of modules (*Figure 5*). It consists of a core common to all features which has fields for comments, citations about the feature, etc. It has a required *Seq-loc* giving the location on the *Bioseq(s)* of the feature. An optional product *Seq-loc* allows the feature to be used to explicitly link two *Bioseqs* through a process defined by the *Seq-feat* (e.g. a coding region feature points to the source DNA *Bioseq* and the product protein *Bioseq*). Different types of features are made by attaching different data structures to the *Seq-feat*. So a *CdRegion* structure has fields for genetic code, reading frame, etc., whereas a citation feature has fields for authors and journal titles. In fact, this design allows extensive reuse of data structures. So a citation feature points to the identical *Pub* structure used for all bibliographic citations anywhere in the data model.

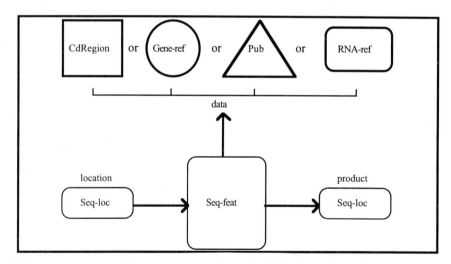

Figure 5. A sequence feature is composed of a core (*Seq-feat*) common to all features, a *Seq-loc* defining its location on a sequence, and optional product defining another sequence (e.g. the protein product of a coding region). Different types of features are made by attaching different data structures appropriate to the type.

Software which operates on *Pubs* will work equally well on a *Pub* from a MEDLINE entry as a *Pub* that is a feature on a GenBank sequence.

A number of the data blocks defining *Seq-feat* types are designed to be both a limited cache of local information, as well as a link to other, non-sequence, databases. For example, a *Gene-ref* (reference to a gene) has fields for locus, allele, description, but also keys to genetic databases. Since it is independent of whether it is used as a feature or alone, a genetic database could produce such simple data structures as an easy way of making summary and linking data available to others. A sequence database could then use those data structures to make *Seq-feats* to link the genetic database to the sequence database.

4. Technical aspects of the NCBI toolkit

4.1 ASN.1 libraries

AsnLib is a library of support functions for using ASN.1. It is built on *Core-Lib* so is portable on all platforms. *AsnLib* can check and validate an ASN.1 specification, produce a parse tree for that specification either at run time or as a C header file that can be included in a program, and read and write data encoded in ASN.1 in either the binary or print format. A supporting program, *AsnCode*, can automatically generate C structures corresponding to higher level ASN.1 data structures, and the code for creating them, freeing them, reading them from an ASN.1 stream and writing them to an ASN.1 stream. The library also provides the capability of exploring the data structures in memory using the ASN.1 defined type names to locate data of interest, to print those data using a template defining its format, to copy any ASN.1 defined type in memory or through a disk intermediate, and to read or write it to a location in memory.

4.2 Object loader layer

On top of *AsnLib* is a layer called the 'object loader' layer. This consists of C structures defined for each higher level type in the ASN.1 specification. Each type has, at a minimum, a set of functions to create an instance of that type, to read it from an ASN.1 stream, to write it to an ASN.1 stream, and to free it (*Figure 1*). There may be additional routines as well, such as a high speed copy, a formatting routine, or to compare it with another type.

4.3 Utilities layer

A large number of utility routines that operate on the C structures of the object loader layer are provided in several source files. Although it is not possible to present the full list of capabilities, a few representative examples can be discussed. *SeqLocCompare()* takes pointers to two *Seq-locs* and returns a value indicating if they overlap, if one is included in the other, or if they do not overlap. This function supports any class of location including multiple exons on more than one sequence. *BioseqFind()* can locate a sequence in memory, given its identifier. A *SeqPort* can be opened on any sequence or any (arbitrarily complex) location on a sequence. The *SeqPort* allows the software to treat the sequence like a file, with functions to seek to locations, get residues, convert them into any alphabet, and so on. *SeqPort* supports sequences composed of parts of other sequences, maps, and compact (multiple base per byte) sequence encodings. *ProteinFromCdRegion()* will produce a protein sequence from a coding region, with support for alternative genetic codes, alternate start codons, exceptions at individual codon positions (such as a suppresser tRNA), and incomplete codons at the ends of the coding region. The sequence editing function library supports functions to cut, copy, or paste regions of sequences, with or without the appropriate changes to their feature tables being made automatically. The editor functions support all classes of sequence including nucleic acid and protein sequences, genetic and physical maps. Many other utilities are also available.

4.4 Access libraries

NCBI distributes a program called *Entrez*, which supports data retrieval of nucleic acid sequences, protein sequences, and MEDLINE bibliographic information from CD-ROM or from an internet client/server. *Entrez* uses a library of access functions which support the construction and evaluation of Boolean queries, the location of entries related to each other by sequence similarity, the location of bibliographic records cited by a sequence entry and vice versa, the location of nucleic acid entries coding for a protein sequence and vice versa, the location of bibliographic records concerning a similar subject area to a given bibliographic record, and the retrieval of sequence or bibliographic records given an identifier (e.g. GenBank accession number, MEDLINE unique identifier, etc.). NCBI distributes the access library as source code so that other groups may incorporate parts, for example to fetch sequences on demand, or the whole spectrum of capabilities from *Entrez*, into software of their own design with minimal effort. Such software can be developed using the CD-ROM or a local file server, then turned into a true internet client/server program just by linking to a different set of access libraries.

4.5 *Vibrant* portable graphical interface

Vibrant is a portable graphical interface building tool. It allows the building of source code identical applications on Apple Macintosh, Microsoft Windows, Windows NT, UNIX X11 and Digital VAX/VMS X11 using Motif. The executable on each platform uses the underlying windowing system native to that platform and maintains the look and feel appropriate to it. *Vibrant* can be used at a number of levels. At the lowest level (from the programmer's point of view) one can call functions which, on a command line system, will process and prompt for command line arguments, but on a windowing system, will lay out a dialogue box automatically to interact with the user. At the next level, the programmer can simply request a number of data elements through function calls, and *Vibrant* will automatically build and lay out the appropriate dialogue box at run time, process the user's input in a manner appropriate for the host platform, and return the data to the program in the supplied variables. A more ambitious programmer may control the layout and appearance of the dialogues to a greater or lesser degree, as desired. Higher level routines support the presentation of documents with variable founts and layout and selectable text. Finally, a graphics segmentation system allows the building and presention of two-dimensional graphics, with automatic scaling from user defined coordinates to screen coordinates, and support for selection of graphical objects by clicking on the screen. An example of such a display is shown in *Figure 4*.

4.6 Network client/server libraries

NCBI has implemented a number of services, including *Entrez* and *BLAST*, as true internet client/server programs using the NCBI Dispatcher. It is a flexible and relatively easy way to offer client/server access to structured data over the internet. The interface is defined in ASN.1 and automatic tools generate code that allows communication by passing C structures through functions. The client/server can make the connection, maintain the connection, and even offer data encryption in a manner transparent to the application. Servers currently must be on a UNIX platform, but clients can be on a large number of platforms, including Macintosh, MS-Windows, UNIX, VMS, and Windows NT, among others.

A server demon (*ncbid*) is launched on the server computer, and it announces to the dispatcher what service(s) it provides. The dispatcher notes that the server and its service(s) are available, listing the latter in an on-line catalogue. A client then connects to the dispatcher and requests the service by name. If a server of that type is available, the dispatcher connects to *ncbid* on the server machine, and *ncbid* connects directly back to the client. Once that connection is made, *ncbid* forks and launches the server program, connecting the server's standard input and output streams to the opened TCP/IP socket with client. The client and server then exchange ASN.1 defined structures until they are finished.

Because the server program communicates with the client through standard input and output streams, it is possible to develop the server and test it just by redirecting files of ASN.1 messages to the server program. A 'brokering' mechanism makes it possible to test a complete client/server combination without hooking them into the dispatcher/*ncbid* system. Once that works, then the real server is made just by launching it from *ncbid*.

5. NCBI toolkit applications

5.1 *Entrez*

Entrez is a molecular sequence retrieval system which provides integrated access to nucleotide and protein sequence information from a number of sources and the MEDLINE citations associated with the sequences. *Entrez* supports a fast and intuitive query system, and also provides ways of easily moving from a nucleotide sequence to the protein it codes for to the MEDLINE records about this and related proteins. *Entrez* also provides indexes based on sequence similarity and the relationships of MEDLINE articles to answer very general queries like 'Find another sequence like this one' or 'Find more papers about the same subject as this one'. Since it is built with the toolkit it runs on many platforms. It can be used from CD-ROM, from a central file server, or directly over the Internet by client/server or World Wide Web.

5.2 *BLAST*

BLAST provides high speed sequence database searching over the Internet. The client software runs on a variety of platforms and supports convenient searching of all the major public sequence databases rapidly and accurately. *BLAST* supports nucleotide, protein, and translation (nucleotide compared with protein) searches.

5.3 *BankIt*

BankIt provides an easy way to submit sequences to GenBank using World Wide Web as the interface and the NCBI toolkit on the back end. *BankIt* gives new meaning to the words 'quick and easy'.

5.4 *Sequin*

Sequin is a stand alone tool for making more complex submissions to GenBank than those supported by *BankIt*. It supports very large sequences, segmented sequences, and more complex annotation problems.

5.5 Others

NCBI is constantly developing new applications. Contact NCBI for a current list of available resources.

6. Summary

NCBI has developed a comprehensive data model and a set of portable software tools which it uses for building and releasing GenBank, *Entrez*, *BLAST*, and other popular databases and applications. The system is used daily to process large amounts of data from many sources, including Gen-Bank itself, SWISS-PROT, PIR, PRF, PDB, MEDLINE, and others. These data are passed through a number of databases, both relational and non-relational, and a variety of analytical tools within NCBI. The data and associated retrieval systems are made available both on CD-ROM and as internet client/servers. All software tools are available to the public as source code with no restrictions. Inquiries about access to GenBank or any of the tools or services described here may be made by e-mail to `info@ncbi.nlm.nih.gov` or by World Wide Web at `http://www.ncbi.nlm.nih.gov` or by mail to: Information, National Center for Biotechnology Information, Bldg 38A, NIH, 8600 Rockville Pike, Bethesda, MD 20850, USA.

Note added in proof
From August 1996 NCBI will no longer be distributing *Entrez* using CD-ROM since the rapid increases in database size makes this medium impractical. The networked version of *Entrez* and the WWW *Entrez* now supersede *Chromoscope*.

3

EBI databases and tools

RAINER FUCHS and GRAHAM N. CAMERON

1. EBI information products

This chapter describes the information products available from the European Bioinformatics Institute (EBI) on CD-ROM and by network access, including the software developed by the EBI and third parties for use with EBI databases.

In 1980, the Data Library was established at the European Molecular Biology Laboratory (EMBL) with the brief to collect, organize, and distribute nucleotide sequence data and related information. Today, its successor, the EBI, located at Hinxton in the UK, is Europe's prime information provider in the area of molecular biology.

The main endeavour of the EBI service group is still the production of the EMBL nucleotide sequence database. This database is maintained in a close international tripartite collaboration, involving GenBank in the USA, the DNA Databank of Japan (DDBJ), and the EBI. These groups jointly collect nucleotide sequence data and exchange new information electronically on a daily basis.

The details of data acquisition have changed considerably over the years (1). In the beginning, most information in the database originated from literature scanning; today, the vast majority of data is directly supplied electronically by researchers prior to publication. Tools such as Authorin facilitate the preparation of data submissions considerably, and high-speed computer networks ensure efficient communication between the databanks and the scientific community. This way, individual sequences—mainly originating from targeted analyses of particular genes—find their way into the database. Realizing the increasing importance of genome sequencing projects (1), the EBI has complemented these traditional data acquisition methods recently by developing new mechanisms for fully automatic incorporation of bulk data from various European genome projects and large-scale sequencing initiatives (2). Another new data acquisition channel was opened by establishing collaborations between the databanks and national patent offices, resulting in the addition of sequence data from patents and patent applications to the sequence data collections (2).

A second major activity of the EBI service group is the production of the SWISS-PROT protein sequence database together with the University of Geneva (3). This database contains extensively annotated protein sequences from various sources, including sequences derived from the corresponding nucleotide coding regions and directly submitted protein sequence information.

Starting out as a sequence database producer, the scope and activities of the EBI have been broadened constantly over the years towards becoming a general information provider for molecular biology. These activities include the collection and redistribution of other important databases, free molecular biological software and more general biological information (2).

Table 1 lists the databases and information products available from the EBI. The rest of this chapter explains the various methods which can be used for accessing these products and describes several software tools to be used in conjunction with EBI databases.

2. Databases and software on the EBI CD-ROM

Only a few years ago CD-ROM (compact disk read-only memory) was introduced into molecular biology as a data distribution medium and since then it has rapidly established itself as a convenient way of providing large databases.

Production cycles of two or three-months make CD-ROM more suitable for periodic releases of databases every few months than for rapid updates. For many researchers this time lag is unacceptable and more immediate access to latest information, for example by using computer networks, is required (see below). However, if access to latest data is not crucial, or can be achieved by other means (such as using on-line services) a laboratory can be well served with regular updates of its data collections on CD-ROM.

New releases of the EBI databases CD-ROM are published quarterly and are available for a nominal fee. Each release comprises a set of three disks. Volumes I and II contain the EMBL and SWISS-PROT sequence databases plus a variety of index files for use with different retrieval systems described below. Volume III is an extensive collection of other biological data collections as listed in *Table 1*.

To further increase the usefulness of CD-ROM, most database distributors have also developed software products that allow database queries and information retrieval directly on the CD-ROM. Although the speed of access to a CD-ROM is inherently slower than to a magnetic disk, the large storage capacity of CD-ROM encourages extensive indexing, which can, for many applications, overcome the limitations of inherent slow disk access. The

Table 1. Databases and other products from the EBI (for detailed citations see ref. 2)

Database or product name		Mag. tape	CD-ROM	E-mail	FTP/Gopher WWW
EMBL	Nucleotide sequence database	•	•	•	•
SWISS-PROT	Protein sequence database	•	•	•	•
ALIGN	Sequence alignments			•	•
ENZYME	Database of EC nomenclature	•	•	•	•
EPD	Eukaryotic promoter database	•	•	•	•
PROSITE	Protein pattern database	•	•	•	•
BLOCKS	Protein blocks database		•	•	•
PRINTS	Protein signature database		•	•	•
SBASE	Protein domain database		•		•
REBASE	Restriction enzyme database	•	•	•	•
RELIBRARY	Restriction enzyme database		•		•
METHYL	Site-specific methylation			•	•
FLYBASE	*Drosophila* genetic map database		•		•
ECD	*E. coli* map database	•	•	•	•
LISTA	Coding sequences from yeast		•	•	•
TFD	Transcription factor database		•		•
PKCDD	Protein kinase catalytic domains		•	•	•
KABAT	Proteins of immunological interest		•		•
HLA	HLA class I and II sequences		•	•	•
TRNA	tRNA sequences		•	•	•
RRNA	Small subunit rRNA sequences		•	•	•
BERLIN	5S rRNA sequences		•	•	•
SMALLRNA	Small RNA sequences			•	•
REPBASE	Human repetitive DNA sequences			•	•
ALU	ALU repetitive sequences			•	•
HAEMB	Haemophilia B mutations		•	•	•
CPGISLE	CpG islands database		•	•	•
SRP	Signal recognition particle database		•		•
TRANSTERM	Translational termination signals		•	•	•
RLDB	Reference library database		•		•
CUTG	Codon usage tabulated from GenBank			•	•
CODONUSAGE	Codon usage tables			•	•
PDB	Brookhaven protein 3D structures			•	
HSSP	Homology-derived protein structures		•	•	•
DSSP	Secondary structure digests of PDB			•	•
3D_ALI	Structure-based sequence alignments			•	•
SEQANALREF	Sequence analysis bibliography		•	•	•
FANS_REF	Functional analysis bibliography			•	•
BIO_CATAL	Catalogue of mol. biol. software		•		•
JOURNALS	Tables of content of various mol. biol. journals				•
LIMB	Listing of mol. biol. databases		•	•	•
P53	p53 point mutations			•	•
Software	UNIX, OpenVMS, DOS, Macintosh			•	•
General information	Technical information, submission and order forms, etc.		•	•	
Database searches				•	(•)

following describes the query and retrieval systems produced by the EBI for use with the databases on their CD-ROM.

2.1 EBI software for DOS computers

The EBI CD-ROM comes with a query and retrieval system for MS-DOS compatible computer systems called *CD-SEQ*. Hardware requirements are two CD-ROM drives or one drive plus a large hard disk (about 180 megabytes) to hold the index files.

CD-SEQ allows the identification and retrieval of specific entries from the EMBL and SWISS-PROT sequence databases. Queries are entered into input screens as shown in *Figure 1*. The full text of database entries can be searched, or queries can be restricted to various database fields, including entry names, accession numbers, author names, or references. From the resulting hit lists, individual entries can be viewed, printed or exported to disk files.

The functionality of *CD-SEQ* is similar to that one would see in a bibliographic database, allowing identification and retrieval of specific entries. Another perhaps more demanding use of the database is in the location of sequences similar to a given query sequence. The EBI CD-ROM can be used for that purpose in a number of ways. Programs such as *FASTA* (Section 2.3) are sensitive enough to detect distant relationships between biologically related sequences at the expense of being rather slow to run. For many applications it is sufficient, though, to look for close similarities only. The program *EMBLSCAN*, supplied on the EBI CD-ROM, is particularly suited for this task (4). *EMBLSCAN* allows very fast comparisons of nucleotide sequences against the EMBL sequence database; a typical search using a 1000

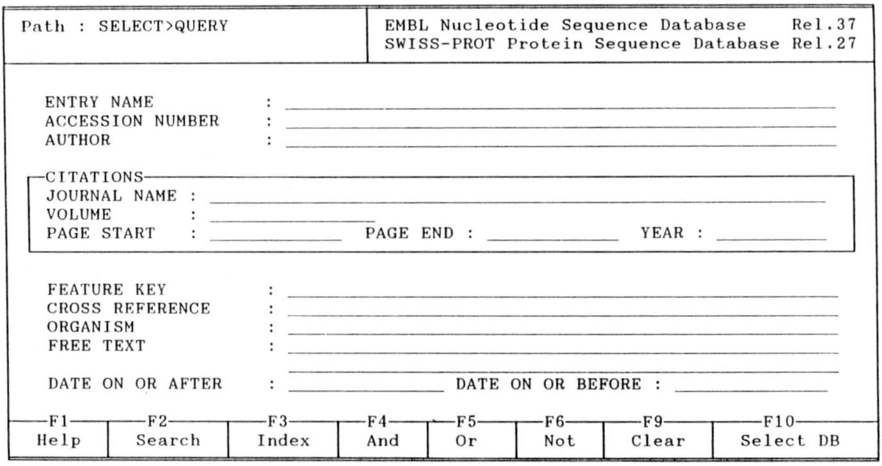

Figure 1. The *CD-SEQ* query software for MS-DOS.

nucleotide query sequence takes only approximately four minutes. This way, quick routine checks of new sequence data are feasible.

2.2 EBI retrieval software for Macintosh computers

Owners of Apple Macintosh computers will find the program *EMBL-Search* (5) useful. It is supplied on the EBI CD-ROM and also available from the EBI network servers. *EMBL-Search* is a general database query and retrieval system for use with the databases on Volumes I and II of the EBI CD-ROM set. It runs on all Apple Macintosh systems with System software version 6 or 7. Two CD-ROM drives are required; one drive is sufficient if the index files used by *EMBL-Search* are moved to a large hard disk (100 megabytes or more).

The present version of *EMBL-Search* supports queries involving the EMBL, SWISS-PROT, PROSITE, ENZYME, and EPD databases. Users can search for entry names, accession numbers, keywords, author names, species names, or free text terms. *Figure 2* illustrates the graphical user inter-

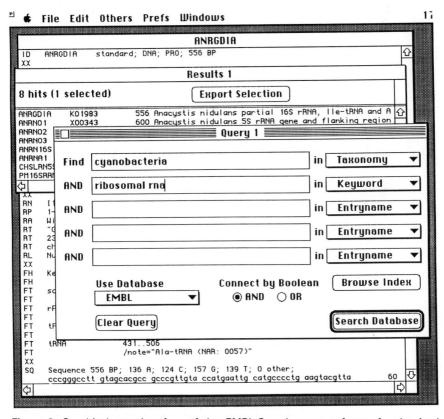

Figure 2. Graphical user interface of the *EMBL-Search* query software for the Apple Macintosh.

face that makes working with *EMBL-Search* easy and intuitive. Any hits found in a query can be viewed on screen, printed, or easily exported in a number of different file formats for further manipulation by other programs.

A powerful feature of *EMBL-Search* is its utilization of cross-references between the supported data collections. This enables researchers to quickly look up related information from several databases. *Figure 3* illustrates the ways in which a user can exploit the cross-reference mechanism. When a database entry is displayed, a special menu contains all pointers from this entry to related information in other databases. In addition to choosing from this menu, the user can also simply double-click on a highlighted cross-references line in the entry text ('HyperLinks'). In either case, a new window shows up displaying the related entry from the other database. This way, researchers can easily navigate between databases and identify related information which otherwise may be overlooked.

EMBL-Search fully supports operation in a networked environment. Thanks to the Macintosh's built-in AppleTalk networking capabilities it is possible to make very efficient use of the EBI CD-ROM by mounting the disks on a central server system so that scientists can access them from their personal Macintosh workstations. This approach is obviously more cost-effective than equipping each individual computer system with CD-ROM drives and buying multiple copies of a CD-ROM product, which is often not feasible for small laboratories. The performance of such a network configuration is dependent on network quality and number of simultaneous users.

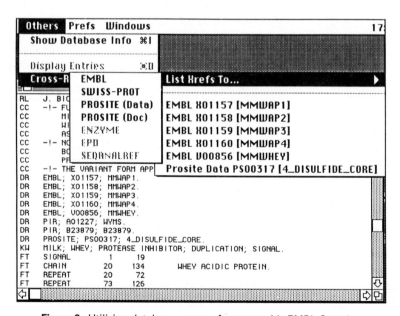

Figure 3. Utilizing database cross-references with *EMBL-Search*.

2.3 Other software

In addition to the software tools produced by the EBI there are a number of third-party products that make direct use of the data on this CD-ROM.

Included on Volume III of the EBI CD-ROM is DOS database management software for use with the *E. coli* Database (ECD) (6) which allows queries of this database on gene names, map location, references, etc., as well as the assembly of map contigs.

The EBI CD-ROM can be used with the popular *FASTA* software (7) to compare a new nucleotide or protein sequence against the EMBL or SWISS-PROT databases in order to identify related sequences (sequence similarity or homology searches). *FASTA* is available for DOS and Macintosh platforms directly from the author (William Pearson, University of Virginia, Box 440, Jordan Hall, Charlottesville, VA 22908, USA).

Other software packages that utilize the EBI CD-ROM for sequence retrieval include the *Staden* package for UNIX systems (Rodger Staden, MRC-LMB) and the *GenMon* package for DOS and OpenVMS platforms (David Lincoln, GBF).

3. Network information services

For many applications, the completeness of a database is essential. Release cycles for molecular biological databases on CD-ROM are typically two or three months, and the rapid growth of today's databases calls for mechanisms which give researchers more timely access to new data. Recent developments in computer networking have enabled database producers to establish database access based on direct computer communication. These methods allow researchers, for instance, to download complete databases at release time as well as retrieve individual entries as soon as they become available (8).

This section briefly describes the network information services available from the EBI. A comprehensive summary of network access to biological information resources is given in Chapter 4 of this book and in ref. 9.

3.1 EBI database and information servers

The EMBL Data Library was one of the first molecular biology database providers offering electronic network access to their products (10). Over time the initial main access mechanism, electronic mail, has been complemented by newer approaches; simultaneously the scope of the Data Library's activities has been extended continuously, and now the EBI archives contain many databases, free software and general biological information (2).

The following describes the network services provided by EBI as well as the information available via these channels. For more information or in case of problems with these services, please contact the EBI at the electronic mail address `nethelp@ebi.ac.uk`.

3.1.1 EBI electronic mail server

EBI's electronic mail server has been in operation since 1985 (10). It is freely available to anybody with access to international computer networks and provides access to most of the databases and free software in the EBI archives. Detailed help files can be obtained by sending an e-mail message to the Internet address `netserv@ebi.ac.uk`, including the word HELP in the body of the e-mail message.

The EBI e-mail server enables scientists to download complete databases, individual entries from nucleotide and protein sequence databases, free molecular biological software, and general information. The EMBL nucleotide sequence database, for example, is updated on a daily basis, and new entries become available on the e-mail server as soon as they are produced by the EBI staff. Combined with the EBIs daily exchange of new sequence data with the collaborating databanks in the USA and Japan, this means that all data from the international nucleotide sequence database collaboration is available in EMBL format this way. For example: a scientist reading about a new sequence of interest in a journal notes that it is cited under the accession number Y12345. A simple e-mail message to `netserv@ebi.ac.uk` with the line `get nuc:y12345` will cause the sequence in question to be returned by e-mail within seconds.

To assist users in identifying database entries of interest, a range of index files, such as author index or keyword index, are updated daily and available on the server. Also, once a week a list of all new or updated entries including a short description of each entry is produced.

In a similar fashion, access to entries from the SWISS-PROT protein sequence database is possible. This database and its associated index files are also updated in regular intervals, typically once a week.

Due to their size, it is not possible to download the complete EMBL or SWISS-PROT databases by e-mail. However, a large number of other, smaller databases of interest to molecular biologists is available this way as illustrated in *Table 1*. In addition to these databases, the EBI e-mail server provides access to general biological information such as files with database growth statistics, lists of network servers, data submission forms, subscription information, etc.

Using the EBI e-mail server it is also possible to access the world's largest collection of free software in molecular biology. More than 200 programs are available for UNIX, OpenVMS, Macintosh, and DOS platforms. These programs range from software for predicting signal cleavage sites to comprehensive sequence analysis packages, from tools for calculating enzyme kinetics to software for linkage analysis, and much more. Listings of software available and instructions for how to download individual programs can be obtained by sending a mail message to `netserv@ebi.ac.uk` with the line `help xxx_software` where xxx should be replaced by unix, vms, dos, or mac.

3.1.2 Anonymous *FTP*

The EBI e-mail server is supplemented by the EBI anonymous *FTP* server, which is available to anybody on the Internet who can utilize the *File Transfer Protocol* (*FTP*). To reach the EBI *FTP* server, researchers should use their *FTP* client software to connect to the address `ftp.ebi.ac.uk`. When prompted for a user name, they should enter the word 'anonymous' and give their e-mail address as the password. New users should check the directory `/pub/help` for further information. Also look at the files `/INDEX` and `/NEWFILES` for a complete list of all files on the anonymous FTP server and a list of all files new or updated within the last two weeks, respectively.

The anonymous *FTP* server gives access to all data available from the EBI e-mail server and more. Complete releases of the EMBL sequence database can be downloaded as soon as they are produced (in `/pub/databases/embl/`). Files with all new and updated entries are provided on a daily and weekly basis, thus providing an easy mechanism for regularly updating local copies of the EMBL database. A new method for retrieving individual database entries by *FTP* is currently being implemented. Similarly, complete SWISS-PROT sequence database releases as well as regular updates are made available for downloading (in `/pub/databases/swissprot/`). *Table 1* shows the range of other databases accessible via the EBI *FTP* server (in `/pub/databases/`). New data collections are constantly being added. Most databases are updated as soon as new releases are published, some are updated more frequently with data between releases.

The anonymous *FTP* server gives easy access to the EBI archive of free molecular biological software (Section 3.1.1). Users should check the subdirectories of `/pub/software/` for software for their particular hardware platform. Each subdirectory contains a `README` file with a description of all programs provided.

3.1.3 Other access methods

The recent development of a new generation of graphically oriented tools significantly alleviates access to network information resources for computer-illiterate life scientists. Recognizing the importance of these new network information retrieval systems, the EBI operates several servers that enable researchers to access the EBI archives with Gopher and World Wide Web client software. See also Chapter 4.

i. Gopher
Recently, the Gopher network information retrieval software (11) has gained popularity among scientists, including molecular biologists. Researchers should direct their Gopher clients to connect to `gopher.ebi.ac.uk`, or use

one of the many existing Gopher 'holes' that point to the 'EMBnet BioInformation Resource EBI'. All data from EBI accessible by anonymous *FTP* are also available under Gopher. Gopher can thus be used as a convenient, user-friendly tool for retrieving databases, software, and related information from the EBI archives.

In addition, the EBI Gopher server is an entry point to a world-wide network of BioGophers. Gopher's capability of providing 'links' or 'Gopher holes' is utilized to guide users from EBI to other places which offer specialized services such as database searches or act as specialized information repositories. An example, access to the SRS database retrieval system, is illustrated in Section 3.2.2.

ii. World Wide Web (WWW)

The latest addition to the EBI network services is a WWW server (12). This WWW server can be contacted using the Universal Resource Locator (URL) `http://www.ebi.ac.uk/`. The EBI's WWW home page offers links to documents describing EBI products and tools, gives access to the EBI anonymous *FTP* and Gopher servers, and provides frontends to EBI sequence analysis services (see below). It also connects to other biological WWW servers around the world.

3.2 On-line database access

While the use of network tools such as e-mail, *FTP*, or Gopher is sufficient for many applications, often researchers find the need for more comprehensive and integrated access to databases and sequence analysis software, which is satisfied by so-called on-line services. Two European examples which are tightly linked to EBI information products are described in this section.

3.2.1 European molecular biology network (EMBnet)

In 1988, EMBL initiated the European molecular biology network (EMBnet) project with a view to establishing an infrastructure of collaborating nodes in most European countries. Each node holds up-to-date databases and provides biocomputing services, including user support and training, to their national academic and commercial constituency. A list of EMBnet national nodes currently in operation is provided in *Table 2*. A national node typically offers access to daily updated sequence databases, a collection of other molecular biological databases, comprehensive sequence analysis software such as the GCG suite (13) or the SRS package (Section 3.2.2), and access to international computer networks and bulletin boards. Many nodes provide services beyond that, often resulting from, and taking into account, local research interests. Most nodes charge only nominal fees. A brochure which describes in more detail the EMBnet project and the activities of the various national nodes can be obtained from any of the sites listed in *Table 2*.

Table 2. European Molecular Biology Network (EMBnet) national nodes

Name	Contact person, fax number, e-mail address
EMBnet node Austria	M. Grabner, +43-1-79515/6108
	`martin.grabner@cc.univie.ac.at`
EMBnet node Belgium	R. Herzog, +32-2-6509762
	`rherzog@ulb.ac.be`
EMBnet node Denmark	H. Møller, +45-86-131160
	`hum@biobase.aau.dk`
EMBnet node Finland	H. Lehuaslaiho, +358-0-4572076
	`heikki.lehuaslaiho@csc.fi`
EMBnet node France	P. Dessen, +33-1-69333013
	`dessen@coli.polytechnique.fr`
EMBnet node Germany	W. Chen, +49-6221-422333
	`dok419@genius.embnet.dkfz-`
	`heidelberg.de`
EMBnet node Greece	B. Savakis, +30-81-231308
	`savakis@myia.imbb.forth.gr`
EMBnet node Hungary	E. Barta, +36-28-330127
	`barta@hubi.abc.hu`
EMBnet node Ireland	A. Lloyd, +353-1-6081969
	`atlloyd@acer.gen.tcd.ie`
EMBnet node Israel	L. Esterman, +972-8-344113
	`lsestern@weizmann.weizmann.ac.il`
EMBnet node Italy	M. Attimonelli, +39-80-484467
	`attimonelli@mvx36.csata.it`
EMBnet node Netherlands	J. Noordik, +31-80-652977
	`noordik@caos.kun.nl`
EMBnet node Norway	L. Akselberg, +47-22958756
	`linda.akselberg@biotek.uio.no`
EMBnet node Poland	P. Zielenkiewicz, +48-2-6584703
	`piotr@ibbrain.ibb.wae.pl`
EMBnet node Portugal	P. Fernandez, +351-1-4431408
	`pfern@pen.gulbenkian.pt`
EMBnet node Spain	J. M. Carazo, +341-585-4506
	`carazo@samba.cnb.uam.es`
EMBnet node Sweden	P. Gad, +46-18-551759
	`gad@perrier.embnet.se`
EMBnet node Switzerland	N. Redaschi, +41-61-2672078
	`redaschi@comp.bioz.unibas.ch`
EMBnet node United Kingdom	A. Bleasby, +44-925-603100
	`bleasby@daresbury.ac.uk`

3.2.2 Sequence retrieval system (SRS)

SRS (14) is a menu driven environment that provides fast integrated access to more than 20 different molecular biological databanks including the daily updated EMBL nucleotide sequence database. A special on-line service enabling researchers to use the SRS software is provided by EMBL and by the Norwegian EMBnet node and is freely available to all scientists on the Internet by using remote login (telnet). Users should open a terminal connection from their local computer to `srs.embl-heidelberg.de` or to `biomed.uio.no`. Instructions and detailed help are available on-line. Gopher and WWW interfaces to the SRS system are currently under development at EMBL and various EMBnet nodes. SRS can also be installed locally under UNIX or OpenVMS. Copies can be obtained from the EBI anonymous *FTP* server. A DOS version is in preparation.

3.3 Remote database searches

Various organizations offer e-mail services for the remote analysis of new sequence data, often using specialized hardware or software not generally available (15; for reviews see Chapter 4 in this book and ref. 9).

The EBI operates three e-mail servers for sequence similarity searches: BLITZ, Mail-FASTA, and Mail-QuickSearch (*Table 3*). Additionally, scientists can use the Mail-PROSITE service for protein function prediction using the PROSITE protein pattern database (16). All services are freely accessible via international computer networks such as the Internet and are activated by sending a normal electronic mail message to a specific Internet address. The message must include the query sequence and can also specify service-specific parameters such as the database to be searched. Since e-mail messages to these servers are processed entirely automatically, special formatting rules must be adhered to. Details of these rules can be obtained by sending an e-mail message containing the single word HELP to the appropriate addresses given in *Table 3*.

As the formatting rules often vary considerably between the different servers, various tools have been developed that assist users in preparing correct service request messages. The EBI product MSU (mail server utility) (17) is available for UNIX and OpenVMS platforms. It prompts the user for all information required for using a particular service, then builds a correctly formatted e-mail message and sends it to the selected server. Results are soon returned into the user's mailbox. A unique virtue of this utility is its use of external service specification files which can be manipulated with any text editor; this way it can be easily extended and adapted to personal needs without changing the program code itself. The distribution contains specifications for more than 25 different services world-wide; copies can be obtained by anonymous *FTP* from `ftp.ebi.ac.uk` as `/pub/software/unix/msu.tar.Z` or `/pub/software/vax/msu.uue`.

4. Contacting the EBI

Detailed information about the databases, software, and other products described in this chapter can be received from the EBI. Communication between information providers such as the EBI and the research community is essential, so for any problems, suggestions, or comments, please do not hesitate to use the contact addresses shown in *Table 4*.

Table 3. EBI remote database search services. (detailed information can be obtained by sending a mail message with the word HELP to the specified Internet address)

Name	Description	Search type	E-mail address
BLITZ	E-mail interface to the MPsrch program of Collins and Sturrock (University of Edinburgh) for ultra-fast sensitive database searches on parallel computers, running on the MasPar massively-parallel computer installed at EMBL. It finds best local similarities of protein query sequences to sequences in the SWISS-PROT protein sequence database. The implementation of nucleotide similarity searches is under way. The method of choice for most applications.	Protein sequence vs. protein sequence database.	blitz@ebi.ac.uk
Mail-FASTA	E-mail interface to the popular FASTA program (7) for finding local similarities of protein and nucleotide sequences against whole databases. Daily updates of the EMBL and SWISS-PROT databases as well as subsets of these databases and other databases such as PIR are available.	Nucleotide sequence vs. nucleotide sequence database or protein sequence vs. protein sequence database	fasta@ebi.ac.uk
Mail-QuickSearch	E-mail interface to an extension of the GCG (13) program QuickSearch. Appropriate for rapidly identifying very similar sequences in the database. Not useful for finding remote similarities. Daily updates of the EMBL database as well as subsets are available.	Nucleotide sequence vs. necleotide sequence database	quick@ebi.ac.uk
Mail-PROSITE	E-mail to the ppsearch program (R. Fuchs, unpublished) for identifying protein motifs and patterns from the PROSITE database in new protein sequences.	Protein sequence vs. PROSITE pattern database	prosite@ebi.ac.uk

Table 4. EBI contact addresses

Postal address

The European Bioinformatics Institute, Wellcome Trust Genome Campus, Cambridge CB10 1SD, UK.
Phone: +44 (1223) 494400
Fax: +44 (1223) 494468

Network addresses

General enquiries	datalib@ebi.ac.uk
Data submissions	datasubs@ebi.ac.uk
Data corrections, notification of data publication	update@ebi.ac.uk
Problems with network services	nethelp@ebi.ac.uk
Problems with software downloading, software submissions	software@ebi.ac.uk

Acknowledgements

Peter Stoehr's contributions were invaluable for establishing CD-ROM and network access to EBI databases. We would also like to thank the EMBnet community for their efforts, in particular Rodrigo Lopez and Reinhard Doelz. The EBI is supported by the European Commission under the BIOTECH funding programme.

References

1. Fuchs, R. and Cameron, G. N. (1991). *Prog. Biophys. Mol. Biol.*, **56**, 215.
2. Rice, C. M., Fuchs, R., Higgins, D. G., Stoehr, P. J., and Cameron, G. N. (1993). *Nucleic Acids Res.*, **21**, 2967.
3. Bairoch, A. and Boeckmann, B. (1993). *Nucleic Acids Res.*, **21**, 3093.
4. Higgins, D. and Stoehr, P. J. (1992). *Comput. Appl. Biosci.*, **8**, 137.
5. Fuchs, R. and Stoehr, P. J. (1993). *Comput. Appl. Biosci.*, **9**, 71.
6. Kröger, M., Wahl, R., and Rice, P. (1993). *Nucleic Acids Res.*, **21**, 2973.
7. Pearson, W. R. and Lipman, D. J. (1988). *Proc. Natl. Acad. Sci. USA*, **85**, 2444.
8. Coulson, A. (1995). *Trends Biotechnol.*, **11**, 223.
9. Fuchs, R. and Cameron, G. N. (1995). In *DNA cloning—complex genomes: a practical approach* (ed. D. Hames and D. Glover), p. 151. IRL Press, Oxford.
10. Stoehr, P. J. and Omond, R. A. (1989). *Nucleic Acids Res.*, **17**, 6763.
11. McCahill, M. (1992). *ConneXions—The Interoperability Report*, **6**, 10.
12. Berners-Lee, T., Cailliau, R., Groff, J., and Pollermann, B. (1992). *Electronic Networking: Research, Applications and Policy*, **2**, 52.
13. Devereux, J., Haeberli, P., and Smithies, O. (1984). *Nucleic Acids Res.*, **12**, 387.
14. Etzold, T. and Argos, P. (1993). *Comput. Appl. Biosci.*, **9**, 49.
15. Henikoff, S. (1993). *Trends Biochem. Sci.*, **18**, 267.
16. Bairoch, A. (1993). *Nucleic Acids Res.*, **21**, 3097.
17. Fuchs, R. (1994). *Comput. Appl. Biosci.*, **10**, 413.

4

Networked services

G. WILLIAMS

1. Introduction

The Internet is a communication network which connects most academic and commercial computers in the world. The available services on the Internet network and their ease of use are increasing all the time. The various genome mapping and sequencing projects have greatly accelerated the use of computers in molecular biology, so that new databases and programs are constantly being released and updated. This chapter examines some of the services available on the Internet and how to access them.

1.1 Logging in to the system

To use e-mail or any of these Internet services, you will probably have to log in to your local institution's computer system. Contact your local computer centre to set up an account. Your account will include two identifying codes: username and password. Once you have an account you must log in to the system from a terminal on the system.

1.2 Computer names

The Internet standard for naming computers and sites is a hierarchical system, naming levels such as computer name, organization, and country, for example: `cray1.mit.edu`. The organizational level becomes more global as you read from left to right. Outside the USA, two-letter country codes are used as the top level, for example: `hgmp.mrc.ac.uk`. Addresses in the USA omit the country name and the following are used as top level names to indicate the type of site:

`edu`	educational institution	`mil`	military site
`com`	commercial organization	`net`	network centre
`gov`	government agency	`org`	all other organizations

2. Electronic mail

2.1 E-mail

Electronic mail (e-mail) allows you to exchange messages with other computer users and to send data to services on the Internet. Electronic mail is one of the most popular uses of the Internet.

Different computers use different software for electronic mail. UNIX systems, for example, may use UNIX mail, *msg*, *elm*, *pine*, or something else. Different software uses different commands, for an example of mail see *Figure 1*. Ask your local computer centre how to use electronic mail on your system.

E-mail addresses are commonly constructed from a site's name preceded by a person's name and an @ (at) symbol, e.g. `jbloggs@jax.org.`

N.B. E-mail addresses on the UK's academic network (JANET) used to be written in the reverse order, so that an address started with the highest level (country or organization) first, e.g. the international address `jenny@oxford.ac.uk` was written `jenny@uk.ac.oxford` when posting from the UK. There may still be historical references to this old system.

2.2 E-mail servers

E-mail can be used as a general purpose tool for submitting simple enquiries to a database or to a public software resource. Known as e-mail or archive servers, these systems are special e-mail addresses that have been established to accept requests in the form of an e-mail message. Any response to the request is then returned to the originator, also by e-mail. These e-mail

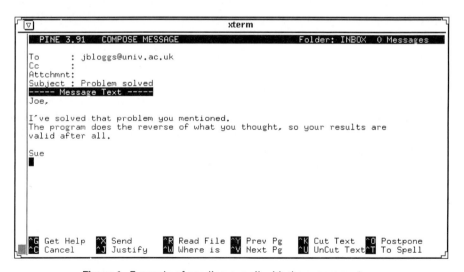

Figure 1. Example of sending e-mail with the program *pine*.

servers can be accessed by most e-mail users even if they do not have direct access to the Internet.

E-mail servers tend to be the results of academic projects that can be subject to funding problems and so may cease to exist. The servers at EBI, EMBL, and NCBI should be stable. A list of e-mail servers is given in *Table 1*.

Before using an e-mail server, you should check its usage and capabilities by obtaining its help documentation. This is usually done by sending a message containing just the word 'HELP' to the server address. The server will then return the documentation via e-mail. It is generally best to leave the 'Subject:' line of the mail message blank.

Some e-mail servers can get overloaded, so do not get impatient with delays of 48 hours or so before receiving a reply.

The program *msu* is recommended if you are going to use e-mail servers a lot. It presents the user with a menu of services and eases the construction of the request e-mail message in the correct format.

Copies of *msu* can be obtained by anonymous FTP from `ftp.ebi.ac.uk` (files: `/pub/software/unix/msu.tar.Z` and `/pub/software/vax/msu.uue`) or from the e-mail server `netserv@ebi.ac.uk` (send `get unix_software:msu.uue` or `get vax_software:msu.uue` in the e-mail message).

3. *File transfer protocol (FTP)*

3.1 *FTP*

The standard program for transferring files across the Internet is called *FTP*. This is used to establish the connection to a named remote computer and then files and programs can be transferred across the network by typing simple commands.

FTP is started by typing '*ftp*' followed by the name of the machine you wish to contact. You will be prompted for a valid account name and password.

Some sites, known as 'anonymous *FTP* servers' provide public access to collections of files or programs. When you use *FTP* to reach an anonymous *FTP* server you should give 'anonymous' as your username and your e-mail address as the password. An example session is shown in *Figure 2*.

Some useful commands when using *FTP* are:

(a) `help` to display help on using *FTP*.

(b) `ls` or `dir` to see a directory listing of files available on the remote system.

(c) `cd` (followed by a subdirectory name) to change directories on the remote machine. Usually the main public files are found under the `pub` subdirectory, but this is not always the case. You may also have to search in other directories.

Table 1. Useful e-mail servers

E-mail address	Description
pythia@anl.gov	Identification of repetitive elements and Alu subfamilies
bioscan@cs.unc.edu	Similarity search on a massively parallel machine
blitz@ebi.ac.uk	Protein similarity search on MasPar machine
dapmail@ed.ac.uk	Similarity searches using the Edinburgh DAP
q@ornl.gov	Similarity searches using *FASTA, BLAST, FLASH* etc.
dflash@watson.ibm.com	Similarity search on an IBM machine
blast@genome.ad.jp	Similarity search using the program *BLAST*
blast@ncbi.nlm.nih.gov	Similarity search using the program *BLAST*
fasta@ebi.ac.uk	Similarity search using the program *FASTA*
mfasta@genius.embnet. dkfz-heidelberg.de	Similarity search using the program *FASTA*
fasta@genome.ad.jp	Similarity search using the program *FASTA*
flat-netserv@ smlab.e g.gunma-u.ac.jp	Server for sequence databases and *FASTA* search
fileserv@ nbrf.georgetown.edu	Server for the PIR databases and *FASTA* search
bioserve@t10.lanl.gov	Server for software, documentation and data
netserv@ebi.ac.uk	Server for databases and software at EMBL
netserv@genius.embnet. dkfz-heidelberg.de	Server for EMBL/SWISS-PROT databases
retrieve@ncbi.nlm.nih.gov	Server for database entries at NCBI
est_report@ncbi.nlm.nih.gov	Server for entries in dbEST database
server@rdp.life.uiuc.edu	Ribosomal RNA sequence software and data
mailserv@gdb.org	Server for the GDB and OMIM databases
genome_database@ genome.wi.mit.edu	Server for the Whitehead/MIT mouse map data
waismail@net.bio.net	Text searching of the BIOSCI WAIS archives
blocks@howard.fhcrc.org	Protein/DNA comparison to blocks of conserved protein regions
sbase@icgeb.trieste.it	Similarity search of database of protein domains
motif@genome.ad.jp	Search for protein motifs in protein sequence
prodom@toulouse.inra.fr	Search for protein domains in protein sequence
cbrg@inf.ethz.ch	Various similarity searches or mass comparison
mowse@daresbury.ac.uk	Peptide mass fingerprint search of protein databases
nnpredict@celeste.ucsf.edu	Protein secondary structure prediction
predictprotein@ embl-heidelberg.de	Protein multiple sequence alignment and secondary structure prediction
netgene@virus.fki.dth.dk	Intron/Exon splice site prediction
genmark@ford.gatech.edu	Finds protein coding regions in DNA of various species
grail@ornl.gov	Finds protein coding regions in human DNA
ftpmail@ ftp.uni-stuttgart.de	Server for *FTP* by mail to any site
ftpmail@sunsite.unc.edu	Server for *FTP* by mail to any site

```
┌─────────────────────────────────────────────────────────────────────┐
│ ▽                              xterm                                  │
├─────────────────────────────────────────────────────────────────────┤
│ Unix % ftp ncbi.nlm.nih.gov                                          │
│ Connected to ncbi.nlm.nih.gov.                                       │
│ 220-Welcome to the NCBI FTP Server (ncbi.nlm.nih.gov)                │
│ 220-                                                                  │
│ 220 ncbi FTP server (Version wu-2.4(2) Mon Apr 18 13:33:40 EDT 1994) ready. │
│ Name (ncbi.nlm.nih.gov:hgcs): anonymous                              │
│ 331 Guest login ok, send your complete e-mail address as password.  │
│ Password:                                                            │
│ 230 Guest login ok, access restrictions apply.                      │
│ ftp> verbose                                                         │
│ Verbose mode off.                                                    │
│ ftp> cd repository/blocks                                            │
│ ftp> dir                                                             │
│ total 14                                                             │
│ drwxr-sr-x    4 4104     506          512 Aug  1  1993 .             │
│ drwxr-xr-x   47 0        0           1024 Jan 23 19:51 ..            │
│ -rwxr-xr-x    1 0        506          158 Aug  1  1993 .cache        │
│ -rw-r--r--    1 4104     506         1291 Nov 24  1993 README        │
│ drwxr-sr-x    5 4104     506          512 Aug 10  1994 dos           │
│ drwxr-sr-x    7 4104     506          512 Aug 10  1994 unix          │
│ ftp> get README                                                      │
│ ftp> quit                                                            │
│ Unix % ▯                                                             │
└─────────────────────────────────────────────────────────────────────┘
```

Figure 2. Example of using *FTP* to get the file README from the directory repository/blocks at the site ncbi.nlm.nih.gov.

(d) binary if you are going to be transferring any files that are not simple printable ASCII files. This includes any form of compressed file or executable program.

(e) put and get (followed by a filename) to send and receive a single file.

(f) mget and mput (followed by a wild-carded filename) when transferring multiple files. You should use * as a wild-card character in the filename you give.

(g) bye or quit to logout of the remote system.

3.2 File formats

There are many programs in common use that compress files or group collections of files together. These programs generate specially formatted files which usually have a standard ending to the filename to indicate what format they are in.

If a file has been compressed in any way, it should be treated as a binary file and you should give the command 'binary' in an *FTP* session before transferring the file.

The following are some common file endings:

(a) .hqx—the file is an Apple Macintosh file compressed with the *binhex* utility.

(b) .tar—several files have been grouped into one file by the UNIX program *tar*. Use the command tar -xvf filename.tar to extract the constituent files.

(c) .uue—the file has been coded using the program *uuencode*. Use the program *uudecode* to restore to normal.

(d) .gz—the file has been compressed with the *gzip* utility. Use *gunzip* to restore to normal.

(e) .z—the file has been compressed with the standard UNIX *compress* command. Use the *uncompress* command to restore to normal.

(f) .ps—the file is a PostScript graphics file and requires a PostScript printer or viewer in order to display it.

In some cases, a series of compression and grouping of files has been performed. The file should be processed in the order of the last part of the file. For example the file `program.tar.z` should first be uncompressed and then *tar* should be used to unpack the component files.

3.3 Archie

If you wish to obtain a file from somewhere by *FTP*, but do not know where to get it from, the service `archie` can be used to search for the file amongst all *FTP* sites.

Give the command `telnet` followed by one of these names:

`archie.au`	Australian server
`archie.doc.ic.ac.uk`	United Kingdom server
`archie.funet.fi`	Finnish server
`archie.internic.net`	AT&T server, NY (USA)
`archie.rutgers.edu`	Rutgers University (USA)
`archie.sura.net`	SURAnet server MD (USA)
`archie.th-darmstadt.de`	German server
`archie.uqam.ca`	Canadian server
`archie.wide.ad.jp`	Japanese server

Use the name of the site closest to you. Give the name `archie` when prompted for a login name and then give the command `help`. The command to search for a file is `prog` followed by the file name.

A list of selected anonymous *FTP* servers is given in *Table 2*.

4. Remote log in

4.1 Telnet

Telnet is a program that allows a person at one site to work on a computer at another site. It is the Internet standard protocol for remote log in.

The command `telnet` followed by the address of a machine on the Internet starts a log in dialogue. If you have a valid account name and password

Table 2. Selected anonymous *FTP* servers

Organization	Machine name	Comments and description
NIC, Finland	`ftp.funet.fi`	General repository of programs. See directory `pub/sci/molbio`.
IUBio, USA	`ftp.bio.indiana.edu`	Molecular biology data and software. Read the `Readme` file.
DDBJ, Japan	`ftp.nig.ac.jp`	Data archive for Japan. Read the `About_DDBJ/Outline` file.
NCBI, USA	`ncbi.nlm.nih.gov`	Major data and software server site.
Houston, USA	`ftp.bchs.uh.edu`	Molecular biology software in directory `pub/gene-server`.
Whitehead Inst, USA	`genome.wi.mit.edu`	Mouse genetic map data and software.
Imperial Coll, UK	`sunsite.doc.ic.ac.uk`	Repository of general software.
EBI, EMBL	`ftp.ebi.ac.uk`	Major data and software server site.
Genethon, France	`ftp.genethon.fr`	Data from Genethon.
TIGR, USA	`ftp.tigr.org`	EST data.
Weizmann Inst, Israel	`bioinformatics.weizmann.ac.il`	Molecular biology databases and programs.

on that machine, you will start a session on that machine. The session may start a database or a general user interface.

The following are some useful addresses to telnet to.

4.2 BIDS

Address: `bids.ac.uk`

This provides access to the Excerpta Medica bibliographic database, the ISI science citation database and to the British Library 'Inside Information' service.

This service is only available to UK users. To register, e-mail: `bidshelp@bath.ac.uk`.

4.3 MEDLARS

Address: `medlars.nlm.nih.gov`

MEDLARS is the MEDical Literature and Retrieval System, an online information service of the National Library of Medicine. This system contains mainly bibliographic medical information on over 20 different databases, including MEDLINE, the Physicians Data Query, and TOXNET.

There is a charge for the use of this service. To register, mail the following address for the 'Online Services Application'.

MEDLARS Management Section, National Library of Medicine, 8600 Rockville Pike, Bethesda, MD 20894, USA
Tel: 800-638-8480
Fax: 301-496-0822

For help in establishing accounts and answering questions, mail: `gmhelp@gmedserv.nlm.nih.gov` or see the WWW address: `http://igm.nlm.nih.gov/`

4.4 MSDN

Address: `bdt.org.br`

The Microbial Strain Data Network (MSDN) is an international non-profit making organization. It provides databases and bulletin boards covering microbiology, biotechnology, genetics, and biodiversity information. Databases describe hybridomas, cell lines, and molecular probes as well as micro-organisms.

For details of MSDN registration and fee, contact either:

MSDN Secretariat, 63 Wostenholme Rd, Nether Edge,
Sheffield S7 1LE, UK
Tel. +44 114 258 3397
Fax +44 114 258 3402
E-mail: `msdn@sheffield.ac.uk`

or

Bioinformatics Department, ATCC, 12301 Parklawn Drive,
Rockville, MD 20852, USA
Tel. +1 301 881 2600
Fax +1 301 816 4363
E-mail: `lynn@atcc.org`

5. Mailing lists and network news

Electronic mail was originally set up to provide communication between two people. It is also used to send messages to lists of people as a means of disseminating news and discussions on a specific topic.

Unfortunately subscribing to a mailing list can lead to a user being flooded with e-mail messages, hiding personal mail. Bulletin boards were developed as a better way for groups of people to discuss ideas. Network news (sometimes referred to as Usenet) is such a system. Whereas mailing lists send e-mail to individual people, network news articles are circulated to computer sites for any user at that site to read.

5.1 Mailing lists

Several hundred special interest groups or mailing lists are available for subscription. They are all free and most are open to anyone interested in the topic under consideration. Some lists have little message traffic on them; others are so active you can be inundated with messages.

Remember to keep a record of mailing lists you have subscribed to and to save any instructions you receive about unsubscribing from a mailing list. It is a good idea to unsubscribe from a mailing list before going on holiday to prevent your mail directory from overflowing and rejecting mail.

Many mailing lists are handled by a program called *LISTSERV*. Commands are sent to *LISTSERV* via e-mail. The following are a collection of some *LISTSERV* mailing lists.

`genetics@indycms.iupui.edu`	Clinical human genetics
`nihggc-l@ubvm.cc.buffalo.edu`	NIH Grants and Contracts Distribution List
`pdb-l@pdb.pdb.bnl.gov`	The Brookhaven Protein Data Bank mailing list
`hum-molgen@nic.surfnet.nl`	Human (Molecular) Genetics Literature

To obtain a directory of academic mailing lists, maintained by Diane Kovacs, send the message `get acadlist readme` by e-mail to `listserv@kentvm.kent.edu`.

Once you know the name and address of a *LISTSERV* mailing list you can e-mail a command to the *LISTSERV* address asking to subscribe to the list. Once you are subscribed to a mailing list mail will be sent to your address automatically each time someone posts a message to that mailing list. This will continue until you remove your address from the subscription list.

To obtain a summary of *LISTSERV* commands, send the message `send listserv refcard` by e-mail to any *LISTSERV* address, for example `listserv@finhutc.hut.fi`.

To subscribe to a mailing list, for example `bnfnet-l@finhutc.hut.fi`, you must e-mail a message `subscribe bnfnet-l your-name` to the address of the listserver, in this case `listserv@finhutc.hut.fi`.

To send messages to people subscribed to a mailing list, send e-mail to the address of the mailing list, for example `bnfnet-l@finhutc.hut.fi`.

To leave this mailing list, e-mail the message `unsubscribe bnfnet-l` to the listserver address, `listserv@finhutc.hut.fi`.

Not all mailing lists are controlled by *LISTSERV* programs. There is a large group of biological mailing lists called BIOSCI which swap messages with the network news `bionet` groups (see Section 5.2).

The BIOSCI and network news messages are identical; you should only subscribe to a BIOSCI mailing list if you cannot obtain access to network news.

Detailed information on BIOSCI mailing lists can be found under WWW at `http://www.bio.net/BIOSCI/biosci.FAQ.html` or can be obtained by anonymous *FTP* from `net.bio.net` in the directory `/pub/BIOSCI/doc` in the files `biosci.FAQ`, `biosci-us.infosheet` and `biosci-uk.infosheet` or you can request the BIOSCI information sheet by e-mail from `biosci-help@net.bio.net` if you are in the Americas and the Pacific Rim countries, or from `biosci@daresbury.ac.uk` if you are in Europe, Africa, or Central Asia.

5.2 Usenet/network news

Network news consists of many thousands of news groups organized in hierarchies under main subject areas like `bionet`, `comp`, `rec`, and `sci`. These news groups cover a wide range of subjects—a small random sample of which gives:

`alt.college.food`	`bionet.molbio.ageing`
`bit.listserv.mednews`	`comp.sources.mac`
`rec.food.cooking`	`rec.humor.funny`
`sci.bio`	`soc.penpals`

When articles are posted to a news group they are passed between computer sites and are stored by them. The articles are then available for reading by anyone at these sites. Articles are expunged after a period of time set by the local news administrator.

There are many software packages available for reading and posting articles (e.g. *News Watcher*, *ANU-NEWS*, *vnews*, *tin*, *nn*, *trn*, and *xrn*; see *Figure 3*). Ask your local computer centre how to use the news software on your system—they are mostly easy to learn and convenient to use.

You may read articles in any news group that interests you. The news-reading software will generally present you with a selection of news groups to read and will then display any unread articles in the groups that you have selected. It will keep a record of which articles you have already read. Some news-reading software allows you to easily follow a thread of discussion on a single subject within a news group.

You may post replies about an article, either privately via e-mail or publicly to the news group. You should be polite when posting to a news group, large numbers of people may read your article. It is generally a good idea to read the messages in a news group for a few days in order to get a feel for the tone of the discussions before posting to that group.

Biologists are well-served by network news. There is a set of about 50 news groups under the `bionet` news group hierarchy specifically intended for use by professional biologists. If the `bionet` news groups are not taken by your local site, urge your local news administrator to add them.

It has been estimated that well over 10 000 people read the `bionet` news-

```
                            xrn – version 6.18
+ 24911 IMPORTANT - BIOSCI MAIL QUEUE BACK TO NORMA   [21] BIOSCI Administrator
+ 24912 5'-RACE PCR for missing cDNA ends?            [16] DREWES@MPASMB.DESY.D
+ 24913 Re: core mills corp                           [35] Martin D. Leach
+ 24914 alfalfa mosaic virus enhancer?                 [8] Bill
+ 24915 Re: y                                         [15] Bill
+ 24916 Re: RT-PCR with Dynabeads                     [17] jm
+ 24917 Re: buying a thermocycler                     [12] jm
  24918 Reference on urea breakdown                   [11] John Altman
  24919 Re: CsCl vs column kits                       [20] Bob Kodrzycki
  24920 Ligation of large plasmids                    [15] Bob Kodrzycki

          Questions apply to current selection or cursor position
```

Quit	Next unread	Next	Scroll forward	Scroll backward
Scroll line forward	Scroll line backward	Scroll to end
Scroll to beginning	Prev	Last	Next group	Catch up	Fed up	Goto article
Mark read	Mark unread	Unsubscribe	Subject next	Subject prev
Session kill	Local kill	Global kill	Author kill	Subject search
Continue	Post	Exit	Checkpoint	Gripe	List old

```
Newsgroups: bionet.molbio.methds-reagnts
Subject: Re: buying a thermocycler
Date: 1 Apr 1995 01:45:33 GMT
Organization: onramp.net
Lines: 12
Message-ID: <joe1-3103951947030001@dal31.onramp.net>
References: <anne.1.00168D58@lmb1.rug.ac.be>
NNTP-Posting-Host: dal31.onramp.net


> Has anyone  experience with the Peltier thermocycler PTC-200 from
> MJ Research?  Since the cycler is not yet very often used in Europe, we
> especially want to know if  the cycler is reliable and PCR will stay
> reproducible throughout the years (meaning that the timeperiod of one PCR
> round should be the exactly the same in years) .

I have just started using it, and so far it is a fine machine. I share it
with another lab which has been doing alot of pcr over the past year. The
manual is good and a quick read.
```
```
Article 24917 in bionet.molbio.methds-reagnts (1046 remaining) (Next group: embnet.ge
```

| Save | Reply | Forward | Followup | Cancel | Rot-13 | Toggle header | Print |

Figure 3. Example of using *xrn* to read the news group bionet.molbio.methds-reagnts.

groups. They provide conference announcements, information, and help on sources of data and software, access to a large number of experts in various fields, and discussions of methods, techniques, and current topics.

A sample of the most read bionet news groups include:

bionet.announce	Announcements
bionet.general	General discussions
bionet.genome.arabidopsis	Arabidopsis genome program
bionet.info-theory	Information theory applied to biology
bionet.jobs	Job opportunities in biology
bionet.molbio.evolution	Discussions of evolution
bionet.molbio.genbank	Genbank database

`bionet.molbio.methds-reagnts`	Discussions & tips on lab methods
`bionet.molbio.proteins`	Molecular biology of proteins
`bionet.neuroscience`	Discussions of neuroscience
`bionet.software`	Software used in biology
`bionet.users.addresses`	Help in locating biologists

6. Information servers

There are many ways of providing information that require you to run a specific program that talks to remote systems. Two of the most common are explained below.

6.1 Gopher

Gopher presents a text-only menu which allows you to go into sub-menus or display files from the menu. This is a browsing tool that allows you to move into unknown areas and find novel information sources. The use of links between the menus presented by gopher servers allows any gopher server to display information that actually exists at other sites, so can easily move from one gopher server to another by selecting items from the gopher menu.

Many sites provide gopher menus describing their work and offering access to data. Some interesting gopher servers are listed in *Table 3*.

The gopher software can be obtained by anonymous *FTP* from: `boombox.micro.umn.edu` in the directory `/pub/gopher`.

6.2 WWW

The World Wide Web (WWW) is a hypertext system for presenting text and graphical information, allowing you to browse widely amongst an extraordinary range of topics. Like Gopher, which it largely supersedes, it can provide links to other sites so that data can be integrated from many sources. It also provides easy access to gopher, network news, and other internet services.

WWW is impressive in its ease of use and is second only to e-mail in its usefulness.

WWW can display mixed text and graphics. It can present forms to be filled in to request services or run remote programs and databases. Displayed graphics can be made clickable to allow maps or diagrams to be context sensitive.

The recommended means of access to the WWW are either the *Mosaic* or *Netscape* programs—these run on X-Windows, Macs, and PCs (MS-Windows). *Figures 4* and *5* show examples of these two programs.

If you can only display text on your terminal, then WWW browsers such as *lynx* can still present WWW information in an acceptable manner.

Table 3. Some gopher menus

Organization	Machine name	Comments and description
Genethon, France	`gopher.genethon.fr`	Maps in Postscript
JHU, USA	`gopher.jhu.edu`	Genome Database and OMIM
HGMP-RC, UK	`gopher.hgmp.mrc.ac.uk`	General gopher
NIH, USA	`gopher.nih.gov`	NIH services and gopher
SEQNET, UK	`seqnet.dl.ac.uk`	General gopher
ANU, Australia	`life.anu.edu.au`	Biological and medical data

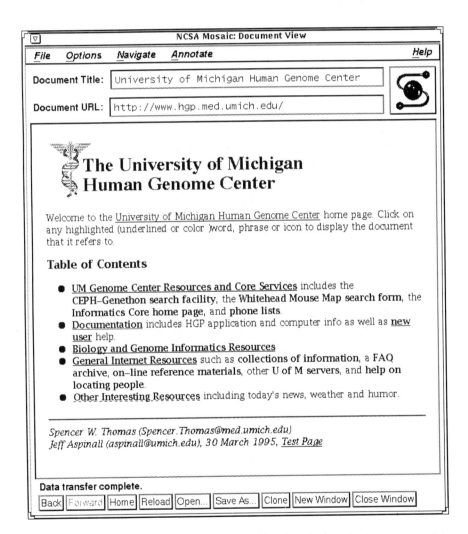

Figure 4. Example of using the WWW browser *Mosaic* to look at the home page of the University of Michigan.

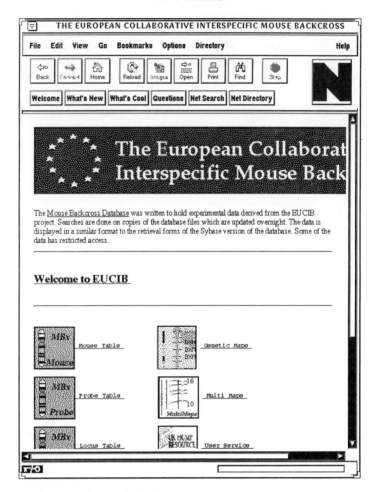

Figure 5. Example of using the WWW browser *Netscape* to look at the European Collaborative Interspecific Mouse Backcross database.

The address used in WWW is known as a URL (Uniform Resource Locator). It has the general form: `type://site_address/path_to_file`, where, among other things, `type` can be one of: `http`—indicating a WWW site, or `gopher`—a gopher site. The `site address` is the Internet address of the machine that provides the WWW information and `path to file` specifies which document or service is being selected. It is common for the `path to file` part of the address to be omitted which, by default, should select the home page for that site.

A list of selected WWW services is given in *Table 4*. WWW services are dependent on resources at the sites providing them and will change with time.

Table 4. Selected WWW servers

WWW address	Comments and description
`http://www.w3.org/pub/WWW/`	The home of the WWW
`http://www.lib.umich.edu/chhome.html`	Subject-oriented Internet guides
`http://www.yahoo.com/`	WWW indexing system
`http://cuiwww.unige.ch/`	Search for topics and sites on WWW
`http://golgi.harvard.edu/`	Large list of genome and sequence data
`http://golgi.harvard.edu/biopages.list`	Large list of biology WWW sites
`http://golgi.harvard.edu/sequences.html`	Search sequence databases
`http://life.anu.edu.au/molbio.html`	Molecular biology databases and sites
`http://www.data-transport.com/`	List of biotechnology companies
`http://www.hgmp.mrc.ac.uk/`	Collection of genome sites and data
`http://www.ebi.ac.uk/srs/srsc/`	SRS database browser at EBI
`http://siva.cshl.org/`	QUEST protein database centre
`http://www.gdb.org/hopkins.html`	Protein databases at JHU
`http://expasy.hcuge.ch/`	Protein sequence analysis
`http://moulon.inra.fr/`	ACeDB-style genome data
`http://www.biotech.ist.unige.it/`	Cell lines database
`http://gdbwww.gdb.org/`	Genome database (GDB)
`http://www.cco.caltech.edu/~mercer/` `htmls/rodent_page.html`	Mouse and rat research information
`http://cancer.med.upenn.edu/`	Oncology information
`http://alces.med.umn.edu/VGC.html`	Sequence analysis programs
`http://bioinformatics.weizmann.ac.il:70/`	Software and databases
`http://atlas.nlm.nih.gov:5700/Entrez/` `index.html`	*Entrez* (search Medline and sequences)
`http://www.bio.net/`	Archives of Bionet news articles
`http://www.ncbi.nlm.nih.gov/`	National Center for Biotechnology Information
`http://www.ebi.ac.uk/`	European Bioinformatics Institute (EBI)
`http://www.chlc.org/`	Cooperative Human Linkage Center (CHLC)
`http://www.cshl.org/`	Cold Spring Harbor laboratories
`http://www.nih.gov/`	NIH grants and information
`http://www.oup.co.uk/`	Oxford University Press
`http://www.netspace.org/MendelWeb/`	Commentaries on Mendel's papers
`http://www.millipore.com/`	Millipore catalogue

It is usual for WWW browsing software to provide a 'hotlist' (also known as bookmarks)—an easy way for you to store and recall the addresses of interesting sites.

The *Mosaic* software can be obtained by anonymous *FTP* from: `ftp.ncsa.uiuc.edu` in the directory `/Web`.

The *lynx* software can be obtained by anonymous *FTP* from: `ftp2.cc.ukans.edu` in the directory `/pub/WWW/lynx`.

The *Netscape* software can be obtained by anonymous *FTP* from: `sunsite.doc.ic.ac.uk` in the directory `/computing/information-systems/WWW/Netscape`. *Netscape* is a product of Netscape Communications Corporation; read the licence before use.

7. Further information

For a thorough overview of the Internet from a biologist's point of view, see *A biologist's guide to internet resources* by Una Smith. This document is available via anonymous *FTP* from `ftp.warwick.ac.uk` in the directory `/pub/archive/news.answers/biology/guide/` split into several files.

A good introduction to the Internet is given in Krol (1).

References

1. Krol, E. (1992). *The whole Internet user's guide and catalog.* O'Reilly & Associates, Sebastopol, CA.

DNA sequencing methodology and software

WILLIAM D. RAWLINSON and BARCLAY G. BARRELL

1. Introduction

Improvements in the ease and speed of sequencing DNA have been the result of simplified methods for DNA template preparation, the use of technically less complex sequencing reactions, simplified preparation of polyacrylamide gels, and computer software that has allowed easier handling of large amounts of sequence data. Improvements in each of these steps have mostly been incremental, and this chapter will describe the changes to some of these stages, concentrating on the currently available software for sequence data acquisition and handling. As the majority of our experience is with the *Staden* package (available from Rodger Staden on e-mail `rs@mrc-lmb.cam.ac.uk`), details of the use of this software will be described here.

Although several other packages are available, we have successfully used the current *Staden* software (and earlier versions) to assemble data from the 230 kilobase (kb) human cytomegalovirus (HCMV) genome (G + C content 57.2%) sequencing project, and also to compare shorter sequences generated from templates made using the polymerase chain reaction (PCR) with the completed sequence. More recently, we have used the software to assemble data acquired from sequencing the 230 kb murine cytomegalovirus (MCMV) genome (G + C content 58.7%), which was done as a single 230 kb 'shotgun' project. Shotgun sequencing is a strategy for DNA sequencing where a (generally large) target DNA (either from subclones in lambda or cosmid vectors (1), or from whole organisms such as purified virion DNA from a DNA virus (2)) is randomly broken into fragments (by sonication, DNase I digestion, or multiple restriction enzyme digestion). These random fragments are subcloned (into sequencing vector such as M13), single-stranded DNA produced, and used as sequencing template. Sequences are then entered into the database and overlapping sequences joined using the sequence assembly software. Entry of progressively more sequences into the database allows production of longer 'contigs', and joining and editing of these contigs results in production of a completed sequence that can be output as a consensus file.

A consensus (sequence) is the final sequence produced by overlapping gel readings. Individual gel readings may differ from the consensus because of:

(a) Gel reading errors due to poor quality data, or stops introduced by the enzyme pausing on the template.

(b) Compressions, usually due to regions of high G + C content that have some secondary structure.

(c) Cloning artefacts (such as from two different but abutting blunt-ended sequences inserted in tandem within the same sequencing vector).

(d) Alteration of inserts by cloning vector enzyme systems.

(e) Variation in the biology of the organism, with corresponding changes in the DNA sequence.

(f) Errors introduced by the sequencing reactions (due to misincorporation of nucleotides by sequencing enzymes).

(g) Human errors introduced by mistakes in editing of the database.

Also, depending on the cause of the mistake, gel reading errors may differ between different strands of the genome, and in some cases between different sequencing reactions (such as (f) and (g) above), or they may remain constant between different sequencing runs (with the same template).

   The technology available to acquire DNA sequence is continually evolving. Recent developments in DNA sequencing technology include machines utilizing fluorescently labelled DNA templates that are visualized using an excitatory laser, with the fluorescent signal being read by photomultiplier tubes. There are three machines currently in common use. The Applied Biosystems (ABI) machine utilizes four different fluorescent dye-labelled primers or dye-labelled dideoxy terminators loaded and scanned in a single lane, and sequencing reactions using the enzymes *Taq* polymerase or T7 polymerase (Sequenase). The machines from Pharmacia (the ALF machine) and from Millipore scan fluorescein-labelled samples loaded in four lanes (ACGT). Alternatively, conventional films can be scanned, and the base calling generated by software on a computer running the Unix operating system. Hardware and software to do this are currently available from Amersham (the Base Scanner), Protein and DNA Imageware (PDI), and Bio-Rad. The autoradiographs are stored as images (of the film), as well as densitometric traces (for each gel reading). Each gel reading scanned in using the Amersham film reader has the raw output from the densitometric traces and the original gel image accessible from within the database.

2. DNA sequencing methods

When determining the sequence of any DNA template, it is important to use the optimum experimental strategy at each step, in order to minimize the cost per base and maximize the efficiency of sequence output from individual workers. For example, the optimum insert length is longer when sequencing using fluorescence-based machines (that are able to generate longer gel reads than conventional radioactive sequencing), and when intending to carry out primer-walking for gap closure. The sequencing strategy will depend on the easiest methods for preparation of the DNA used to produce the sequencing template, the nature of the target DNA, the size and form of the target DNA (PCR product, genomic, or cloned), and the available sequencing technology. Consideration must also be given to the final accuracy of the sequence (that is whether the entire sequence will be studied, or if only a part of the sequence is of interest, such as where specific point mutations are sought), as well as the technology available locally for the sequencing project.

The dideoxy sequencing reactions utilized in most commonly used protocols remain much the same as those originally described (3). DNA is primed with an unlabelled primer, and a DNA polymerase (such as *Taq* polymerase or Klenow fragment of DNA polymerase I) is used to produce complementary copies of the DNA, which incorporate a label that can be visualized in some way. The incorporation of dideoxy nucleotides (ddNTPs) randomly terminates the elongating DNA, and these labelled fragments can be separated on a polyacrylamide gel (1). Changes in the way the reactions are performed (such as the use of fluorescence or radioactively labelled primers), and the use of new enzymes (such as *Taq* polymerase) have made significant improvements in the length and quality of gel reads available from the newer sequencing machines (4). The use of thermal cyclers able to take 96-well microtitre plates, has made these reactions easier to perform for large numbers of samples (5). Other changes (such as the use of robots, multichannel electronic pipettors, and single tube sequencing reactions) have also helped to speed the process of performing the sequencing reactions. Sequencing reactions are most easily performed in microtitre trays (rather than tubes) as we have described elsewhere in detail (1, 5, 6). This allows more samples to be sequenced at one time, easier integration with other hardware (such as robots and multiple-channel pipettors), and simplifies sample storage and record-keeping (see *Figure 1*).

The basic equipment needed for sequencing is either a fluorescence-based sequencing machine, or apparatus for performing conventional ^{35}S radiolabelled sequencing (1). The addition of a film scanner allows the user to avoid much of the time-consuming task of gel reading, and permits display of the raw data from within the assembly software (see Section 4.1). There are a number of emerging sequencing technologies such as capillary electrophoresis (7), and hybridization with arrays of oligonucleotides (8) that are currently being developed commercially. These techniques provide significantly different ways of dealing with difficult sequences (such as compressions), and may in the future be integrated with or replace some forms of sequencing, particularly if they output data in a format recognized by the current soft-

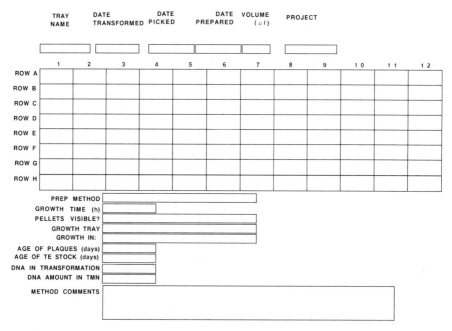

Figure 1. This form is used to record information relating to DNA templates stored in microtitre trays at −20 °C.

ware. As these techniques are not currently in widespread use, they are not discussed further here.

Throughout this chapter, several terms are used frequently when referring to sequencing software. Some of these are defined in *Table 1*, and others are defined in the main body of the text.

3. Sequence handling software and sequence project design

The design of any sequencing project depends on the ultimate aim of the study. If a large genome is being sequenced, then a strategy such as the 'shot-gun' approach (see Section 1) may be used (1). For shorter sequences (such as single genes, where the length is of the order of 1–5 kb), primer-walking may be a more appropriate strategy. The primer-walking strategy involves designing oligonucleotide primers congruent with sequences at the known ends of the template DNA, then performing sequencing reactions using these primers and appropriate templates, to determine the unknown sequences adjacent to the ends of the known sequence. This procedure is then continued, using the new sequence information gained.

Sequence assembly requires hardware (currently we use computers

Table 1. Definitions of some terms used when referring to sequence assembly packages

Clipping	The hiding of certain parts of sequences (such as poor quality data and vector sequences) that are able to be viewed by the user, but which are not used in the calculation of the consensus sequence
Compression	A region of DNA sequence that shows abnormal spacing between bases, usually as a result of secondary structure within the sequencing reaction products
Contig (11)	A set of gel readings whose sequences overlap
Contig editor	A scrollable window in which are displayed the sequences of the gel readings, their names and numbers, and the consensus sequence. Also displayed are the clipped data, which the user can view, but which are not used to produce the consensus sequence. The sequence can be edited within this window, and tags (see below) can be added
Padding characters	Non-specific insertions (indicated by *) that the program introduces during assembly to improve the alignment of the gel readings being entered (as shown in *Figure 2*)
scf files	Used to store traces, can be produced for various on-line and off-line gel scanners and are accepted by the *xbap* program (9)
A trace	A densitometric display of the intensity of labelling of each base that is output from the sequencing machine (or the film scanner). A trace contains significant peaks, each of which represents a base call (as shown in *Figure 2*)
Uncertainty codes	Used to represent probable base calls for a given base, where a certain base call cannot be made. The uncertainty codes in current use are Staden, IUPAC (generally not used for sequencing), and Intelligenetics
Unused data	The sequences of poor quality (or vector sequences) that are clipped, that are not used in calculating the consensus, but which can be viewed by the user

running Unix and X-windows) with a colour display, a three button 'mouse', and storage facilities for raw data from the sequence traces (each of which is 140–170 kb in size). Software for sequence assembly (preferably integrated with analysis software) is required, and ideally the software should accept data from a number of different sources (9). *xbap* is the sequence assembly program (10) that is run using the X-windows interface. Previous versions of the program were called *sap*, *SCREENV*, *SCREENR*, *DBAUTO*, and *DBUTIL* (11).

3.1 Conventions

Certain conventions are used to indicate actions by the user (on the keyboard) when using the sequence assembly software (*xbap*).

- `Select` indicates the user should select the alternative (described in the inverted commas) from those listed in the menus of *xbap*.

79

- `Option` indicates a message generated by the sequence analysis software that requires a response from the user.
- `Choose` indicates that a particular option (usually `Yes`, `No`, or `Cancel`) should be checked from the boxes that *xbap* generates.

Also, the sequencing software uses certain text to indicate information relating to the sequence assembly. The symbols most commonly used in the *xbap* program are:

* (asterisk) inserted by the program to indicate a padding character to improve the alignment.

\- (dash) indicates a base is indeterminate

, (comma) indicates padding characters that are inserted only during assembly to improve the alignment between the new gel reading and the consensus.

/ (slash) is used in front of gel reading names when searching for gel reads by name within *xbap*.

3.2 Display of trace data from within the database

The accurate editing of any sequence data is best done while viewing the original data. The display of fluorescence traces from the sequencing machines (ABI and Pharmacia), as well as the densitometric traces from one film scanner (the Amersham Base Scanner) is currently possible from within the contig editor of the *Staden* software package. The use of confidence values assigned to each base call is not included in the currently available version of *xbap*, although the future releases may include this feature. This will allow more accurate editing of the bases both automatically by the software, and by the user. *Figure 2* shows a typical display of the contig editor from *xbap*.

Each trace will generally contain some sequence information from the sequencing (or in some cases the cosmid) vector at the 5′ end. If the resolution of the sequencing gel is shorter than the length of the DNA insert in the sequencing vector, then the 3′ end of the trace will also contain sequence from the sequencing vector. *VEP* (or the Vector Excision Program) is used for marking vector (or cosmid) sequences at the 5′ and the 3′ ends of the DNA insert. These sequences are clipped (see *Table 1*) to the last base of the vector sequence, and tags are added at that point. These tags are displayed within the *xbap* contig editor, and prevent the vector sequence being used in any calculation of the consensus sequence.

3.3 Software created to make design of sequencing reactions easier

The selection of optimal oligonucleotides for use as PCR or sequencing primers is possible using a section of the program invoked by choosing

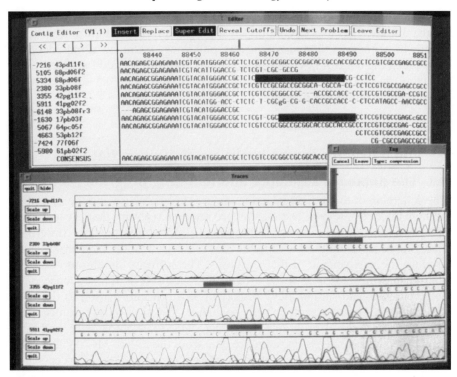

Figure 2. The X-terminal window as it appears when using the contig editor of *xbap*. The figure shows in the Editor window the gel readings names, the gel readings sequences, the consensus sequence (labelled CONSENSUS), the tags on the gel readings (shown as dark regions, in real-life they are red for a compression), the tag editor window (to the right of the figure), and in the window Traces the trace files associated with four gel readings (numbers − 7216, 2380, 3355, and 5911) with the accompanying original base calls (above the trace files).

Select oligos from the pull-down menu that is available when using the contig editor (see Section 4.1). The selection of an oligonucleotide is based on parameters that can be altered by the user (including primer location constraints; primer composition constraints such as length, G + C content, 3′ nucleotide, and melting temperature; and annealing constraints such as self-annealing characteristics). This software chooses templates that are suitable for primer-walking, and attaches a tag containing information about the primer to the template. Currently the user (and not the software) identifies which templates contain short inserts, and are therefore not suitable for primer-walking.

Tags are annotations added to the gel readings within the database (but in *xbap* not added to the consensus) that can be viewed from pull-down menus available from within the contig editor. Each tag has a type (see Section 4.1),

81

a position, a length, a space for comments, and an associated colour shown on the display. Tags can be created (by using the left mouse button to select a length of sequence, then invoking the pull-down menu by simultaneously depressing the control key and the left mouse button), edited (as before, but selecting the `edit tag` option), or removed (by using the `delete tag` option on the pull-down menu).

4. The software for assembling sequence data

4.1 The database assembly and handling program (*xbap*)

The collection of the individual gel readings into a database and the assembly of them into a single final DNA sequence requires software that is capable of accepting the new gel readings, comparing them with the previous (intermediate) consensus, and then linking the new readings with the old consensus to produce the new version of the consensus. The software should also be capable of allowing the user to determine how accurate is the sequence data being entered, and should allow display of as much information as the user requires to continue assembling the sequence. We have most recently used the *Staden* package of programs (including *xbap*) to assemble, edit, and perform preliminary analysis on sequence data from the 230 kb MCMV genome. The programs consist of *xbap*, and a number of scripts we use to process and enter data into the database. *Xbap* incorporates the use of coloured `tags` that can be displayed from within the contig editor which allow regions of the sequence with specific problems (such as compressions) or features (such as short inserts) to be identified (*Figure 2*). Comments can also be written into pop-up boxes (tags labelled as `comments`), that relate to the sequence being assembled.

When starting to assemble the data from a sequencing project, it is first necessary to construct a project database (see *Protocol 1*). This database will be added to by entering new gel readings into it, and all assembly will be performed using this database.

(a) Assembly is the act of entering gel readings into the project database and subsequent comparison of each gel reading (and its complement) with a consensus of all gel readings already stored in the database.

(b) It is important to always edit, view, and work with version 0 of the database. If it is necessary to create backup copies of the database, name these versions 1 to 9 or A to Z using a single character only.

The sequences that are obtained (from fluorescence-based or radioactivity-based sequencing methods) can be entered into the database either one at a time, or (in larger scale projects) as batches of around 50–200 sequences. The assembly of larger numbers of sequences at any one time is more

---

Protocol 1. Starting a project database

1. Log on to the computer, change directory to the directory where the database and all trace files will be stored.[a]

2. Open *xbap* by typing xbap. Three boxes will appear (named Dialogue, Graphics, and Output).

3. From the Dialogue box, select Open existing database?, accept option No.

4. Option New project name?. Type in a short, logical name of up to 12 characters.[b]

5. Option Database size?. Type in a size appropriate to the memory requirements to be expected during the assembly. For small projects, this will be around 2000, for large-scale projects (with sequence lengths over 100 kb) this size will be around 4000 to 8000. The size can be altered at a later date if necessary.

6. Option Maximum reading length?, can be from 512 to 4096 characters (currently). Accept 1024 characters.

7. Option Database is for DNA?, accept Yes.

8. *Xbap* will respond with the message that the database has been started, copy 0, and that it is currently empty.

[a]The sequence files can be moved out of the subdirectory containing the database files and the trace files once the sequence files are assembled. This reduces the size of the subdirectory containing the database to the smallest size necessary.

[b]The name of the database must not contain a full stop or reserved characters (either / or \), and it is recommended that it only contains alphanumeric characters (avoiding the characters {} () : ; * ! @ £ $ % ^ & * ,< > + | " or other unusual characters).

---

convenient, as procedures such as comparison with vector sequences, vector excision (using *VEP*, see Section 3.2), and assembly itself can be performed on larger numbers of sequences in a similar amount of time as it takes to perform the same procedures on smaller numbers of sequences. Furthermore, it is possible to make assembly simpler by writing scripts that perform the functions involved in entering new gel readings into the database. These scripts are easily written and can eliminate a number of keyboard steps, using the version of *xbap* (known as *bap*) written for a non-X-windows interface. For example, in the assembly of the MCMV sequence, in order to prevent entry (into the database) of some of the contaminating sequences from the mouse fibroblast DNA, we used the mouse Alu equivalent B1, and LINE 1 repeat sequences of *Mus domesticus* obtained from the EMBL database (that cover a total of 7–12% of the *Mus domesticus* genome) to screen out some of the murine (non-MCMV) DNA sequences.

Protocol 2. Summary of the steps required to enter sequences
into the database

The procedures are described in more detail in *Protocols 3* to *9*.

1. Read in the sequences, in a form (as *scf* files) allowing display of the
   original data from within the sequence assembly program (see *Table
   1*).

2. Clip the sequences to remove inaccurate data from the 3′ end of the
   gel read using the program *vep*.

3. Generate a list of filenames (e.g. called `fofn`) for all gel reads. In
   Unix, this can be done (e.g. for gel reads with names ending in `read`)
   by typing `ls *read > fofn`.

4. Compare these gel reads with sequences of the vector used in this or
   previous subcloning (such as M13 sequencing vector or cosmid vector) to remove vector sequences, and to clip the 3′ end of the
   sequence if the insert is shorter than the resolution of the sequencing gel ('short inserts').

5. Generate a second list of filenames to assemble, that only include
   non-vector and clipped sequences.

6. Tag the 3′ ends of short inserts using the script `tagscript`, to identify sequencing templates of no use for primer-walking experiments.

7. Compare these gel readings with sequences of possible contaminating (non-vector and non-target) DNA.

8. Assemble these gel readings using *xbap* (*Protocol 3*), which compares them with each other and those gel readings already in the
   database.

9. Repeat these steps for each new batch of gel readings to be assembled.

10. Use the function, Find internal joins (*Protocol 5*) to find joins
    between contigs not previously found during automatic assembly
    (because of poor alignments), utilizing the clipped (hidden) data.

    ● Steps 4, 7, 8, and 9 are easily run using a script that non-
    interactively compares the gel readings with sequencing vector, com-
    pares the gel readings with any other screening sequence, automati-
    cally assembles the data, and then performs a second assembly (of
    reads that have failed on the first assembly) once the first assembly is
    completed.

The creation of a project database generates five files (described in *Table
2*) that contain all the information relating to the database. The five files consist of the user-defined database name (in *Table 2*, * represents the name of

Table 2. Summary of the five files created within xbap that store all the information relating to the database

Filename	Details	Contents
`*.ARn`	File of archive names	List of each of the original gel reading files in the database.
`*.CCn`	Comments file	Comments relating to specific readings.
`*.RLn`	Relationships file	All the information needed to assemble the gel readings into contigs during processing by the computer. The information in this file relates to each gel, and its relationship to other gels (stored on gel descriptor lines), about the length of each contig and readings at the end of each contig ('contig descriptor lines') and the number of gels and contigs in the database ('general data').
`*.SQn`	Sequences file	All the individual sequences stored in the database .
`*.TGn`	Tag file	All the different tags relating to sections of the sequence. Currently this contains tags denoting the beginning and end of the used sequence, and internal tags (viewable from within the contig editor) with information relating to comment (for general comments), oligo (for PCR or sequencing primers), compression (for regions of sequence with altered base spacing due to sequence compression), stop (for regions of the gel with all four bases overlapping), repeat (for repeat sequences), cosmid vector, ignore cutoff – sequencing vector (for inserts shorter than the gel resolution), and ignore cutoff – cloning vector (for inserts at the end of cosmid, lambda or other cloning vectors).

the database), and the last letter in the database name (in *Table 2* indicated by n) represents the copy number of the database (from 0 to 9).

Sequences derived from individual gel readings are stored unassembled, and relational information is used to assemble the sequences during processing. Currently no gel reading within the database can exceed 4096 characters.

Once a project database is created, randomly generated gel readings are entered. In the early stages of a large-scale sequencing project, these gel reads will usually not overlap, and the number of contigs will increase rapidly (as shown in *Figure 3* region a). As the number of reads entered into the database increases, the contigs will begin to overlap, and the gaps of unknown sequence between the contigs will decrease. At this stage the number of contigs will generally plateau (shown in *Figure 3* region b) and then fall fairly rapidly (*Figure 3* region c). The number of characters required to produce a single contiguous sequence (contig), that represents the entire sequence of interest will depend on the length of the original sequence being determined, the pres-

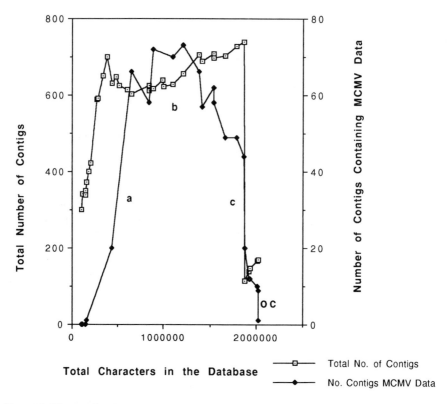

Figure 3. The totalled lengths of contigs over 800 nucleotides long (presumed to represent only MCMV data), and totalled lengths of all contigs (including contaminating murine eukaryotic sequence) are shown plotted against the total number of characters accumulated in the database. 'oc' represents the point at which a single contig was achieved. The plot was generated with data from the examine quality option of *xbap* (10).

ence of regions that are difficult to sequence and/or difficult to assemble (such as tandem repeats, regions of secondary structure, or regions that are difficult to clone) and the techniques used to produce the sequence. As shown in *Figure 3*, around 2.2 megabases of random data were required to determine the sequence of a 230 kb, G + C rich (58.7%) DNA virus (MCMV).

In all aspects of handling files relating to the database, logical, short names should be used. The file of filenames is most easily named using the prefix fn and the date in the form 930907 (for the 7 September 1993), with no spaces between the text (i.e. as fn930907). This means that at a later date, the nature and use of the file of filenames are easily identified, and when listed by numerical value, the latest date will be last.

The output from a typical auto assembly is shown in *Figure 4*. The entering gel reading (on the lower line) is shown aligned against the consensus

Protocol 3. Assembly of gel readings

1. With *xbap* open, select `Auto assembly`.

2. Option `Permit entry?`, choose `Yes`.

3. Option `Hide alignments?`, choose `No`.

4. Option `Use file of filenames?`, respond `Yes`, and enter the name of the file that contains a list of the gel reading names (i.e. the file of filenames). Alternatively, if single gel readings are being entered, respond `No`, and enter the filenames individually.

5. Option `File for names of failures`. This file will contain the names of all sequence files that do not enter the database.

6. Option `Perform normal shotgun assembly`. The other options `Put all sequences in one contig` or `Put all sequences in new contigs`, are used for special purposes (usually later in a project).

7. Option `Permit joins?`, accept `Yes`.

8. Option `Minimum initial match?`, enter the minimum number of exact matches contained in overlaps between gel readings and contigs that will be used to enter the gel readings (the default value is 15). Any overlaps with a lower number of consecutive identical characters will not be entered. The range allowed is from 14 to 4097 matches.

9. Option `Maximum pads per gel`, the default is 8. This is the maximum number of padding characters (shown as commas) the assembly program will introduce into the gel reading sequence before failing the gel reading (and hence not allowing it to enter the database).

10. Option `Maximum number of pads per reading in contig`, the default is 8.

11. Option `Maximum percent mismatch after alignment`, the default is 8.00%.

   • Options 9, 10, and 11 determine the stringency used to enter gel readings into the database.
   • If a gel reading being entered does not overlap any existing contig, then it enters as a new contig.
   • Alignments that have more than the defined maximum number of padding characters in the contig or the gel reading (generally set to 8, with a maximum of 25 characters), or alignments exceeding the maximum mismatch (generally set between 8.00% and 15.00%) will be displayed, but the gel readings will not be entered into the database. These represent 'Failed reads', and a list of the filenames will be made for subsequent re-entry after editing the gel readings (such as by shortening the length of used data, thus removing poor quality data).

sequence. The alignment has been optimized by insertion of padding characters (commas) into the consensus sequence and the sequence of the new gel
reading. The mismatch of the new reading with the consensus (7.5%), the
number of padding characters, the location of the gel reading overlap with
the consensus sequence, and the result of the assembly are all shown. Information relating to the status of the database (such as the total contig length,
total characters in the database, and the average 'depth' or number of gel
characters per consensus character) are shown in the summary at the top of
the output window in *Figure 4*.

Towards the end of the initial random stage of a sequencing project, some
regions of the genome will usually be represented on one strand only, and
other regions may still be compressed, meaning a completely accurate consensus cannot be generated. At this stage other techniques, e.g. custom
oligonucleotide primers, heated gel plates or other sequencing reagents such

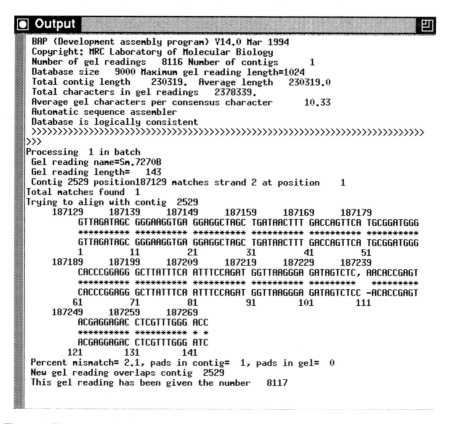

Figure 4. The output window of *xbap* (within the X-terminal window) as it appears after
assembly of new gel readings in the database, using *xbap*. The top sequence line represents the current consensus, the lower sequence line the gel reading being entered, and
the commas represent pads introduced into either sequence to improve the alignment.

as dITP or C^7-deaza dGTP (12) are used in order to accumulate data in regions with difficult sequences (such as compressions).

Minimal editing is required during assembly of the sequence if good quality data are used. If a sequence is determined to a depth of 8, with an error rate in base calling of 1%, then a theoretical cumulative error rate (before any editing of the database) would result in an average of 2 miscalls every 25 bases of raw sequence (not of consensus sequence) (14). However, neither the errors nor the sequence depth are random, and padding is used to produce the alignments. Thus the practical error rate is usually very different from the theoretical rate. This means that at the end of all projects some editing of the data will be required to produce the correct consensus sequence. It is especially at this editing stage that the raw (trace) data are needed to determine the final base call, or to direct the person editing to the need for further sequence data. Within *xbap*, raw data from the sequencing machines (and from the Amersham film scanner) can be displayed as traces (shown in *Figure 2*), and therefore all editing can be performed on-screen by the user in an interactive way. When editing, traces can be displayed from readings derived from fluorescent sequencing machines (currently those from ABI and Pharmacia) and from film scanners (currently the machine obtained from Amersham) by double-clicking the middle mouse button on the sequence of interest. The centre base of the trace displayed will then be the base clicked on. In all cases, editing should take account of the highest quality data on each strand, and use this data to determine the final consensus.

Protocol 4. Editing in *xbap*

1. With the database open, select Contig editor.

2. Option `Contig identifier?`, as the default shows the longest contig. If another contig is required, the name can be entered (as /gel reading name) or the line number (which can be found by using the `Show relationships` option).

3. Option `Replace` is checked as the default, and allows the user to replace any character in the gel readings. Alternatively if option `Insert` is used (by checking this box), editing can be performed (on the consensus only) to remove or insert padding characters.

4. Reads can be extended (into good quality data that has been clipped) by first displaying the unused data (by checking the option `Display cutoff`), locating the cursor on the extreme end of the used data, then depressing the meta key and the left or right arrow keys.

5. Tags (see Section 3.2) can be added to particular parts of any gel reading by using the tag editor menu (available by depressing the Control key and the left mouse button simultaneously).

Protocol 4. *continued*

6. Items can be found within the database using the `Search` option on the pull-down menu (that also contains the tag editing functions), accessible by depressing the Control key and the left mouse button simultaneously.

● When the mouse pointer is in the scroll bar, the mouse buttons function to set the editor to the position (middle button), move forward one screen (left button) and scroll backwards one screen (right mouse button). The screen can also be moved using the four scroll buttons displayed on the screen (shown in *Figure 2*).

● The `Super Edit` function has the potential to cause havoc with the alignments within the database if used incorrectly. Therefore this function should ONLY be utilised when the user is familiar with the software, and then used as infrequently as possible. Always make a copy of the database before editing

● Edits on the consensus will delete all characters to the left of the cursor, and will shift all gel readings so that the alignments (to the right of the deletion) remain in-register. This feature can be used to bring poor alignments into register by hand. Thus it is possible to insert padding characters into each gel reading, which brings the reads into alignment, and then remove any additional pads (from the consensus) generated during this procedure.

● Equally it is possible to insert an additional base into the consensus (where this base is correct, but present in the minority of gel readings) by inserting padding characters into the consensus and then adding characters to the gel readings to align them with the altered consensus. Both these procedures should be done with extreme caution, but are useful where gel reading differences occur between different strands, and between different quality reads.

The contig editor provides a useful search feature (see *Protocol 4*, step 6). Searches can be for *position*, *reading name* (that is the gel reading name preceded by /), *tag type* (comment, oligo, compression, stop, repeat, cosmid vector, ignore cutoff-sequencing vector, or ignore cutoff-cloning vector), *annotation* (searches for strings within the tag comments), *sequence*, *problem* (searches for a consensus character that is a * or a -), or by *quality* (searches for regions where the two strands disagree or where the consensus for either strand is a * or a -). The search feature can therefore be used as a method of locating individual gel readings once they are entered into the database. It can also be used during assembly to assess the quality of the sequence data within the database. This directs the user's attention to any significant reduction in the quality of sequences entering the database, and allows this problem to be addressed.

It is usually in the later stages of the sequencing project that the small number of contigs (often with only small gaps between them) can be joined interactively, using the find internal joins function (*Protocol 5*), or using the joins editor (*Protocol 6*). The find internal joins function allows use of the clipped or hidden data (that are generally of poorer quality) at the ends of the gel readings, and the number of contigs usually falls fairly rapidly once these gaps are filled (region c in *Figure 3*), often with the addition of relatively little new sequence data.

Protocol 5. Searching for internal joins in the database

1. With the database open, select Find internal joins.

2. Option Minimum initial match?, usually this is set at a small number (20) of consecutive characters that must match before joining will be attempted. The range that can be used is from 14 to 4097 characters (currently).

3. Option Maximum pads per sequence?, usually this is set at a large number (20) of pads that are allowed in the initial overlap (the range is from 0 to 50).

4. Option Maximum percent mismatch after alignment?, this usually is a large mismatch initially (75%) to allow for the poor quality data at the ends of the reads. The available maximum is 100% mismatch.

5. Option Probe length, is the length of the sequence from the two ends of each contig that is compared with the sequence along the total length of all other contigs.

6. Option Use clipped data?, choosing No, does not utilize the hidden data in attempting to perform the joins. This should be performed first, as it is usually fairly fast, and identifies any joins not made with the 'used' data.

7. Option Use clipped data?, choosing Yes, utilizes the clipped data (see *Table 1*) in attempting to perform the joins.

8. *Xbap* then responds by sequentially listing the alignments between potential joins, and the contig numbers. It is possible at this stage to accept the joins, or reject the (large number of) spurious overlaps.

● The majority of overlaps will be false (in a long sequence), because of the loose definitions of overlap used in steps 2 to 8.

It is also possible to join contigs interactively using the contig editor to create potential alignments, then editing these (using the raw trace data) to determine if a real join exists between the two contigs. The idea of the find internal joins and the interactive joins option is use poorer quality data at the ends of the gel readings (that have been clipped) to design experiments allowing joins to be made, that would not be made otherwise. The sequence data may be of sufficient quality to read the overlapping sequences, but may be inadequate to allow generation of an accurate consensus. Secondly, other information (available from restriction maps) may be used to indicate two contigs are adjacent. In both these cases, the 'unused' data from gel readings at each end of the contigs may be used to design experiments (such as primer-walking) that may confirm (or contradict) the putative join. In some cases, the unused data may be utilized to join overlapping contigs, only if at least two accurate gel reads (one from both strands of the DNA) are present.

Protocol 6. Joining contigs interactively using the contig editor

1. With the database open, select `Join contigs`.
2. Option `Contig identifier?`, type the name or line number of the left hand contig that you wish to join.
3. Option `Contig identifier?`, type the name or line number of the right hand contig that you wish to join. The window then displays the left (top) and right (bottom) contigs, and a box containing exclamation marks where the bases disagree between the two sequences. Agreement between the two contigs is indicated by blank space.
4. The upper and lower contigs can be moved relative to one another using the on-screen scroll buttons ($<<$, $<$, $>$, $>>$) or by positioning the cursor in the scroll bar and moving the cursor in the direction of travel desired.
5. The optimum join position can be stabilized by checking the box `Lock`, which fixes the positions of the two contigs relative to each other.
6. The alignment of the two sequences must then be checked over the length of the whole overlap. If the two sequences are truly overlapping, editing of each sequence can then be performed, using the trace displays (invoked by double-clicking on the sequence of interest) to verify the sequence.
7. The join can then be completed by checking the box `Leave editor`, and saving the join (by replying `Yes` at the prompt `Save the join?`), or discarding the join (by replying `No` at the prompt `Save the join?`) or cancelling (`Cancel`) and returning to the join contigs editor.

● Once the join is completed, it is saved within the database. If an incorrect join is made, this may be corrected using the option `Break contig?`.

On completion of the sequencing project (or at an intermediate stage) the consensus DNA sequence can be calculated, and output to a file. The user is able to calculate a consensus from either the entire sequence or a subset of this sequence.

Protocol 7. Calculating and outputing a consensus file from the database

1. Select `Calculate a consensus`.

2. Option `Name for consensus file?`, name the consensus.

3. Option `Use clipped data?`. This allows the user to include data that have not been used to calculate the consensus viewed within the contig editor (i.e. data that have been clipped, as described in *Table 1*), but which may be of use for other purposes.

4. Option `Make consensus for whole database?`, accept `No` to select a single contig, or `Yes` to concatenate all contigs into a single consensus.

5. Option `Contig identifier`, choose the contig of interest, either by typing in the line number of the contig or a / followed by the name of a gel reading within the contig.

6. Option `Start position in contig?`, choose the beginning of the region of interest.

7. Option `End position in contig?`, choose the finish of the region of interest. The software will respond with the message `working` until the calculation is finished.

8. Option `Select another contig?`, choose `No`.

9. Option `Staden` or `FastA`, allows the output of the consensus in either type of format.

The consensus file can be directed to the screen or to a new file (by using the `Redirect output` option). This redirection option is particularly useful if the consensus is to be compared with other sequence files.

4.2 Alternative packages

A number of other sequence assembly programs are also available, such as GCG (13), and the ABI sequence editor (*SeqEd*). We use the *Staden* package because it allows display of the raw data from within the contig editor, it accepts data from more than one source using the *scf* files (9), and it can be used to assemble the large amount of data needed to determine the sequence of longer DNA (of 230 kb and above) using random shotgun techniques. The package can also be used to manipulate data from shorter DNA sequencing projects (such as derived from sequencing PCR products). The software for sequence assembly is continually evolving, and it is to be expected that additional (and different) features to the ones described here will be available in the future.

5. Assessment of sequencing projects

5.1 Recording information about the sequencing templates

When gel readings are first entered into the database, information is compiled using a proforma, into which data are entered by hand (although this may be automated in the near future). The form represents the outline of a microtitre tray (corresponding to a microtitre tray containing the template DNA stored at −20°C), and data entered into each cell (shown in *Figure 1*) indicate for each well the nature of the DNA (sequencing vector, non-target DNA, or target DNA), the sequence read length (after clipping poor quality data), whether the insert is shorter than the gel resolution, whether forward and/or reverse primers were used in the sequencing reactions, and the amount of template remaining. The form also contains information about the location of the microtitre tray, and some information about the methods used to prepare the template DNA. This allows later appraisal of an individual tray, if subsequent poor quality sequence data are obtained from sequencing reactions performed using templates derived from that tray.

5.2 Assessment of the sequence data during assembly

The sequence quality is judged during assembly using data generated by *xbap*. These data relate to the total characters in the database, the total contig length, the number of contigs, the number of gel reads, and the mean depth of the gel reads across the genome. Additional information not available from within the software (but which can be determined by the user) is the range of depth of gel reads across the genome, and the depth of the gel reads at the ends of the sequence (where these ends are known from other studies of the organism).

The 'examine quality' option of *xbap* may be used to find the percentage of the sequence in the database that is 'well-determined' on both strands of the sequence (as outlined in *Protocol 8*, part A).

In the example presented here of 'shotgun' sequencing a complete 230 kb genome, the ends of the sequence were apparent because of the presence of two sets of readings, with the majority of gel reads in the same orientation, each terminating at or close to an identifiable end. Readings from the ends were over-represented in the final sequence, when compared with the mean depth of the whole genome. The majority of the terminal clones were encoded by the same strand (as would be expected), as the length of the inserts (1 to 2 kb) was greater than the resolving power of the sequencing gel systems used (200 to 300 bp for the radioactive sequences and 400 to 550 bp for the fluorescent sequences) (14).

Protocol 8. Examining the quality of data in the database

A. *Using the* show relationships *option of xbap to display information about all of the contigs in the database*

1. With the database open, select Show relationships.

2. Option Select contigs?, choose No.

3. Option Show gel readings in positional order?, choose No.
   The output window then shows the gel name, gel number, position in
   the contig of the gel, length of the gel (which is negative if the gel has
   been complemented to enter it into the database), and the line
   numbers of the gel readings that are the right and left neighbours of
   the gel. This listing can be viewed by scrolling using the left-hand
   mouse button and the scroll bar on the left side of the output window.

B. *Using the* examine quality *option of xbap to show which regions of
the sequence are 'well-determined'*

1. Select Examine quality.

2. Option Contig identifier?, write the name of the contig (/ followed
   by a gel reading name or the line number of a gel reading from the
   contig).

3. Option Start position in contig (n – nn)?, choose the region to
   be examined.

4. Option End position in contig (n – nn)?, choose the region to be
   examined. *Xbap* then displays (in the output window) the percentage
   of sequence:
   OK on both strands and they agree
   OK on plus strand only
   OK on minus strand only
   OK on both strands but they disagree.

5. Option List codes or Plot codes, allows the display (using characters or graphically) of the quality of the consensus at each position.
   The analysis assigns a code at each position indicating the base is:
   well determined on both strands and they agree (code 0), well determined on plus strand only (code 1), well determined on minus strand
   only (code 2), not well determined on either strand (code 3), or well
   determined on both strands but they disagree (code 4).

---

6. Discussion

This chapter presents the recommendations for the use of sequence assembly
software based upon our use of the *Staden* package (*xbap*), and accompanying scripts on a computer using the Unix operating system. The design of

any sequencing project should consider a number of factors as well as the assembly software.

(a) The use of longer inserts (of 1–2 kb rather than 0.5–1 kb) allows more efficient reverse priming and primer-walking.

(b) Template preparation should be done using the most efficient techniques available. Single-stranded DNA templates may be prepared using a semi-automated procedure (15) or using standard phenol preparation. In our hands, the templates produced using techniques involving phenol extraction of DNA took considerably more effort (two workers could prepare 96 phenol templates in one day, whereas to one worker was able to prepare 384 templates in one day with the semi-automated procedure) but produced sequence read-lengths around 10% longer than from sequencing reactions using single-stranded DNA prepared using the semi-automated procedure as template.

(c) Performing the sequencing reactions in microtitre trays (1, 5) allows easier handling of material for multiple repetitive steps.

Closure of gaps (between contigs) present towards the end of a sequencing project can be done in a number of ways. Reverse primed reactions can be performed using templates from the ends of all remaining contigs. Also, custom oligonucleotide primers (usually 16 to 20-mers) can be made for the ends of contigs not joined using reverse priming, and sequencing reactions performed using these primers (either with radioactive techniques or with fluorescence-labelled ddNTPs). In a small number of sequencing projects (such as some large-scale sequencing projects), other information (such as map location or homologies with other organisms) may be used to position contigs relative to their expected position on the complete genome. Subsequent experiments (such as primer-walking) can be designed to determine if the two contigs are adjacent, overlapping or separated on the final sequence.

There are an increasing number of new techniques available to produce single-stranded DNA sequencing templates (6, 15); to perform sequencing reactions (4, 5); to acquire sequence data (16); to assemble the sequence (10); and to analyse (17) and organize the final sequence information. The overall result (for large-scale sequencing projects) of the incremental advances in sequencing techniques and technology has been to allow a single worker to sequence around 100 kb per year (2). Improvements in the efficiency of sequence data acquisition are likely to result both from incremental improvements in the currently available sequencing technology, as well as from the contribution of completely new sequencing technologies. Future developments likely to be important in increasing the speed of acquisition of sequence data include:

(a) The use of techniques that allow more rapid and straightforward acquisition of data from reverse primed, random templates (18).

(b) The automation of a number of the procedures involved in preparation of the sequencing templates (15).

(c) The ability to read further from a single sequencing template.

(d) The use of simple techniques to determine the sequence of double-stranded DNA templates (4).

(e) The ability to accurately sequence multiple templates from a single lane (19).

(f) The introduction of new techniques that can be auto-mated (8).

Acknowledgements

The authors wish to thank Rodger Staden, Alan Bankier, and Helen Farrell.

References

1. Bankier, A. T. and Barrell, B. G. (1989). In *Nucleic acids sequencing: a practical approach* (ed. C. J. Howe and E. S. Ward). IRL Press, Oxford.
2. Davison, A. J. (1991). *DNA Sequence*, **1**, 389.
3. Sanger, F., Nicklen, S., and Coulson, A. R. (1977). *Proc. Natl. Acad. Sci. USA*, **74**, 5463.
4. Craxton, M. (1991). *Methods*, A companion to *Methods in enzymology*, **3**, 20.
5. Smith, V., Craxton, M., Bankier, A. T., Brown, C. M., Rawlinson, W. D., Chee, M. S., *et al.* (1993). In *Methods in enzymology*, Vol. 218 (ed. R. Wu), p. 173. Academic Press, London.
6. Rawlinson, W. D., Chee, M. S., Smith, V., and Barrell, B. G. (1991). *Nucleic Acids Res.*, **19**, 4779.
7. Kambara, H. and Takahashi, S. (1993). *Nature*, **361**, 565.
8. Southern, E. M., Maskos, U., and Elder, J. K. (1992). *Genomics*, **13**, 1008.
9. Dear, S. and Staden, R. (1992). *DNA Sequence*, **3**, 107.
10. Dear, S. and Staden, R. (1991). *Nucleic Acids Res.*, **19**, 3907.
11. Staden, R. (1982). *Nucleic Acids Res.*, **10**, 4731.
12. Mizusawa, S., Nishimura, S., and Seela, F. (1986). *Nucleic Acids Res.*, **14**, 1319.
13. Gribskov, M. and Devereux, J. (1991). *UWBC biotechnical resource series*, 279. R. R. Burgess. Stockton Press, New York.
14. Edwards, E. and Caskey, C. T. (1991). *Methods*, A companion to *Methods in enzymology*, **3**, 41.
15. Smith, V., Brown, C. M., Bankier, A. T., and Barrell, B. G. (1990). *DNA Sequence*, **1**, 73.
16. Hunkapillar, T., Kaiser, R. J., Koop, B. F., and Hood, L. (1991). *Science*, **254**, 59.
17. Karlin, S. and Altschul, S. F. (1990). *Proc. Natl. Acad. Sci. USA*, **87**, 2264.
18. Smith, V. and Chee, M. (1991). *Nucleic Acids Res.*, **19**, 6957.
19. Church, G. M. and Kieffer-Higgins, S. (1988). *Science*, **240**, 185.

6

Molecular biology software for the Apple Macintosh

M. GINSBURG and M. P. MITCHELL

'To find a world in a grain of sand and heaven in an hour', William Blake

1. Introduction

The notion of having sufficient power in a personal computer to perform almost any sequence analysis task has come much closer to reality in the last few years, as powerful RISC-based systems become affordable. Furthermore, operating systems have become more friendly and accessible for even the most computer-naive scientist, through the advent of the graphic user interface. As a result of such developments, programmers in both the commercial and public sector have responded by writing molecular biology software packages apparently able to perform anything from the most trivial to the most complex task. Programs vary from HyperCard stacks to programs written for another hardware platform and then re-compiled for the Macintosh. In the latter case they may retain the text based feel of the original. At the other end of the scale, programs may be complex offerings with a full range of windows, drop-down menus, and impressive use of colour and graphical displays. These may also require large amounts of hard disk space and RAM, use copy protection, and may be expensive. The quality and reliability of programs varies enormously and it is not always the case that commercial offerings are the best in terms of ease of use, freedom from bugs, and providing continuing support.

The programs described in this chapter are all available for the Apple Macintosh computer. The selection we discuss illustrates the variety of capabilities and user interface styles rather than implying endorsement of any particular system. Limitations of space prevent us from covering all software packages. This chapter covers both an overview of three general purpose programs (*GeneWorks*, *MacVector*, and *DNAStar LaserGene*) and a selection of more specialist commercial, public domain and shareware programs.

All the programs have been used on an Apple Macintosh IIci with 20 Mb of RAM, an extended keyboard, a 160 Mb hard disk, and a 13 in. colour screen, a Centris 660AV with 16 Mb RAM and a 500 Mb hard drive, and a PowerBook 180c with 4 Mb RAM and 140 Mb hard disk. Some applications have also been tested on a PowerMac 7100/80 with 8 Mb RAM, 32 Mb of virtual RAM, and a 700 Mb hard disk. The programs have been tested using system 7, system 7 Pro, and system 7.5. An Apple CD drive has been used where appropriate. The IIci, although not a contemporary system, is a medium power machine representative of many in use today.

2. *GeneWorks*

2.1 Overview

2.1.1 System requirements

GeneWorks requires at least 2 Mb of RAM and 5 Mb of hard disk space as a minimum configuration. A hardware copy protection device or dongle is required in order for the program to run. The dongle is inserted between the keyboard and the main chassis of the computer and without it the program will only run in demonstration mode. *GeneWorks* can be installed from floppy disks or CD-ROM, and while colour is an advantage it is not essential to have a colour screen.

2.1.2 Program layout and structure

The program (at the time of writing version 2.4) is wholly integrated, that is, one program performs all analyses. When starting the program for the first time the user sees a program banner which is then replaced by the *GeneWorks* desktop and the on-line help window. The desktop has four icons on the top window bar representing Stop, Go, the Analysis control panel, and Help. The start-up views may be customized to a limited extent using the Preferences option in the File menu. A new document can be created by selecting the appropriate type from the New option in the File menu. There are a number of types that may be selected, each type having its own icon. The main types are new sequence document (DNA or protein), a new database search, new sequence alignment, new dot matrix, new text alignment, or sequencing project.

   GeneWorks operates around a set of windows or views, each of which contains either graphic, tabular, or textual information. Highlighting a region of a sequence in one window, such as the sequence editing window, will also highlight the appropriate sequence region in other windows. Within the graphic views the resolution may be varied from 512 residues per pixel to one residue per character, where the actual sequence will be displayed in the associated graphic.

   Individual documents are represented on the *GeneWorks* desktop by appropriate icons, as shown in *Figure 1*. Double clicking on an icon will open that document. The user may opt for sequence documents to open automati-

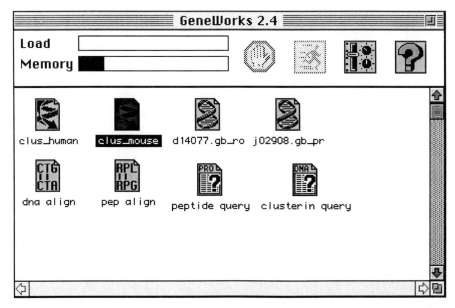

Figure 1. The *GeneWorks* desktop has different icons for the different data file types.

cally on start-up using the `Preferences` option. The `Help` window provides context sensitive help and may be called up via an icon, from the menu by using a keyboard command (cmd H), or using the extended keyboard help key.

2.1.3 Network versions
A network version now exists which relies on key server technology.

2.2 DNA analysis
2.2.1 Entering and editing data
The sequence is entered directly into a sequence editing window which opens with three icons representing a digitizer control, a panel for customizing the appearance of the editing window, and an icon for calling up the on-line help. The font, font size, colour, style, line spacing, and number of characters and character blocks per line may all be set after selecting the `Format` icon.

Sequences are entered by typing directly into the window or by the usual cut, copy, and paste features. Sequences may be imported from `Intelli-Genetics` format files directly as long as the files have a PEP or SEQ extension to the file name, otherwise these will be opened as simple text. Other formats are also supported. Pre-existing files are imported via the `Open` option from the `File` menu whereupon a new window with two boxes appears. The top box contains the available files which are then added to the bottom box for importing in to the program by double clicking on the sequence name or alternatively using a single click followed by the command `Add`. Pressing an `OK` button completes the operation, the window disappears and the sequences are added to the *GeneWorks* desktop. The editor will only accept standard IUPAC code characters and alerts the user accordingly when errors are made. The sequence can be verified by highlighting a segment and selecting `Read` from the `Edit` menu. The sequence is read back in a female voice as single characters rather than as base names.

2.2.2 Annotating the sequences

The sequence can be annotated in a number of ways. The simplest is to high-light a segment of sequence in the editor view and select the `Make feature` option (cmd M) from the `Edit` menu. This calls up a new window which con-tains a drop-down menu from which the user selects a feature key such as *allele*. Boxes highlight the range of the feature selected; a large text box is provided for the description of the feature as well as check boxes indicating how the feature will be displayed in the graphic window. The user may select from these and specify whether the feature will be shown in the sequence or graphic views. This latter is especially useful where many features cluster in a small region or where a large sequence is being displayed on a compact scale. Alternatively, the user may select the `Feature Table` option from the `Views` menu. In this instance the features are entered manually, starting with the range of the sequence which is being annotated.

2.2.3 Import/export formats

GeneWorks can import a number of sequence data formats including those used by *PC/Gene*, IntelliGenetics, GCG, *DNA Strider*, and *MacVector*. Other formats are opened as text documents which can then be copied and pasted into a new sequence editor window. *GeneWorks* will, by default, change any U it encounters to T. The sequence can be exported (using the `Save As` com-mand) in IntelliGenetics, EMBL, *FastA*, *FastA* server, *BLAST* server, and *Phylip* formats. In addition, any views may also be saved as text or as PICT format.

2.2.4 Pattern analysis

Restriction enzyme recognition sites are identified automatically when either the `Restriction Sites Table`, `Restriction Sites Map`, or `Circu-lar Map` is selected from the `View` menu. The user can set up the required restriction enzyme set using the `Master Control Panel` from the *GeneWorks* desktop. Choosing this icon brings up a new window, the control panel, which enables other pattern analyses to be selected including DNA and protein motif sets, DNA and protein analysis algorithms, ORF criteria, protease, and restriction enzyme sets.

The pattern sets are displayed as groups, such as all enzymes, or four, five, six cutters. The pattern sets can be edited by selecting a group and then pressing the `Edit` button. This brings up a new window with left- and right-hand panels. The active enzyme set is in the right-hand panel. Highlighted selections from this panel can be edited or removed from the list. Furthermore, if the `Select by criteria` box is checked, then enzyme salt concentration may be used to define further the group in use. The user may select to see either separate maps or one map of all the selected enzymes.

Once the criteria have been established the user selects `OK`, and is returned to the previous window, and eventually to desktop level by repeatedly selecting `OK/Done` as appropriate for each window. At this point any DNA sequences will be mapped according to the established criteria. The results can be viewed as a spreadsheet, a simple listing, a graphic (in the `Restriction Map` view), or an agarose gel as shown in *Figure 2* (from the `Restriction Table` view), or the sites may be displayed in the sequence editing window. The display is controlled by using the `Format` icon to open the format control panel. Within this panel is a check box and menu which allows the user to select between restriction sites or motifs to display with the sequence.

A number of DNA sequence motifs are provided. These can be edited to form new subsets or further motifs may be added. The results of motif mapping can only be displayed as a map or marked above the DNA sequence in the sequence editor window. New motifs are added by selecting `DNA Motifs...` from the `Master Control Panel`, then selecting `New` or `Edit` to create a new set of motifs or to edit an existing set. Finally `New Motif` is selected and the name, description, and pattern are entered by filling out a dialogue box.

2.2.5 Gene finding

Open reading frames (ORFs) can be viewed either as an `ORF` map, as shown in *Figure 3*, or via a `Reading Frame` table from the `Views` menu. Using the `Master Control Panel` the ORF criteria should be selected and altered accordingly before running the analysis. The user may select from a number of different translation tables. The default is the universal code but the user may also select from the mitochondrial code, the *Drosophila* code, etc. The codes may be edited and new tables established.

Other criteria for defining an ORF may be combined, for example specifying that an ORF must begin with an ATG, or is preceded by a particular sequence which falls within a specified range of bases from the putative initiator, or that the open reading frame must be a minimum number of base pairs long.

DNA composition may be examined by using one of the analyses found among the `DNA algorithms`. These include di- and trinucleotide composition and codon composition. The required analyses are copied to the active

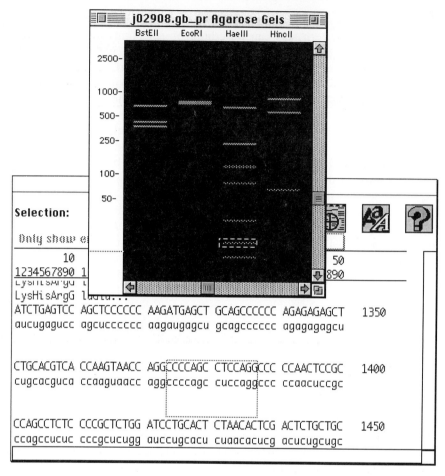

Figure 2. When a band in an agarose gel is highlighted the fragment is highlighted in the sequence window.

algorithm window and the results displayed as a map or table of results. The user will have to switch to the plot layer or the analysis table in order to see the results. The results may be further manipulated in the table view and a histogram, peak finder, or scatter gram can be used for display.

2.3 Protein analysis

2.3.1 Entering and editing data

This has already been covered in the DNA section. A protein sequence document has a different icon and upon opening the document the digitizer icon is greyed-out. As before the sequence can be verified by voice and again the sequence is read back as characters, rather than residue names.

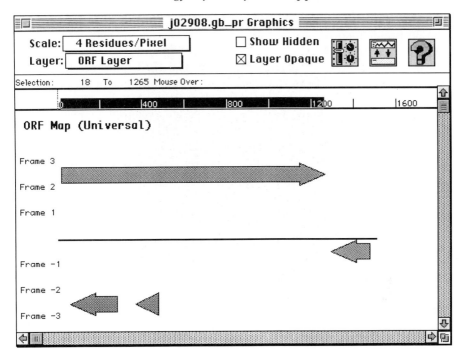

Figure 3. A six frame ORF analysis reveals four ORFs in a nucleotide sequence.

2.3.2 Annotating

The procedure for making a segment of a sequence into a feature is the same as for DNA. The list of keys that may be used is specific to protein sequences. Thus the first feature in the Key drop down menu is ACT_SITE.

2.3.3 Structural analysis

A number of protein sequence analyses can be performed which facilitate a structural interpretation and these are selected via the Protein Algorithms of the main master control panel. The analyses include secondary structure prediction, hydropathicity, surface probability, signal sequence analysis, and an opportunity to add new algorithms. The parameters for any analysis may be edited by initially selecting the parameters button and highlighting the parameter of choice in the new window that appears.

The analyses are initiated by selecting the analysis plots or analysis table. The default analysis is the Kyte and Doolittle (1) hydrophobicity plot. Further plots are added by selecting the control panel icon on the top part of the menu bar and selecting a plot from the left-hand listing and copying it to the right, active analysis list. Selecting an analysis in the right-hand panel allows the user to look at and edit the parameters for that analysis. In addition, new analyses can be made by using a combination of existing analyses and from the 30 pre-defined operations. Definitions of operations may be found via help.

The user may select from Chou and Fasman (2) or Garnier (3) protein structure predictions. In the plot view the results for alpha helix, beta sheet, turn, and coil are displayed as separate plots. The values are likewise represented in different columns in the analysis table. As before, graphs may also be used to represent the results. The Argos *et al.* (4) algorithm, for prediction of transmembrane regions, may also be chosen.

The remaining analyses in the group are Eisenberg's (5) hydrophobic moment, surface probability calculations, chain flexibility, DeLisi (6) amphipathic indices, and eukaryotic or prokaryotic signal peptide cleavage sites which may also be displayed as plots or tables.

pI analysis is new to version 2.4 of *GeneWorks* and is available from the `Views` menu. A graph of the variation of the overall charge of the whole protein, or region, with pH is displayed. The left-hand side of the window shows information about the N and C termini and the number of charged amino acids in the whole protein, or region. Placing the mouse over the pH axis displays the relevant pI and pH values. Four buttons give access to help, setting the pH scale of the plot, editing the pK values of amino acids, and post-translation modification parameters. The plot may be saved as a PICT file.

Helical Net analysis is also new to *GeneWorks* version 2.4 and likewise is available from the `Views` menu. The Helical Net window contains a two-dimensional representation of a helical structure of a whole protein (*Figure 4*) or a previously highlighted region. The diagonal lines represent a 360° turn of the helix with the vertical separation representing the pitch. Each amino acid is represented by a circle with a diameter proportional to the size of the residue, and the colour from the table on the left of the window. The analysis is limited to a maximum of 2000 amino acids. Four buttons give access to help, the helical parameters, the amino acid scale used for the colour scheme, and editing of the colour scheme. The colour scheme may be sorted by amino acid single letter codes or by the value of the scale. Colour editing is only possible after sorting by amino acid. The helical net may be saved as a PICT file.

2.3.4 Pattern analysis

Pulling down the `Views` menu allows the user to select either `Peptidase Map` or `Peptidase Sites Table`. Selecting either will enable peptidase sites for the sequence to be calculated. The results can then be displayed as a spread-sheet-like table or as a graphic display similar to that for restriction enzyme analysis. The table contains the number of cut sites for the peptidases and the position of the cut site.

Enzymes or reagents used for digesting polypeptides are called agents and the list used may be customized or new agents added by selecting the `Pep-`

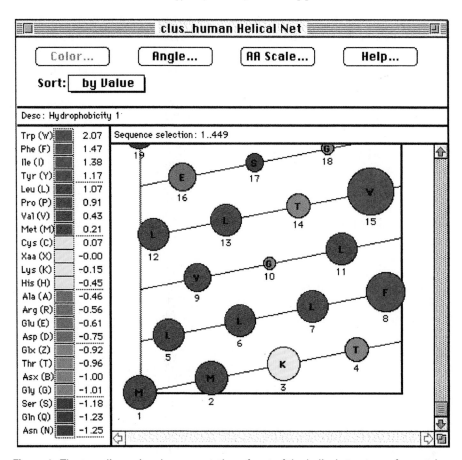

Figure 4. The two-dimensional representation of part of the helical structure of a protein.

tidase Sites control panel. Selecting this icon takes the user through to a new window containing the three default protease sets. There are four buttons called New, Edit, Delete, and Done. Selecting New allows the user to select from the list of all proteolytic agents and create a new subset. Within this new window there are two panels; the left-hand one contains the pre-existing set and the right-hand one is empty. Agents are copied across, removed or created in the right-hand listing and become the new subset. A similar protocol can be followed to edit or add agents to pre-existing subsets. When a new agent is added the details such as sequence, name, and description are entered in an editor window. The actual cut site is set using an arrow slider which is above the recognition sequence.

The spread-sheet window may be sorted by first, last, fewest, and most sites. Additionally, a graphic representation of an idealized sizing gel of the fragments may be displayed. Selecting a band on this gel will also highlight the respective segment of the sequence in the editor view and place a highlight in the associated views

Motif analysis is performed in a similar fashion. The default motifs are taken from the Prosite database and subsets may be selected and edited similarly. The documentation for individual entries is available by selecting the question mark icon in the window listing the entries.

2.4 Special analyses

2.4.1 PCR

The parameters for setting up the search for polymerase chain reaction (PCR) primers are accessed through the master control panel. The program locates the forward primers first and then the reverse primers. The results are displayed in the Primer View table. A primer pair is displayed on each row. Some of these will be designated optimal primers, that is to say they differ by a small margin from the desired characteristics.

The user can restrict the search region within the sequence, or specify a mandatory residue range. Additionally, the length of the amplified sequence or its T_m may be specified. The minimum, maximum, and optimum values for length, T_m, and GC content of the primers can all be set by the user in order to customize the search for PCR primers.

In the graphics view the amplified region appears as a thick line, the ends of which are the primers. On colour screens, optimal primers appear in red, whereas non-optimal ones are designated by blue lines. Double clicking on a primer brings up a new window displaying the primer sequences and characteristics such as length, T_m, or annealing temperature. The primers can also be tested against the GenBank database by clicking on the appropriate button in this window. In addition the user may enter pre-existing primer sequences and test these on the currently active sequence.

2.4.2 Gel assembly

The sequencing project document type has its own icon on the *GeneWorks* desktop. Projects are initiated by selecting New Sequencing Project from the New option in the File menu. This immediately opens a new window containing a number of buttons and two boxes. In a new project only one button, Add Gel, is not greyed-out. This is used to add gel readings to the project from existing sequence files within *GeneWorks*. Thus sequences must have been entered previously, e.g. either by typing into the sequence editor window or by using a digitizer.

Once the readings have been imported into the project they appear in the left hand of the two boxes in the sequencing project window. Length information is also displayed at this point. Clicking on a single reading activates a number of buttons such as Delete Gel, Edit, Auto, or Manual.

If the `Manual` button is pressed a new window appears which asks if the user wishes to align the selected sequence. The user may select to do this one gel reading at a time or all readings at once. Two buttons remain greyed out at this stage, these represent the choices for alignment once one or more contigs have been built. The user then selects the stringency for the alignment using a drop-down menu. Once the `OK` button has been pressed the program tries to align the sequences and when this is successful an editor window appears which shows the aligned gels as arrows where the arrowheads are the sequence orientation. In addition, a bar appears above the gels showing the quality of the alignment. If a resolution of one character per pixel is selected the lines then become the sequences ready for individual editing and may be viewed using a colour coding for the bases.

2.4.3 Multiple sequence alignment

The sequences to be aligned must first be loaded into the *GeneWorks* program. Multiple sequence alignments can then be performed by selecting a new document of the type `New Alignment`. There are separate types for protein or DNA. A new window opens, and the sequences must be added to the document. The alignment parameters are set by selecting the `Parameters...` button from a panel of seven buttons at the top of the alignment window. The parameters include two penalties for gaps, a gap opening and a gap lengthening penalty, and maximum and minimum diagonal offsets. Colour can be used to indicate the quality of the alignment. Prior to performing the actual alignments the designation of subsequences to be aligned can help reduce the time needed and avoid memory restrictions.

New sequences can be aligned either in pairs or in a group. By clicking on the `Align Next` button *GeneWorks* will choose the pair of sequences with the best matching regions and align them; they will then be displayed in the editing window. Having made changes the user then returns to the strategy window and clicks on the `Align Next` button and repeats this strategy until all the desired sequences are aligned. If the sequences are to be aligned in one go then the `Align All` button is selected. The alignment proceeds until all the sequences have been aligned. A strategy window is opened and the sequence names are listed on the left. This window also displays a dendrogram of the alignment. By clicking on any boxed number at the branch of the tree, the alignment for that branch is presented in an editing window. The window allows a number of editing functions to be performed directly on the sequences. A palette of tools is presented which contains a grabber hand for moving regions of contiguous sequence left or right. There are two tools which handle gaps, one inserts them and the other removes them. By clicking on the mouse button, gaps are added or deleted one at a time. If the button is held down and the deletion tool is selected, then a segment of sequence to be deleted may be selected by 'rubber banding'. If the mouse button is held down while the insertion tool is active, gaps are added continually. A grabber hand is also used to move subsequences left and right relative to each without disturbing the overall alignment. This is useful when trying to determine where edits need to be placed in order to improve the alignment.

The consensus may be saved as a separate sequence by choosing `Extract Consensus` from the `File` menu. The final alignment can be formatted before printing. Characteristics such as font, style, and residues per line are chosen using the `Format` menu. The aligned residues can be boxed or shaded and the cost of the alignment is represented by the value chosen for boxing or shading from drop-down menus.

2.4.4 Dot-matrix comparison

This is performed by selecting the appropriate document type (there are separate types for DNA or protein) from `New` in the `File` menu. The sequences to be compared must either be open within *GeneWorks* or both selected on the *GeneWorks* desktop. The analysis makes use of colour to indicate the quality of the alignment. To the left of the displayed matrix there is a percentage scale which uses colour to represent the percentage identity. Clicking on a colour will filter the plot and remove anything with less than that percentage identity.

An alignment can be performed on a region of the diagonal simply highlighting that region. `New Alignment` is then selected and the highlighted region is then automatically displayed.

2.4.5 Database searching

GeneWorks supplies GenBank, EMBL, and SWISS-PROT on CD-ROM. Initially a query document of the desired type (protein or DNA) is opened. The user is presented with two windows, one of which contains the database entries, the other contains the query tools. Performing the search may require up to 12 Mb of RAM. The user may need to find the database `Names` file if the program does not know where the files are located. It is best if the size of the database can be limited in some way prior to performing the search, for example by using `Sequence Criteria`. This allows the user to select, for example sequence length or a feature such as `Active Site` as a filter. The search may also be narrowed by using keywords or species or phylogeny. Using the query tools entries may be `Added`, `Subtracted`, or `Restricted`. Once the final set has been chosen the user sets up the query by clicking on the `Create Query` button. A number of different query types such as `Pattern` or `Similarity` are supported and these can be combined with modifiers such as `Precedes` or `Contains` to provide quite complex searches. The search is initiated by clicking on the `Go` (running man) button. The results of the similarity search, for example, are displayed as a spreadsheet with the result `True` or `False` beside each sequence searched. The results may be sorted and individual sequences examined. The results of an actual alignment can be displayed by clicking on the `Create Query`

`Summary` button or for a pair of sequences by selecting `New Protein Dot Matrix` for example.

3. *MacVector* suite

3.1 Overview

3.1.1 System requirements

The *MacVector* suite consists of two parts, the main analytical program *MacVector* and a separate sequence assembly program, *AssemblyLIGN*. *MacVector* is distributed on floppy disks and the version at the time of writing was 4.5, with version 5.0.2 available now. The software requires a minimum of 2 Mb of RAM for system 7 but 4–8 Mb is recommended. If all the files on the distribution disks are installed, at least 3 Mb of hard disk is required.

Sequence searching on CD-ROM is supported, although in contrast to other products *MacVector* does not have its own proprietary formats. Instead the user may choose from the EMBL, NCBI/GenBank, or *Entrez* CDs. Program copy protection is provided by a dongle installed on the Apple desktop bus connected between the keyboard and main chassis. Without this in place the program will not run and a warning message will appear.

3.1.2 Program layout and structure

On starting the program, the user sees a menu-bar containing six choices. These are `Files`, `Edit`, `Options`, `Analyze`, `Database`, and `Windows`. The user should select `Open` from the `File` menu and then select a file to bring up the sequence editor. The top bar of the editor window contains nine icons. Alternatively, selecting `New` and then a file type from the submenu will bring up the appropriate file editor.

The general style of operation is first to select an analysis, then use the window which appears to select or modify the parameters for that analysis. When the analysis is complete a new window appears in which the user may define the display parameters. The results of the analysis are then displayed. In some instances it may be necessary to open an additional file, containing, for example sequence motifs, and select these before running the analysis.

3.1.3 Network version

Network versions of *MacVector* are available which enable a number of users on the same AppleTalk or Ethernet network to share *MacVector* licences. Rather than use a system whereby copies of the program are shared from a central file server, the *MacVector* approach uses a central licence database which can allocate licence tokens to users when they start up their local copy of *MacVector*. The advantage of each user having their own installation of *MacVector* is that network traffic and contention for the central file server is minimized. Network *AssemblyLIGN* usage is also managed in this manner.

3.2 DNA analysis

3.2.1 Entering and editing data

MacVector uses a number of different icons to designate the different file types and these are visible on the Macintosh desktop. There are also special icons for files containing restriction enzymes, sub-sequence (pattern) files, and scoring matrix files, as well as protein and DNA sequence files.

Sequences are entered and edited by typing directly into the editor window, shown in *Figure 5*, or by the usual cut, copy, and paste. This latter feature is active only if the sequence is not locked.

Protein or DNA sequence data may be imported from text files on the hard disk or the CD-ROM databases in GCG, *Staden*, EMBL flat file, Pearson, *FastA*, CODATA, or *DNAStrider* ASCII using the `Open` command. Using the `Save As` option the user can select the export format from *MacVector*, GCG, *Staden*, BioNet, or GenBank styles. If the program is unable to determine the sequence type, the user clicks on a DNA or protein check-box to force *MacVector* into accepting the sequence type. If a DNA sequence is in the currently active window all protein analyses in the `Analyze` window are greyed-out and vice versa.

Correct entry of the sequence can be checked by using the proof-reader. A new window appears, which resembles a video cassette player control panel. Using this, the volume and speed of playback are set by slider controls and the gender of the narrator is chosen by clicking on an icon. The sequence or a portion must be selected before playback can commence. However, by clicking at any point within the sequence and selecting the 'fast forward' or 'fast rewind' icons the sequence can be automatically highlighted. This highlighting continues until the stop button is pressed. Read-back commences after the playback button is selected.

Sequences are annotated by clicking on the book icon in the window bar above the sequence. A new window appears which bears only two icons. The first, a plus sign, defines the type of annotation to be attached, for example definition or keyword. The second icon, a triangle, brings up an editor window into which the comments may be entered. If the annotation is being added for the first time then the plus icon will bring up this second window automatically.

3.2.2 Pattern analysis

First the user must select the enzymes to be used in the analysis. This is done by opening an enzyme file. A new type of window with four icons on the top window bar is displayed. Each line has the name of an enzyme, its recognition sequence with the cut site marked, and the names of the known

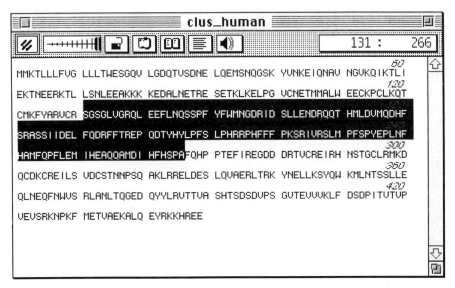

Figure 5. The sequence editor window of *MacVector* with a region of sequence highlighted.

isoschizomers. The user selects individual enzymes by clicking on each one and the enzymes are highlighted by a bullet on the left-hand side of the window. Having selected the enzyme set to be used, the analysis is performed by choosing `Restriction Enzyme` from the `Analysis` menu.

When the display options appear check boxes are used to select listings of restriction sites, showing non-cutters, displaying the map, and fragment prediction. Two of the check boxes can activate extra drop-down menus for showing linear or circular maps and displaying single or double digests. Once the selections are complete, pressing the OK button activates the selections and further windows appear with the results of the analysis. In the map display window, as shown in *Figure 6*, a new cursor, a magnifying glass, appears which can be used to select portions of the sequence by pressing the mouse button and dragging. This activates the zoom-in feature to increase the scale of the map; the final zoom remains a graphical map display. It is not possible to zoom in to view the actual sequence. A double click returns the view to the full scale display.

DNA sequence motifs are known in *MacVector* as sub-sequences. The analysis is performed in a very similar way to restriction enzyme analysis. In this case a nucleic acid sub-sequence file is opened and the required sub-sequences selected. Display options are set and the final results displayed in a number of windows.

A new sub-sequence may be added by clicking on the 'plus sign' icon which invokes the sub-sequence editor. Into this window the user adds

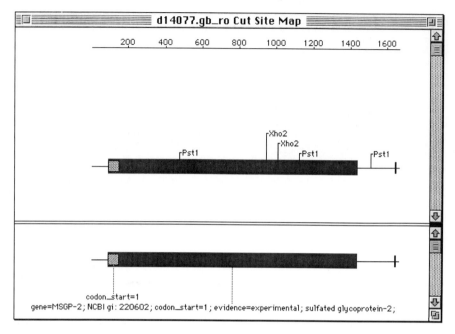

Figure 6. The map display window of *MacVector* showing *Pst*I and *Xho*II sites; the lower panel displays feature table information.

a name for the pattern and can define up to three parts for the pattern which may be combined with the OR logical operator. If the logical AND operator that combines pattern elements is used, then the user can specify the range of the gap between each element in the pattern. In addition, perfect matches and the number of mismatches allowed may be specified prior to initiating the search. Comments may also be entered in to a comments box.

3.2.3 Gene finding

An open reading frame analysis is initiated from the Analyze menu by selecting the Open Reading Frame menu item. A window appears containing the analysis options. The top half of this window contains the analysis type, while the bottom contains the options for setting the genetic code, the region to be analysed, and which strands are to be analysed. The criteria are set using a number of check boxes. These tell the search to look for ORFs which have start and stop codons at the 5′ or 3′ ends of the sequence and which satisfy a minimum protein length. A second type of search uses Fickett's (7) method, where the user may set the minimum length of DNA to be analysed.

When the analysis is complete a display option window appears. Three

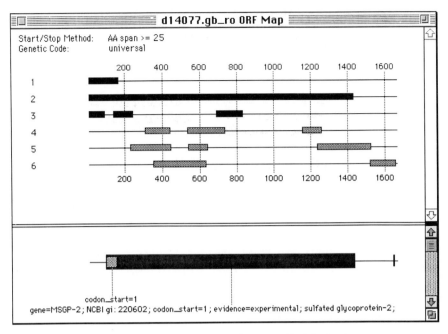

Figure 7. A six frame ORF map with associated feature table information.

check boxes are displayed to select display by ORF map as shown in *Figure 7*, annotated sequence or a list.

Codon bias analysis is selected from the Options menu and is only used with sequences chosen from the GenBank or *Entrez* CDs. The option allows the user to set the sequences to be analysed and then uses the feature table to set the region on which to base the analysis in those sequences. The results can then be saved to disk as a codon bias file.

A codon preference plot can be used to locate regions of sequence with high coding probability. This requires the use of an appropriate codon bias file for the sequence in question. Codon Preference Plot is chosen from the Analyze menu. The options for this analysis allow the user to select the codon bias file, the minimum window size and the genetic code to be used. Once the OK button has been clicked the analysis starts. When it is complete a window containing the preference plots appears.

3.3 Protein analysis

In order to perform any protein analysis the Protein Analysis Toolbox must first be selected from the Analyze window. A window is displayed bearing a number of check boxes representing the different analyses. The user must click on these and may choose between plotting and/or listing the results.

3.3.1 Entering and editing data

The approach used by *MacVector* for entering DNA and protein sequence data has been outlined above. When sequences are being typed the computer will make a tone when invalid characters are entered. Sequences are annotated using a special editor invoked by clicking on the annotations button in the header portion of the sequence editor window. The annotations window contains three buttons bearing a plus, a minus symbol and a delta or triangle symbol. Pressing on the plus symbol displays a submenu of annotation types which the user may select. Once a choice has been made the user fills in the template that appears. The minus button is for removing annotations and clicking on the delta symbol allows the user to change annotations. The program assigns a locus name to the sequence automatically and this cannot be altered.

3.3.2 Structural analysis

The most commonly used plots for hydrophobicity and hydrophilicity are Hopp and Woods (8), and Kyte and Doolittle (1). To this has been added a third, the Goldman, Engelman, and Steitz (9) plot. In order to choose a particular plot the user must select one from a pop-up menu in the Protein Analysis Toolbox. The user should specify the range and window size before running the analysis.

Two types of secondary structure prediction can be performed using *MacVector*. These are Chou-Fasman (2) and Robson-Garnier (3). The results are displayed graphically. In the case of Chou-Fasman, there is no attempt to resolve conflicting predictions. Thus each prediction is displayed as a separate plot. The user clicks on the secondary structure check box in the plot selections of the Protein Analysis Toolbox to activate this analysis.

MacVector can calculate amphiphilicity, pI, surface probability, chain flexibility, and antigenic index. It does not do helical wheel plots. pI, molecular weight and amino acid composition are selected by clicking on the Show option in the Analysis Toolbox.

3.3.3 Pattern analysis

Using this analysis option both chemical and enzymatic cleavage agents may be selected, up to a maximum of six agents. The option Proteolytic Enzyme must be chosen from the Analyze menu. A dialogue box appears which allows the user to select the file containing the cleavage agents and whether all or just selected agents are required. The user may also specify the range of sequence to be analysed. The results of the analysis are displayed after setting the options on the Proteolytic Enzyme Analysis Display. Within this box the display is controlled by selecting the maximum number of cuts the agent may perform, and how the cut map will be displayed, whether by separate or combined maps. The user may also opt to see fragment size predictions as well as listing non-cutters.

Motifs and patterns are referred to in *MacVector* as sub-sequences, and both simple and complex patterns may be used in searches. Complex patterns are built up using AND/OR logic. One can also specify the number of mismatches allowed and whether any residues should match exactly. The results can be displayed as separate or combined maps, lists of motifs found or annotated sequences.

3.4 Special analyses

3.4.1 PCR and primer analysis

Sequencing or hybridization primers may be predicted using the `Sequencing Primers/Probes` option from the `Analyze` menu. A dialogue box appears which allows the user to set the primer selection parameters and the region of interest. The parameters available allow the user to specify the size range of the predicted primers, the desired range of T_m, the range of percentage G + C content, a specified 3′ dinucleotide and the hairpin forming or dimer forming conditions to be excluded. Other parameters define the regions to be analysed, the strand to be used, and primer and salt concentrations. After the analysis has been performed a second dialogue box appears which allows the user to choose a text list and/or a graphical map of the primers, and to filter the reported primers by T_m and/or percentage G + C content. This second dialogue box also displays the number of screened primers, and the reasons why primers are rejected.

Least degenerate hybridization probes may be found using the `Reverse Translation` option from the `Analyze` menu while working in a protein sequence window. The region of interest is specified by amino acid range or from a region in a feature table. The genetic code is chosen from a pop-up menu and alternative genetic codes can be created by using the `Modify Genetic Codes` option from the `Options` menu. A list of the least ambiguous oligonucleotide probes of a specified size range may then be produced.

PCR primers may be found using the `PCR Primer Pairs` option from the `Analyze` menu. The region of interest is specified by either product size or by flanking regions. The primer parameters are selected in the same manner as the sequencing primers described above. After the analysis has been performed a second dialogue box appears which allows the user to choose a text list and/or a graphical map of the primers, and to filter the reported primers by the maximum pair T_m difference. This second dialogue box also displays the number of screened primers, and the reasons why primer pairs are rejected.

3.4.2 Multiple sequence analysis

MacVector allows the user to search a folder of nucleic acid sequence files with a nucleic acid file and create multiple alignments of those sequences that are found to be similar. The user must have a nucleic acid scoring matrix open or available on the hard disk and the nucleic acid query sequence, or a region of interest, cannot be longer than 8000 bases. The folder may contain *MacVector* files and text files; however, the text files must all be either GCG format or one of the IUPAC formats. Files using the *Staden* ambiguity codes should not be used. The `Align to Folder…` option from the `Database` menu brings up a dialogue box which allows the user to set the folder to be searched, to specify whether text files are in GCG format, to set the minimum number of consecutive matches before an alignment is scored (hash value), the number of scores to keep and the scoring matrix to be used. Alignments may be carried out on-the-fly, which is slower, or at the end of the search. Once the search has been completed a second dialogue box appears which displays some statistics of the search and allows the user to filter the results by quality or region of alignment. This dialogue box also allows the user to specify how the matches are displayed, either as an ordered list, a graphical map or as an alignment.

3.4.3 Dot matrix analysis

MacVector allows the user to identify regions of similarity (23) between two sequences of the same type (protein or nucleic acid) or of differing type. The `Pustell DNA Matrix`, `Pustell Protein Matrix`, and `Pustell Protein & DNA` options form the `Analyze` menu and bring up an initial dialogue box which allows the user to select the two sequences and the region of interest, with a limit of up to 8000 bases for the horizontal sequence and 32000 for the vertical sequence. The comparison parameters are also set using this dialogue box, including the scoring matrix to be used. Once the comparison is complete, a second dialogue box appears which displays some statistics about the comparison and allows the user to filter the results by sequence region and similarity score. The comparison may be displayed either as a graphical dot plot with lines, a text dot plot with coding characters and a text alignment. The `Pustell Protein & DNA` option first translates the nucleic acid sequence in either three frames or six frames and these translations are then compared against the protein sequence.

3.4.4 Database searching

MacVector does not provide a proprietary format CD-ROM but instead allows the user the flexibility to use the GenBank or *Entrez* CD-ROMS. The user may either align a sequence to the database (database search) or browse the database (locate and extract a sequence). Either of these may be performed by selecting from the `Database` menu. When browsing, for example GenBank, a new window, called the GenBank browse opens. This has a number of icons down the left-hand side of the window representing the database divisions. There are also three pop-up menus containing the categories to search. Single or combined categories may be used to search the database.

3.5 *AssemblyLIGN*

AssemblyLIGN is distributed on floppy disks and the version at the time of writing was 1.0.7. The software requires a minimum of 1.2 Mb of RAM for system 7 but 2–4 Mb is recommended. If all the files on the distribution disks are installed at least 1 Mb of hard disk is required. Program copy protection is provided by a dongle installed on the Apple desktop bus connected between the keyboard and main chassis. Without this in place the program will not run and a warning message will appear. Network versions of *AssemblyLIGN* are available which enable a number users on the same AppleTalk or Ethernet network to share *AssemblyLIGN* licences as described above.

On starting the program the user sees a menu-bar containing six choices. These are `File`, `Edit`, `Project`, `Contig`, `Sequence`, and `Windows`. To start work the user should select `New Project` or `Open Project` from the `File` menu. New sequence data may be typed into a window of a simple editor using the `New Fragment` option from the `Sequence` menu. Existing data in TEXT, *MacVector*, GCG, *GeneWorks, GeneJockey, Gene Construction Kit, DNA Strider*, or *DNAStar* formats may be imported using the `Import Fragment...` option. Specific sequences such as restriction sites, vectors, or primers are entered or imported as `Templates`, and can be used to screen the fragments in a project automatically on entry or manually by dragging a fragment onto the template icon.

The parameters for sequence assembly are set using `Assembly Options...` from the `Project` menu. These include the minimum overlap length, the per cent stringency, the consensus cut-off, and if fragments are to be aligned with or without gaps. Once these parameters are chosen, some or all fragments can be assembled, either using the `Assemble` command from the `Project` menu, or by dragging a group of highlighted fragments onto another fragment icon in the project window.

Once a number of fragments have been assembled the fragment icons in the project window are replaced by contig icons. Double clicking on a contig or using the `Open Selection` command from the `Project` menu opens the Contig Map window. The window displays the fragments and their orientations in the contig, the plus and minus strand consensus sequences and has a tool palette down the left side. The contigs may be edited using either a text editor or a colour editor that displays the sequence as colour bars (or patterns on black-and-white monitors). Both editors highlight ambiguous bases, as shown in *Figure 8*.

4. *DNAStar*

4.1 Overview

DNAStar requires at least 4 Mb of RAM and 14.8 Mb of hard disk space as a minimum configuration. DNA and protein databases are distributed on a set

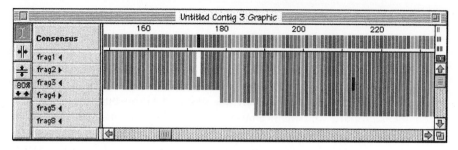

Figure 8. The colour editor of *AssemblyLIGN* displays the contig in coloured bars highlighting ambiguous bases.

of two CD-ROMs which include the software for the programs. A separate installation (floppy) disk is required to install from the CD.

Due to its modular nature *DNAStar* is different from the other programs discussed so far. There are a number of separate programs which can be used individually or accessed via a central program called the *LaserGene Navigator*. Six buttons activate sequence editing and analysis, restriction analysis and mapping, multiple sequence alignment, sequence project management, and Biological Database Resource and protein analysis. Two further buttons give access to on-line help (there is no printed manual) and quitting the *Navigator*.

No hardware copy protection device is required. The user may require a CD-ROM drive for use of the Biological Database Resource.

4.2 Sequence editing

The *EditSeq* program is used for entering and editing protein or DNA sequence data and is accessible from the *LaserGene Navigator* as Sequence Editing & Analysis. Starting the program opens a split pane window. The user can select from New DNA or New protein file types. An icon in the top left-hand corner indicates the sequence type. The top pane is used for entering the sequence, the bottom pane is used for entering sequence annotations. Along the bottom of the window there are five buttons, a hand icon to terminate proof-reading, a mouth icon to initiate proof-reading, and three buttons associated with searching the sequence. The main search button brings up a dialogue box to specify the nature of the search, and the other two buttons are to used to indicate the direction of the search from the current insertion point. There is also a pull down Search menu. The program supports proof-reading of the sequence either on entry or read back by selecting Proof Read from the Digitizer menu. To obtain proof-reading on entry the user must select Macintosh Voice or Tone from the Digitizer menu prior to entering the sequence. The user can also perform a basic compositional analysis as well as translations using the Goodies menu.

4.2.1 Annotation of sequence data

A sequence may be annotated using free text entry. When the sequence is created a minimum annotation of time and date is automatically entered. If a sequence is imported, database entry annotations are retained. When new sequences are generated, for example from translations or backtranslations, the annotations include information about the genetic code used.

4.2.2 Import formats

The following formats can be imported into *EditSeq*: *DNA Strider*, ASCII, BIONET, GCG, Mount Conrad, *Gene Jockey*, *MacVector*, IntelliGenetics, and ABI trace files.

4.2.3 Translations and backtranslations

The program can perform translations or backtranslations of sequences using either the standard genetic code or a number of other codes such as mito-chondrial genetic codes. The user must select the requisite code from a cus-tomizable menu. The options are found in the `Goodies` menu. The translated or backtranslated sequence appears in a new window complete with basic annotations.

4.3 Pattern analysis

The *MapDraw* program is used for creating restriction enzyme maps of DNA sequences and is accessible via the *LaserGene Navigator* as `Restriction Analysis & Mapping`. The user may select either to work on a `New` sequence or to `Open` an existing map. If the user selects `New` a dialogue box prompts the user to find the sequence file, if `Open` is chosen the user must locate a map file. Once this has been selected the resulting maps can be dis-played showing the sequence, restriction sites, and translations. The transla-tions can be three frame from either strand or a full six frame translation. A variety of map types are available, including a schematic map (or worksheet), a series of linear charts (or minimap) and linear or circular illustrations. The bottom of the map window contains three buttons associated with searching activities. Two control the direction of the search from the current insertion point while the third is used to specify the search type. The options include literal or ambiguous as well as open reading frame or restriction site searches.

The program comes supplied with a file of restriction enzymes. Additional enzymes can be added by the user, and further customization of the enzyme lists is possible using filters to create subsets of enzymes. These can be based on criteria such as overhang type or frequency. Information about restriction sites found on the sequence may be sorted and displayed by position, enzyme, or frequency. Choosing any one of these options will open a new window showing the information in list form.

DNA sequences can be translated to protein using a number of different genetic codes, and ORFs may be determined as described in Section 4.2.3.

Annotations or features may be added to the display. The options for this are found in the `Features` menu, however, most of these options are greyed-out until a range of the sequence is highlighted and a feature created with the option `New Feature`. This brings up a new window which includes text areas, check boxes, and buttons where the feature can be given a name and a title. It can also be given a `Description` and `Style`. The description includes the standard feature table key words and the style controls the graphical appearance, for example, colour or shape. The check boxes control whether the feature will appear in the different forms of map display.

4.4 Protein analysis

The *Protean* program is used to analyse protein sequences and is accessible via the *LaserGene Navigator*. On starting the program the user may select from `New` or `Open`. If a new sequence is selected the display switches to a window of graphical displays with a button toolbar on the left-hand side of the screen. Analyses may also be displayed in a tabular form. In the top left- and right-hand corners of the main window are two small round objects. These are the curtain pulls. Grabbing one of these with the mouse and pulling it across will display the list of available analyses. The left-hand curtain displays the analysis list in a manner resembling the `View by name` option of the Macintosh Finder. Double clicking on an analysis brings up a parameters dialogue box.

The analyses available cover physical properties such as pI titration curves, prediction of regions of positive or negative charge, compositional analysis, and the Jameson–Wolf (10) antigenicity prediction. Other analyses cover secondary structure prediction using the Chou–Fasman (2), Robson–Garnier (3), and Roux–Deleage (11) methods, and alpha helices may be predicted using the Goldman–Engelman–Steitz (9) method. Both helical wheel and helical net analyses are available. The helical wheel shows the arrangement of residues in an alpha helix as if the observer is looking down the core of the helix. The helical net view displays the helix as if it has been cut open and unfolded. In addition a beta net shows how the amino acids are arranged on opposite sides of a beta strand.

Other analyses include the Ariadne system (12) of hierarchical secondary structure patterns that facilitate the search for similarities at a level higher than primary sequence, and displays of linear space filling models and carbon skeleton backbones using specially designed fonts with each amino acid represented according to the single letter code.

Any combination of analyses may be saved as a method outline. If the outline is saved as the default outline. New analyses will be analysed with this set of analyses and associated parameters and display formats.

4.5 Special analyses

4.5.1 PCR

The *PrimerSelect* program is used to find and test PCR and other primers and is accessible via the *LaserGene Navigator*. On starting the program the user may select from New or Open. If a new analysis is selected a window opens with a button toolbar on the left-hand side of the screen. Enter Sequence... from the File menu allows the user to choose the sequence to be analysed. The sequence is displayed in a window that shows either the T_m or ΔG profiles of the top and bottom strands.

Using the Conditions menu the user can define physical conditions such as salt (Na^+) concentration, initial primer concentration and the temperature standard used for computing free energy (ΔG). Primer characteristics may also be defined, including primer length, melting temperature, stability, 3' terminal stability, the uniqueness of the 3' terminal sequence, and primer location. Template composition, length of product, and mismatching between the template and primers can also be used. The Options menu sets the conditions for *PrimerSelect* to calculate thermodynamics based on the data for DNA–DNA, DNA–RNA, or RNA–RNA base pairing, whether repetitive sequences should be avoided, and what sort of backtranslation code to use. The Locate menu is used to identify sequencing Primers and Probes or PCR Primer Pairs using the specified conditions and both strands of the template. Sequencing primers and probes are displayed in a new window that lists the location, length, T_m, ΔG, and ΔG profile for each primer in a tabular form. PCR primer pairs are displayed in the upper half of a graphical window and are sorted by quality, the lower half of the window displays either alternative pairs or alternative products for a pair selected from the upper display. If a primer pair or probe is double clicked the corresponding sequence or region is displayed in the sequence display window.

The Report menu allows the user to generate a number of reports, such as Primer Self Dimers, Primer Pair Dimers, and Primer Hairpins. The Log menu includes a Project Notebook, a Primer Catalogue, and an Oligo Order Form. A user supplied catalogue of primers can be included in the analysis. The identified primers are searched to find compatible pairs for PCR amplification, excluding those with products outside the desired range or those that form primer dimers.

To help design primers Work on Upper Primer or Work on Lower Primer from the Edit menu opens a Primer Workbench Window which can display backtranslations and restriction sites. Any edits in the primer sequence are reflected in changes in the displayed sites and backtranslations.

4.5.2 Gel assembly

The *SeqMan* program is used to assemble contigs and manage data from sequencing projects and is accessible via the *LaserGene Navigator*. On starting the program the user may select from New, Open, or Import. If a new project is selected an untitled window opens. The `Add` option of the `Sequences` menu is used to enter sequences into a project. This opens the Unassembled Sequences window that has five buttons, `Assemble`, `Add Sequences`, `Set Ends`, `Trim 3' End`, and `Options`. The options button controls scanning for vector or contaminating sequences and optimizing the order of sequences in a contig. The trim 3' button is used for automatic sequencer trace file data. The set ends button is used to specify which region of a sequence is to be used in the assembly. The assembly button is used to assemble contigs from the list of entered sequences. Vectors can be associated with a particular sequence so that these can be screened and trimmed off. The user can search for vectors from a vector catalogue.

As the sequences are assembled, a new `Report` window opens that displays which contig each sequence is added into and the parameters used. These are set using the `Parameters` option from the `Project` menu. The details of the contigs are displayed in the upper half of the project window. Clicking on a contig displays its component sequences in the lower half of the window. Double clicking on a contig or one of its sequences opens the editor window, displaying the sequence of the fragments and the consensus with any discrepancies highlighted in red. The user can edit individual sequences or the consensus. Raw trace data can be displayed and the amplitude of the display changed. This is done by choosing `Show Trace Data` from the `Sequence` menu and resizing the window.

The program uses the Martinez algorithm (13) to align newly entered sequences with existing contigs. The Needleman–Wunsch algorithm (14) is used to refine the alignment. If no alignment is found the reading becomes a new contig.

There are a number of special options. For example, the option `Contaminating sequence` allows the user to ensure that any known contaminant DNA is removed from the project and there is a similar option for repetitive sequences. In both instances the user stores the offending sequences in a special folder. Sequences containing such DNA are assembled last.

The user may force two contigs into one by selecting the `Force joins` option from the `Contig` menu. There are two alternatives: either to have manual joins or to have gapped joins. The contigs to be merged are selected in the first instance from the Project window. Once a contig has been assembled and edited to the user's satisfaction the consensus may be saved using the `Save Consensus` option of the `Contig` menu, while the project as a whole is `Saved` from the `File` menu.

4.5.3 Multiple sequence alignment and dot-matrix comparison

Within the *DNAStar* package there are two programs that will produce sequence alignments and dot-matrix comparisons. *MegAlign* is used for multiple sequence alignments and dot-matrix comparisons whereas *Align* is used for pairwise alignments and dot-matrix comparisons.

The *MegAlign* program can be used to create multiple sequence alignments, and alignments or dotplots between pairs of sequences. The program can be launched from the *LaserGene Navigator*. On starting the program the user must specify New to open a new alignment view. Sequences are then added by selecting the Enter sequences... option from the File menu.

A number of methods are available for comparing sequences and the display of the aligned sequences can be controlled by the user using a method termed decoration. Decorations are rules for highlighting matching or mismatching residues and are displayed in the Alignment report.

The multiple sequence alignment method is selected from the Align menu. Sequences are displayed in a window, to the left of which is a toolbar. This contains such tools as a button for translating DNA sequences to the corresponding protein or toggling between coloured and black residues and bases.

The Jotun Hein method (15) is a phylogeny based technique which builds a graph of all possible alignments based on pairwise comparison of sequences and uses this to guide the final alignment. The Clustal method uses the Clustal V algorithm (16). Sequences are grouped into clusters by examining the distances between all sequence pairs. This is also a phylogeny-based method.

There are some methods which are specifically for pairs of sequences only. These are the Wilbur and Lipman method (17), the Martinez/Needleman–Wunsch method (13, 14), and the Lipman–Pearson method (18). The Wilbur–Lipman method, which is for DNA only, constructs tables of sequence words (*k*-tuples) which are used to find regions of similarity. The Martinez/Needleman–Wunsch, which is also for DNA only, initially finds perfect matches. These regions are then optimized using the Needleman–Wunsch algorithm. The Lipman–Pearson method constructs tables of sequence words (*k*-tuples) which are used to find regions of similarity. The alignments are then optimized using a version of the Needleman–Wunsch algorithm. This method is intended for use with protein sequences.

Dotplots are used to give a graphic display of regions of similarity by placing dots on a 2D matrix. The parameter settings for the dotplot method are window and percentage match. The density of information shown on the plot can be set by filtering the display. The aligned pair of any diagonal may be displayed by clicking on the diagonal to select it, then choosing a button from the tool bar located at the side of the dotplot window.

The *Align* program is used to create alignments between pairs of sequences. It is not available through *LaserGene Navigator*. The user can create dotplots of the sequence pairs as well as viewing alignments with matching residues or bases marked out. The methods used are Wilbur–Lipman (17), Martinez (13)/Needleman–Wunsch (14) and Lipman–Pearson (18) as discussed above in Section 4.5.3. A novel feature of this program is its batch processing capability. Lists of alignments may be created and saved for later processing.

4.5.4 Database searching

The *GeneMan* program is used for database searching and manipulation and is accessible via the *LaserGene Navigator*. The databases supported are a merged set of GenBank and EMBL, GenBank EST and the databases on one CD. The protein database is found on a separate CD and contains PIR, Swiss-Prot, and GB (GenBank) Translated merged set. Medline Genetic abstracts and Prosite are also included. The NRL_3D database is included once a year after an annual update. If the databases are unavailable the program will display a warning and request that the CD be inserted into the drive. It displays a further message asking the user to shutdown and restart or use files from disk. If portions of the databases are copied onto the hard disk, searches will be speeded up. The databases can be browsed, or searched using either text to search through the feature table fields or sequence similarity, keywords or patterns. Searches may be combined using AND/OR/NOT logical operators in a drag and drop query building interface.

The results of searches are displayed in a new window. For text searches the results may be displayed as either single line or expanded to show the sequence and feature table or comments fields.

4.5.5 Miscellaneous analyses

The *XrayViewer* program is usually launched directly or from within *GeneMan* to display 3D structures using the data in PDB entries. These structures may be displayed as wire frames or space-filled models and with or without the amino acid side chains. Stereo views are available for the wire frame models, however special glasses are required. Each view may be freely rotated using either the 3D axes or a square grid. These are selected from the X-ray menu by selecting either the option `Virtual Sphere Controller` or `Sliders Control` the view point may be zoomed in or out, and all views may be exported as Macintosh PICT files.

5. *Sequencher*

5.1 Overview

The program, which is installed from floppy disks, requires a hardware copy protection device. The total installation space required is about 3.8 Mb (PowerMac version) and the minimum RAM requirement is 3.5 Mb. The program works with Apple system 7 and supports drag and drop. The current version is 3.01. The basic program is network-aware and the simultaneous use of the program is managed by network management software. In short the program may be installed on a number of machines but can only be run by a number up to and including the number of available licences.

The program is started in the usual way by double clicking on the *Sequencher* icon and a new untitled project window appears. Seven menus and five buttons are associated with this window. The buttons include a

Parameters, Assemble Interactively, and Assemble Automatically options. The menus include the usual File and Edit menus as well as Select, View, Fragment, Contig, and Special menus. Old gel readings are imported into a project, whereas new gel readings are added as New Fragments from the Sequence menu. The program supports voice readback for sequence verification.

There is a comprehensive set of user definable options which can be set from within a User Preferences dialogue box. This is invoked from the Windows menu. The user can select from a number of preferences such as the minimum time between auto saves (from 1–15 min), and the number of minutes the keyboard is idle before an auto save (1–4 min). Balloon help has been implemented and the user may select a time delay before the balloon appears. There is a also a features preference which controls the display style.

5.2 Entering sequences

On selecting the New Fragment option a new window appears containing four buttons, including Cut Map, Find, and Ruler. The latter controls the formatting of the editing window as well as whether a restriction map should be shown. A fourth button is the Assembly Parameters. When this is selected a window is displayed which contains slider controls for controlling the minimum match and overlap and the maximum loop size, and two radio buttons for controlling the accuracy of the data. This is of interest to users whose data are derived from fluorescent sequencers, where the quality of the data degrades towards the end of the trace but may still be of value in the final assembly. Furthermore, users who are building large projects may create sub-projects by highlighting items in the project window and using the Export Selection as Project.

There is also a button for displaying a map of start and stop codons which updates as the sequence is typed in. A number of different import formats are supported that allow the user to import gel readings from TEXT files, GCG sequence files, and contig files, popular molecular biology software application files, and automatic sequencer trace files.

5.3 Assembling the data

Data are assembled automatically or interactively. Assembling automatically brings up a status window which displays messages as the aligning process occurs. The final display shows the number of comparisons made, the construction performed, and the time taken.

Double clicking on a contig icon brings up a new window which is an overview map of the gel readings comprising the current project. There are four buttons associated with this window, an Overview button which toggles between the overview map and the editing window, a Sort, Option, and Find button.

When assembling interactively the user is presented with a new window which has three major boxes in it. One contains the name of the candidate sequences. On selecting one from this window it is then compared with the other gels or contigs. A scored match window which shows overlaps, gaps, and mismatches between the candidate and target sequences. Thirdly, a window with information about the sequences being built shows the assembled contigs and the unaligned sequences. Having accrued this information the user can then alter the match parameters and try to align the remaining sequences. Readings may also be removed from the project and contigs disassembled by using the `Gel Disassemble` option. Where the user is working on very large projects with potentially hundreds of gel readings, new fragments are added in a different way. The command `Add Selected Items to Others` is chosen from the `Assemble Contigs` in the `Contig` menu. This compares only new gel readings, which have been highlighted, to the existing items in the project

At this stage the user can perform a restriction site analysis selecting either unique, double, triple, or four or more cutters. The user should first select the enzymes to be used. A window counting the enzymes listed in alphabetical order also shows the cut site and salt concentration information when the enzyme is highlighted. The enzyme can be included in the analysis by clicking on the left-hand side of the window. The results can be displayed as a cut position list with sizes or as single or multiple lines.

5.4 Editing the data

When editing the sequences, the altered bases can be highlighted in colour. The bottom of the screen contains a status bar which shows the quality of the consensus. Two symbols are used to indicate gaps or contig ambiguities. These change as the editing process progresses. Using a command key combination during editing, the user may jump from one base ambiguity or contig disagreement to the next very rapidly. This feature speeds up the editing process considerably. If automatic sequencer trace files are incorporated into a contig the original chromatograms may be viewed as four colours or black and white shading, and the base calls can be edited. Sequence data may be exported in a number of formats including ASCII, GenBank, GCG, IntelliGenetics, *DNAStrider*, and SCF (Standard Chromatogram File).

Other options include searching for vector contamination, viewing protein translations, or restriction maps.

6. *Amplify*

6.1 Overview

Amplify, a freeware application, is usually installed from a self-extracting archive and requires at least 400 Kb of hard disk space and 1 Mb of RAM. At the time of writing the current version was 1.2, with version 2.0 now available.

6.2 Running the program

The program is launched by double clicking on the application icon, or a previously saved primer list or sequence file. Six menus will be visible on the menu bar: `File`, `Edit`, `Window`, `Primers`, `Sequence`, and `PCR`. Sequences and primer lists are first opened using `Open Target Sequence` and `Open Primer List` from the `File` menu. A sequence file is a text file that can have an optional header that ends with two periods ('..'), anything following this is assumed to be part of the sequence. The `Preferences...` command from the `Edit` menu can be used to change the header delimiter to any two-character string if you wish to include '..' as part of the header text. Characters in a sequence after the header that are not A, C, G, T, or N (upper or lower case) are removed during analysis, and the data is reformatted to 80 characters to a line. Using the `Save As...` option from the `File` menu will create an *Amplify* sequence file. If the `Save As...` option from the `File` menu is used and the file is given a `.pri` extension an *Amplify* primer file will be created. Both the primer list and the sequence data may be edited and saved from within their respective windows.

Once a target sequence and a primer list have been opened, the user selects the primers for analysis by clicking on primers in the primer list window and selecting `Use This Primer` from the `Primers` menu. The selected primer is added to the `Primers in use` window, and a PCR simulation can be performed using `Run PCR` from the `PCR` menu. The primers are then checked against themselves and each other for potential primer dimers, the sequence is then searched for matches, in both orientations, to the primers. A graphical map of the target sequence is presented displaying arrowheads in the positions of the primers that bind the sequence as shown in *Figure 9*. Potential PCR products are displayed as horizontal bars, the vertical thickness represent the likelihood of amplification. Any potential primer dimer pairs are also displayed.

The graphical map elements have additional information available that may be viewed by clicking on them. An Info window then appears with these details. For example, clicking on a primer arrowhead displays the quality of the match and clicking on a PCR product displays the size and some composition information. The graphical map can be exported as a Macintosh PICT file by `Selecting All` from the `Edit` menu, and then using `Save As...` from the `File` menu. The Info window may be saved to a text file.

Version 2.0 of *Amplify* has a number of new features: the T_m of a primer can be calculated, the target sequence can include N as well as A, C, G, and T. Primers may contain degenerate characters such as Y or R and the on-line help window no longer needs to be closed before continuing use of the software.

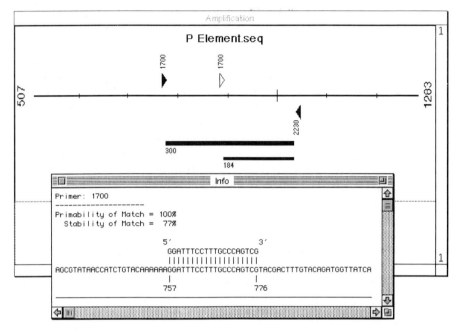

Figure 9. The amplification map showing primer matches and products and the info window displaying primer match details.

7. *MacPattern*

7.1 Overview

MacPattern is a public domain program. It is well designed and conforms to the standard Macintosh interface guidelines. The program can be run on all machines from Macintosh Plus and upward. The program runs under Apple system 7, and supports balloon help. An on-line manual is provided, which can also be printed out. The program package is generally installed from a self-extracting archive and requires about 540 Kb for the basic program (version 3.4), the on-line manual, and a Read-Me document. At least another 5 Mb is required if the user wishes to use databases such as Prosite and BLOCKS. The minimum suggested memory size is 1500 Kb. Although a maths co-processor is not essential it improves program performance.

7.2 Running the program

In its simplest form, the program can be loaded with a DNA or protein sequence, presented in a wide variety of sequence formats. A user-supplied motif or pattern database can scan the input sequence for matches. The program reports the number and location of any identified matches. The results can be sorted by pattern or location. Pattern databases can be Prosite (19), BLOCKS (20), or TFD (21).

The user starts the program in the normal Macintosh way and is presented with a menu bar. If this is the first time, then indices must be built for the pattern database(s). This is accomplished simply by selecting the menu command `New Index` from the `File` menu (there is a different submenu for Prosite or BLOCKS). Once the indices have been built (and this must be done for each new release of a database) the user loads the index file by giving the command `Open Database File` from the `File` menu. The sequence is treated similarly by using the command `Load Sequence File` from the same menu. At its most basic, the analysis is performed by having the database index file window active and then choosing the appropriate search command from the Prosite or BLOCKS menu. The results are then displayed in a new window. Although only one sequence file may be analysed at a time, the file may contain many entries. In addition the user may also enter custom motifs following the Prosite pattern syntax. The large number of patterns in Prosite or BLOCKS may be reduced to more manageable proportions or sets, as they are known, by selecting the desired members and saving them to a new set.

In the current version of *MacPattern* (3.4) a number of statistical analyses for identifying unusual regions are also presented. Maximum segment scoring (MSS) is useful for identifying charge clusters, hydrophobic regions, cysteine clusters and regions such as DNA-binding or transmembrane domains. It is also possible to set up your own scoring scheme for MSS providing you are adept with *ResEdit*. New to release 3.2 of *MacPattern* is an implementation of an algorithm by Eguchi and Seto (22) which identifies the most dissimilar region in a sequence, that is to say the most information-rich region. This may therefore represent the region most likely to have a functional or structural role.

The results from a Prosite search, as shown in *Figure 10*, are relatively easy to interpret: either a pattern is found or it is not. In some cases, such as kinase sites, many apparent matches can be reported (hence the relegation of these patterns to the skipped set). In the case of a BLOCKS search, the block is converted into a scoring matrix where an allowance is made for contributions from multiple closely related sequences. Compensation is also made for any amino acid bias using a frequency table calculated from the latest release of Swiss-Prot. This is then used to evaluate the input sequence. The user may set the number of highest scoring results to be reported. A score is given for each reported BLOCK. In general terms, if the score is above 1000 then the match is interesting. Matching residues are shown in capitals, unmatched are in lower case.

Although we have said that it is relatively easy to interpret Prosite Searches, it should be remembered that a pattern will only be as good as its definition. This will have been based on known members of a sequence fam-

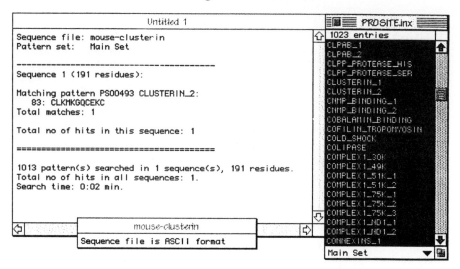

Figure 10. The three windows of a Prosite search; the database window, the sequence window, and the results window.

ily and use of such a pattern may give a positive or negative match. It could be argued that to allow mismatches might include distant relatives but in actual fact may also degrade the pattern's usefulness. However, using a protein BLOCK as the search 'pattern' takes into consideration the type of sequence drift that the most closely related protein families exhibit. It is probably the next best thing to a region-specific score matrix for every family of proteins. It is possible therefore that where Prosite will fail to show any matches, BLOCKS will identify something.

By adding in analyses such as the Eguchi and Seto algorithm and the maximal segment score analysis the program becomes a powerful analytical tool. It is an application that fits well within the Macintosh philosophy, it is easy to use and is highly recommended. The program is well supported, and frequently updated. The pattern databases such as Prosite and BLOCKS are also regularly updated, making this a set of tools no molecular biologist should be without.

8. Other programs

This section includes useful programs which we have not had space to review fully (*Table 1*). Where we have indicated that a program is available by FTP the user should note the following.

All programs may be obtained from the *File Transfer Protocol* (*FTP*) and World Wide Web sites mentioned in Sections 8.2.1 and 8.2.2, although not all sites may hold copies of all the programs (see also Chapter 4). You may

Table 1. Other useful programs

Type	Name	Sources
Commercial sequence analysis	DNASis	National Biosciences Inc.
	GeneJockey II	BioSoft
	Gene Construction Kit	Text Co Inc.
	DNA Strider	Christian Marck
Alignment/phylogeny	ClustalW	FTP
	DottyPlot	FTP
	SeqVu	FTP
	Phylip	FTP
	MACAW	FTP
Sequence editor	DNAid	FTP
	ReadSeq	FTP
	SeqSpeak	FTP
	Speakquencer	FTP
Analytical	MacP12	FTP
	TopPred II	FTP
	PlotA	FTP
	MitoProt	FTP
	MacStripe	FTP
Structure	LoopDLoop	FTP
	MulFold	FTP
	MacMolecule	FTP
	RasMac	FTP
Genome mapping	MacAce	FTP
	MapManager	FTP

also need some basic utilities before attempting to retrieve the programs. Most of the programs may be stored in a compressed and encoded format. If this is the case you will need, for example, *BinHex4.0* to decode files ending in .hqx, *Stuffit* to decompress files ending in .sit, or *CompactPro* for files ending in .cpt. Occasionally files will be self-extracting, this is sometimes designated by .sea (stands for Self Extracting Archive). A very useful utility is *Stuffit Expander* which can handle .sit, .sea, and .cpt files. Some sites use *GnuZip* to compress files further, this is usually indicated in the log-in message or a README file. For those files which generally have the extension .gz, you will need *MacGZip*. Note that most terminal emulation programs support some sort of *FTP* protocol. However, a utility such as *Fetch* which only performs *FTP* may be useful.

8.1 Suppliers

Commercially available software is available from the following suppliers (addresses given in the Appendix): *GeneWorks* (Oxford Molecular Ltd (UK), IntelliGenetics Inc. (USA)); *MacVector* (Scientific Imaging Systems (UK), Eastman Kodak Co. (USA)); *DNAStar* (DNAStar Ltd); *Sequencher* (Gene Codes Corporation); *DNASis* (MedProbe A.S.; National Biosciences

Inc.); *GeneJockey II* (BioSoft); *Gene Construction Kit* (Textco Inc.); *DNAStrider* (Christian Marck).

8.2 Internet sources

8.2.1 FTP sites

```
ftp://ftp.ebi.ac.uk/pub/software/mac/
ftp://ftp.bio.indiana.edu/molbio/mac/
ftp://sumex-aim.stanford.edu/info-mac/sci
```
(mirror sites across the world)

8.2.2 WWW sites

These sites are mentioned as examples of sources of software only. They are not intended to be a comprehensive list of services or service providers. There are many hundreds of molecular biology World Wide Web sites many more of which are described in Chapter 4.

```
http://pubweb.nexor.co.uk/public/mac/archive/data
/misc/molbio/index.html
http://www.umich.edu/~archive/mac/misc/molbio/
```

References

1. Kyte, J. and Doolittle, R. F. (1982). *J. Mol. Biol.*, **157**, 105–32.
2. Chou, P. Y. and Fasman, G. D. (1978). *Adv. Enzymol.*, **47**, 45–148.
3. Garnier, J., Osguthorpe, D. J., and Robson, B. (1978). *J. Mol. Biol.*, **120**, 97–120.
4. Argos, P., Rao, J. K. M., and Hargrave, P. A. (1982). *Eur. J. Biochem.*, **128**, 565–75.
5. Eisenberg, D., Weiss, R. M., and Terwilliger, T. C. (1984). *Proc. Natl. Acad. Sci. USA*, **81**, 140–4.
6. Cornette, J. L., Cease, K. B., Margalit, H., Spounge, J. L., Bersofsky, J. A., and DeLisi, C. (1987). *J. Mol. Biol.*, **195**, 659–85.
7. Fickett, J. W. (1982). *Nucleic Acids Res.*, **10**, 5303–18.
8. Hopp, T. P. and Woods, K. R. (1981). *Proc. Natl. Acad. Sci. USA*, **78**, 3824–8.
9. Engelman, D. M., Steitz, T. A., and Goldman, A. (1986). *Annu. Rev. Biophys. Chem.*, **15**, 321–53.
10. Jameson, B. A. and Wolf, H. (1988). *Comput. Appl. Biosci.*, **4**, 181–6.
11. Deleage, G. and Roux, B. (1987). *Protein Eng.*, **1**, 289–94.
12. Lathrop, R. H., Webster, T. A., and Smith, T. F. (1987). *Commun. ACM*, **30**, 909–21.
13. Martinez H. M. (1988). *Nucleic Acids Res.*, **16**, 1683–91.
14. Needleman, S. B. and Wunsch, C. D. (1970). *J. Mol. Biol.*, **48**, 443–53.
15. Hein, J. (1989). *Mol. Biol. Evol.*, **6**, 669–84.
16. Higgins, D. G., Bleasby, A. J. and Fuchs, R. (1992). *Comput. Appl. Biosci.*, **8**, 189–92.
17. Wilbur, W. J. and Lipman, D. J. (1983). *Proc. Natl. Acad. Sci. USA*, **80**, 726–30.

18. Lipman, D. J. and Pearson, W. R. (1985). *Science,* **227**, 1435–40.
19. Bairoch, A. (1991). *Nucleic Acids Res.*, **19**, 2241–45.
20. Henikoff, S. and Henikoff, J. G. (1991). *Nucleic Acids Res.*, **19**, 6565–72.
21. Ghosh, D. (1992). *Nucleic Acids Res.*, **20**, 2091–3.
22. Eguchi, Y. and Seto, Y. (1992). *Genome Informatics Workshop III*, 14–15.
23. Pustell, J. M. and Kafatos, K. C. (1982). *Nucleic Acids Res.* **10**, 4765–82.

Further reading

Sequence comparison and alignment

STEPHEN F. ALTSCHUL

1. Introduction

DNA and protein sequences are compared and aligned in order to elucidate common structure or function. A great variety of alignment tools have been developed over the past 25 years, appropriate to different biological questions. These tools can be divided roughly into **global** and **local** algorithms, which in turn separate into **pairwise** (two sequence) and **multiple** (more than two sequence) alignment methods. For the user of these methods, three central questions arise:

(a) *Definitions*. Precisely what sort of alignment is sought? How is the quality of an alignment measured?

(b) *Algorithms*. How is an optimal or near optimal alignment found efficiently?

(c) *Statistics*. How good must an alignment be before it may be considered surprising? What can be expected to arise purely by chance?

There is a certain tension between the first of these questions and the other two, for a more complicated definition may better reflect biology, but it is generally easier to find efficient algorithms and rigorous statistics for more simply stated problems. The user of sequence comparison tools should therefore be aware of the trade-offs implicit in the choice of any given approach. This chapter will consider the main classes of sequence comparison methods, discussing for each when it is appropriate, what are its computational costs, how to choose appropriate parameters, and when to consider its output statistically significant.

2. Global sequence alignment

2.1 Algorithms

The standard method for globally aligning two protein or DNA sequences,

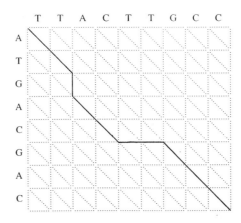

Figure 1. An alignment path graph.

dynamic programming (1–4), is best illustrated with the aid of a **path graph** (*Figure 1*). One sequence is written down the left of the graph, another along the top. Every possible forward path through the graph starting at the upper left-hand corner and ending at the lower right corresponds to a unique alignment. Diagonal edges represent the alignment of a letter from each of the two sequences, while horizontal or vertical edges represent the insertion or deletion of a letter. For example, the path shown in *Figure 1* corresponds to the following alignment:

$$
\begin{array}{ccccccccc}
\text{T} & \text{T} & - & \text{A} & \text{C} & \text{T} & \text{T} & \text{G} & \text{C} & \text{C} \\
\text{A} & \text{T} & \text{G} & \text{A} & \text{C} & - & - & \text{G} & \text{A} & \text{C}
\end{array}
$$

In order to evaluate the quality of different alignments, scores are usually assigned to all possible pairs of aligned letters, as well as to letters in one sequence aligned with null characters inserted into the other. These scores can be thought of as associated with the edges of a path graph. The score for a path through the graph (i.e. an alignment) is then taken to be the sum of its edge scores. For example, if each aligned pair of identical letters receives a score of 2, each aligned pair of non-identical letters a 0, and each letter aligned with a null a -1, the alignment above has score 7. One may consider high scores desirable, and thus seek the alignment(s) with the greatest possible score, or conversely consider low scores preferable and seek to minimize the alignment score. These two approaches are equivalent, and the former convention is followed in this chapter.

The dynamic programming algorithm is based on the observation that an optimal path ending at a given node must pass through one of the three immediately preceding nodes. Therefore, if the optimal scores for reaching all the relevant preceding nodes are already known, one need maximize over only three numbers to calculate the optimal score for reaching a node. This allows one to 'fill in' scores for all the nodes of a path graph in order, with only a fixed amount of calculation performed for each node. The total time required is proportional to the product of the lengths of the two sequences compared. Refinement of this basic algorithm can reduce substantially the number of nodes that need to be considered to produce an optimal alignment (5–7). In order to generate not only the optimal alignment score, but an optimal alignment as well, 'traceback' information must be stored at each node. This can be done with minimal demands on computer memory if only a single, as opposed to every, optimal alignment is desired (8).

2.2 Substitution and gap scores

The user of global alignment methods needs to specify, in addition to the sequences, a set of **substitution scores** for aligning various letters with one another, and a set of **gap scores** for aligning letters in one sequence with nulls in the other. A given program will generally provide default scores, but no scores are ideal for all purposes, and different scores can produce very different results. Substitution scores have been based upon a wide variety of rationales, including the genetic code, physical-chemical properties, and studies of molecular structure and evolution (9–21) (*Figure 2*). Although a statistical theory is available that describes 'optimal' substitution scores for certain local alignment problems, the theory does not extend to global alignments. Nevertheless, the same scores have frequently proved useful in both global and local alignment contexts. The popular log-odds scores, originally pro-

```
A    4
R   -1   5
N   -2   0   6
D   -2  -2   1   6
C    0  -3  -3  -3   9
Q   -1   1   0   0  -3   5
E   -1   0   0   2  -4   2   5
G    0  -2   0  -1  -3  -2  -2   6
H   -2   0   1  -1  -3   0   0  -2   8
I   -1  -3  -3  -3  -1  -3  -3  -4  -3   4
L   -1  -2  -3  -4  -1  -2  -3  -4  -3   2   4
K   -1   2   0  -1  -3   1   1  -2  -1  -3  -2   5
M   -1  -1  -2  -3  -1   0  -2  -3  -2   1   2  -1   5
F   -2  -3  -3  -3  -2  -3  -3  -3  -1   0   0  -3   0   6
P   -1  -2  -2  -1  -3  -1  -1  -2  -2  -3  -3  -1  -2  -4   7
S    1  -1   1   0  -1   0   0   0  -1  -2  -2   0  -1  -2  -1   4
T    0  -1   0  -1  -1  -1  -1  -2  -2  -1  -1  -1  -1  -2  -1   1   5
W   -3  -3  -4  -4  -2  -2  -3  -2  -2  -3  -2  -3  -1   1  -4  -3  -2  11
Y   -2  -2  -2  -3  -2  -1  -2  -3   2  -1  -1  -2  -1   3  -3  -2  -2   2   7
V    0  -3  -3  -3  -1  -2  -2  -3  -3   3   1  -2   1  -1  -2  -2   0  -3  -1   4

     A   R   N   D   C   Q   E   G   H   I   L   K   M   F   P   S   T   W   Y   V
```

Figure 2. The BLOSUM-62 amino acid substitution matrix (19, 90).

posed by Dayhoff and co-workers (10, 11), are described in some detail below. Here it is noted only that closely and distantly related sequences warrant different scores (16, 22, 23).

No good theory exists for the selection of gap scores, and they have therefore been chosen by trial and error. Because a single mutation can delete or insert several nucleotides or amino acids at once, it has been argued that the penalty for a gap of length x should be less than x times the penalty for a gap of length one (24). If **affine** gap scores of the form $a + bx$ are employed, a simple modification of the basic algorithm described above can still find optimal alignments in time proportional to the product of the lengths of the sequences compared (8, 25, 26). More complex, but essentially equally efficient algorithms have also been described for **concave** gap scores that have, for example, a logarithmic term (27). Few programs in common use, however, employ anything more complicated than affine gap scores. Recently, methods have been described for finding optimal alignments simultaneously for all possible affine gap scores (28, 29). These allow the user to investigate the robustness of 'optimal' alignments to different scoring regimes.

2.3 Statistics

To assess whether a given alignment constitutes evidence for homology, it helps to know how strong an alignment can be expected from chance alone. This requires an idea of the distribution of optimal scores when unrelated or 'random' sequences are aligned. Unfortunately, under even the simplest random models and scoring systems, very little is known about the random distribution of optimal global alignment scores (30). Monte Carlo experiments have provided rough distributional results for some specific scoring systems and sequence compositions (31), but these can not be generalized easily. Therefore, one of the few methods available for assessing the statistical significance of a particular global alignment is to generate many random sequence pairs of the appropriate length and composition, and calculate the optimal alignment score for each (32, 33). Although it is then possible to express the score of interest in terms of standard deviations from the mean, it is a mistake to assume that the relevant distribution is normal and convert this z-value into a p-value; the tail behaviour of global alignment scores is unknown. The most one can say reliably is that if 100 random alignments have score inferior to the alignment of interest, the p-value in question is likely less than 0.01. One further pitfall to avoid is exaggerating the significance of a result found among multiple tests. When many alignments have been generated, e.g. in a database search, the significance of the best must be discounted accordingly. An alignment with p-value 0.0001 in the context of a single trial may be assigned a p-value of only 0.1 if it was selected as the best among 1000 independent trials.

3. Global multiple alignment

Non-uniform evolution, large insertions and deletions, exon shuffling, and other mutational events can result in DNA or protein molecules that share only isolated regions of similarity. Methods to detect such local patterns will be discussed at length in a later section. Sometimes, however, a set of sequences will be known to be related across virtually their entire lengths, and the question of aligning them correctly arises. Multiple alignment presents certain novel problems that are absent from pairwise alignment. The first and in many ways central problem is that of alignment score.

3.1 Scores

The rationale for most biological sequence alignment is the theory of evolution, which implies that the relationship among homologous sequences can in general be represented by an evolutionary tree. This has important implications for how the score of a multiple alignment is defined.

A single 'column' of a multiple alignment places into correspondence letters or nulls from each sequence in question. A column from the alignment of five DNA sequences might, for example, contain three As and two Cs. Perhaps the simplest way to evaluate such a column is to compare all pairs of letters and sum their pairwise scores. We call an alignment whose columns are evaluated in this manner an **SP alignment**, for 'sum-of-the-pairs' (*Figure 3*). An approach with a better biological rationale is to describe the evolutionary tree that gave rise to the observed sequences, with reconstructed 'ancestral' sequences at the internal nodes of the tree. The score for a column in such a **tree alignment** is taken as the sum of pairwise scores for letters adjacent within the tree (*Figure 3*). A simpler case arises when the tree is taken to have only one internal node. A column in such a **star alignment** is evaluated by summing the score of a consensus letter compared to each observed letter (*Figure 3*).

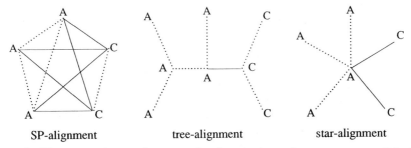

SP-alignment   tree-alignment   star-alignment

Figure 3. SP, tree, and star alignments for five one-letter input sequences. Pairwise alignments between identical letters are indicated by dotted lines, and those between non-identical letters by solid lines.

The choice of a multiple alignment scoring system is an example of the sort of trade-off alluded to in the introduction. Although tree alignments make more sense biologically, they present many algorithmic complications, including the specification or reconstruction of an evolutionary tree and ancestral sequences. Most multiple alignment algorithms therefore employ SP scores. For the user, the selection of a particular alignment program generally entails the choice of a particular type of alignment score. It is best, therefore, to be aware of the different possible approaches.

Finally, it should be noted that SP scores have a fundamental conceptual difficulty. Because the sequences one seeks to align usually are not 'independent observations', but rather are related by means of an evolutionary tree, giving all pairwise alignment scores equal weight allows a group of closely related sequences to 'outvote' a smaller set of more distant but more informative sequences (34). Various schemes for mitigating this problem by weighting individual sequences or sequence pairs have been described (34–39), and could be employed fruitfully by most multiple alignment programs that use SP scores.

3.2 Algorithms

There are several basic approaches that have been taken to the global multiple alignment problem. Perhaps the most obvious is a direct extension of dynamic programming to higher dimensions (40–48). The main advantage of this approach is that, because it is a true optimization method which takes account simultaneously of information from all input sequences, it is more likely to produce correct alignments. Its main disadvantage is that its execution time and memory requirements grow exponentially with the number of sequences. In practice, this limits full-blown dynamic programming to three average length proteins. The 'volume' of the dynamic programming path graph that needs to be considered can, however, be vastly reduced by employing inequalities based on optimal pairwise alignment scores (45–47). This can extend the applicability of dynamic programming to as many as six sequences. These ideas have been implemented in the program MSA (47).

The second, and most widely used global multiple alignment method is **progressive alignment** (49–57). The central idea is that a multiple alignment may be treated as a generalized sequence. Two such generalized sequences may be aligned with one another by standard pairwise dynamic programming methods. Therefore a multiple alignment may be constructed 'progressively' by coalescing ever greater numbers of sequences into a single alignment.

The main difficulty with this approach is that alignments formed early in the process are constructed in ignorance of most of the available data, and therefore may easily freeze in a mistake that can not be corrected later. One way to mitigate this failing is to align first those sequences that are most closely related, and which are therefore easiest to align accurately (49, 51–55,

57). This requires the prior knowledge or reconstruction of a phylogenetic tree relating the input sequences. Methods for inferring such trees are described in Chapter 15. A way to correct mistakes nevertheless made early in the multiple alignment process is to iterate the procedure: sequences are systematically removed from the composite alignment and realigned (50, 51, 56). Because of the ability of the progressive strategy to produce alignments for large numbers of sequences, it is used by all the most popular global multiple alignment programs, such as *Clustal V* (57) and the Genetic Computer Group's (GCG) *Pileup* (58).

A final approach to global multiple alignment is based on finding short, relatively highly conserved regions shared by all the input sequences, aligning these, and then repeating the procedure on the remaining unaligned regions (59–61). Whereas the goal of these methods may be to produce global alignments, their basic algorithmic strategy is that of local multiple alignment. Discussion of such approaches is, therefore, deferred to a later section.

4. Local sequence alignment

While early work on sequence comparison concentrated on methods for aligning complete sequences, it became apparent over time that the more important question was how to find isolated regions of similarity between two protein or DNA sequences. Because of large insertions and deletions, exon shuffling, and the evolutionary conservation of relatively short protein motifs crucial to structure or function, biologically important similarities can often be confined to small regions within large sequences. Programs that demand complete alignments can easily miss such similarities.

4.1 Algorithms

An early approach to seeking local sequence similarities was based on the method of **sliding windows** (9). Using an amino acid substitution matrix, all segments of a fixed length within one protein were compared with all such segments in a second. One problem with this approach is that the width of the 'window' must be specified a priori. If the similarity of interest is substantially longer than the chosen width, it may be that no segment pair will have a score sufficiently high to appear of interest. If the chosen width is too long, however, the scores for extra, unrelated amino acid pairs may degrade the signal. The solution is to let the data themselves determine the extent of the similarity reported. Given two sequences, one simply finds their respective segments, of whatever lengths, whose alignment score is maximized. The Smith–Waterman algorithm (62) for this purpose, and its variations (63–66), have become the standard for pairwise local sequence comparison. These algorithms require computation time proportional to the product of the lengths of the two sequences compared. 'Dot plots' (67) based on the earlier sliding windows methodology are still frequently employed because they are somewhat simpler to program. However, they virtually never have any advantage to variations of the Smith–Waterman approach, either computationally or in the quality of their output (*Figure 4*), and the user is generally advised to eschew this outmoded methodology.

One important sidelight on the local alignment problem is that frequently two sequences will contain several distinct regions of local similarity. For example, the yeast Ste6 protein (68) and the yeast Cdc25 protein (69) share both a rasGEF domain and an SH3 domain (70) (*Figure 4*). It is therefore a mistake to seek only the highest scoring pair of matching segments. To formalize the notion of when two local alignments are really distinct, and not just variations of the same theme, Sellers (63) introduced the concept of local optimality, variations of which have since been proposed (64, 65). The practical user generally need not be concerned with the precise definition of local optimality embodied in a given program. However, he/she should beware that programs which report only the single highest scoring local alignment between two sequences may easily miss similarities of biological interest.

4.2 Local alignment statistics

Fortunately statistics for the scores of local alignments, unlike those of global alignments, are well understood. This is particularly true for local alignments lacking gaps, which we will consider first. Such alignments are precisely those sought by the *BLAST* database search programs (71–73), which will be discussed in Section 5.

4.2.1 Ungapped alignments and the extreme value distribution

A local alignment without gaps consists simply of a pair of equal length segments, one from each of the two sequences being compared. Given two sequences and a set of substitution scores, a simplification of the Smith–Waterman algorithm (62) will find the maximal segment pair (MSP) score. To analyse how high a score is likely to arise by chance, a model of 'random' sequences is needed. For proteins, the simplest model chooses the amino acid residues in a sequence independently, with background probabilities $p_1,\dots, p_{20}$. Additionally, one constraint on the scores s_{ij} used for aligning amino acid pairs is crucial. Namely, the **expected score** $\sum_{ij=1}^{20} p_i p_j s_{ij}$ for aligning two randomly selected residues must be negative. Were this not the case, long alignments would tend to have high scores independently of whether the segments aligned were related.

Just as the *sum* of a large number of independent identically distributed (i.i.d.) random variables tends to a normal distribution, the *maximum* of a large number of i.i.d. random variables tends to an extreme value distribution (74). (We will elide the many technical points required to make this statement rigorous.) The MSP score falls into the latter category (75–77). The

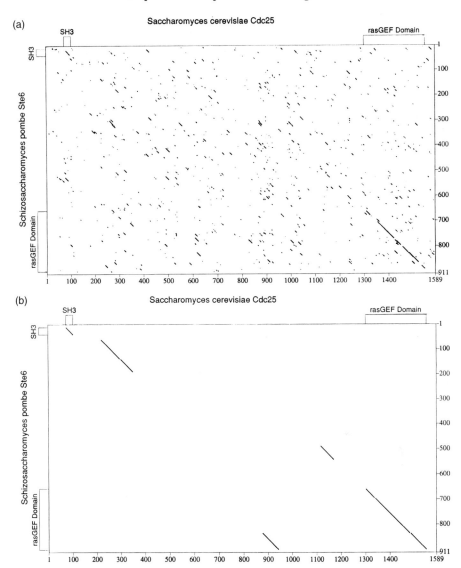

Figure 4. Comparison of yeast Ste6 (68) and yeast Cdc25 (69) proteins. These molecules, members of the superfamily of GTPases, share rasGEF and SH3 domains (70). Panel (a) shows a comparison of the two proteins using a dot plot or sliding window algorithm (9, 67) from the GCG package (58). Panel (b) shows the same two sequences compared using a modification (65, 66) of the Smith–Waterman algorithm (62) that returns the highest scoring locally optimal (63, 64) alignments; the display tool employed is described in ref. 163. The Smith–Waterman algorithm requires only a constant factor more time than the dot plot algorithm and returns much cleaner and more easily interpreted output. In this example, it places the shared SH3 domain among the highest scoring local alignments; the same domain is buried in the noise of the dot plot.

extreme value distribution is characterized by two parameters: the **characteristic value**, u, and the **decay constant**, λ. The probability that a random variable with this distribution has value at least x is given by the formula

$$1 - \exp[-e^{-\lambda(x-u)}] \qquad [1]$$

The probability density of the extreme value distribution with $u = 0$ and $\lambda = 1$ is shown in *Figure 5*.

To calculate p-values from Formula 1 one needs estimates for u and λ. For MSP scores, analytical expressions for these parameters are available (75–77). λ is the unique positive solution of the equation

$$\sum_{i,j=1}^{20} p_i p_j e^{s_{ij}x} = 1, \qquad [2]$$

and u is given by the equation

$$u = (\ln Kmn)/\lambda \qquad [3]$$

where m and n are the lengths of the two sequences being compared, and K is a calculable constant dependent upon the scores s_{ij} and the background residue probabilities p_i (75–77). More generally, the number of distinct (i.e. locally optimal (63, 64)) segment pairs with score at least x is approximately Poisson distributed (75–77), with parameter

$$Kmn\, e^{-\lambda x} \qquad [4]$$

From this formula one sees most clearly that λ serves simply as a natural scale for the scoring system, and K as a natural scale for the size of the 'search space'. Similar analytic results are also available for Markov dependent sequences (78).

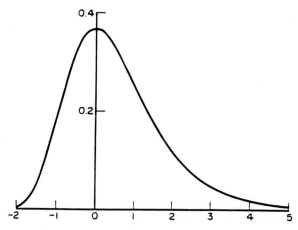

Figure 5. The probability density function of the extreme value distribution (74), with characteristic value $u = 0$ and decay constant $\lambda = 1$.

It is always possible to express a given score in terms of number of standard deviations from the mean (z-value). Unfortunately, the use of z-values frequently is assumed to imply normality for the underlying distribution, and may therefore be misleading. Because the right-hand tail of the extreme value distribution decays exponentially in x, as opposed to x^2 for the normal distribution, the invalid assumption of normality can yield grossly exaggerated claims of statistical significance. It is best, therefore, to avoid discussing extreme value distributed random variables in terms of means and standard deviations, and use instead the more appropriate parameters u and λ. For reference, however, the mean of the extreme value distribution is $u + \gamma/\lambda$, where $\gamma \approx 0.577$ is Euler's constant, and the standard deviation is $\pi/(\sqrt{6}\,\lambda)$ (74).

4.2.2 Applicability of the extreme value distribution

Although a formal proof is not yet available, experiment (22, 79–82) and analogy (23, 75, 83–85) give good reason to believe that the scores for local alignments with gaps will also be extreme value distributed. A necessary condition is that the substitution and gap scores be on average sufficiently negative (86). Unfortunately, no precise (i.e. calculable) condition on these scores can be stated. However, when comparing random sequences of length n, any scoring system will yield optimal local alignments with average score proportional either to n or to log n (86). It is only scoring systems of the latter form that are amenable to analysis with the extreme value distribution.

Analytic formulas for u and λ are not available for gapped alignments, but these parameters may be estimated (80–82) by collecting the optimal alignment scores from many pairs of 'random sequences' (32, 33). Scores may alternatively be collected from pairs of actual but unrelated proteins (22, 79, 81). An advantage of this latter approach is that the subtle residue dependencies of actual molecules, which no statistical model can capture completely, are retained. A disadvantage is that for validity all pairs of related sequences should be excluded, but this can never be done with full confidence.

The extreme value distribution is likely to apply as well to measures of local sequence similarity more complicated than Smith–Waterman scores (23, 83). The underlying reason is that optimizing most truly local similarity measures involves maximizing over a large number of effectively i.i.d. random variables. Values for the two relevant parameters may be determined as before through simulation with random or real but unrelated sequences.

There is one frequent case where the random models and therefore the statistics discussed in this section break down. As described in Chapter 8, as many as one quarter of all residues in protein sequences occur within regions with highly biased amino acid composition. Alignments of two regions with similarly biased composition may achieve very high scores that owe virtually nothing to residue order but are due instead to segment composition. Alignments of such 'low complexity' regions have little meaning in any case: since these regions most likely arise by gene slippage, the one-to-one residue correspondence imposed by alignment is not valid. While it is worth noting that two proteins contain similar low complexity regions, they are best excluded when constructing alignments (87–89). Filtering methods for the purpose are described in Chapter 8.

4.3 Local alignment scoring systems

The results produced by a local alignment program depend strongly on the scores it uses. No single scoring scheme is best for all purposes, and an understanding of the basic theory of local alignment scores can improve the sensitivity of one's sequence analyses. As before, the theory is fully developed only for scores used to find ungapped local alignments, so we start with that case.

As mentioned above, a large number of different amino acid substitution scores, based upon a variety of rationales, have been described (9–21). However, the scores of any substitution matrix with negative expected score can be written uniquely in the form

$$s_{ij} = (\ln \frac{q_{ij}}{p_i p_j})/\lambda, \qquad [5]$$

where the q_{ij}, called **target frequencies**, are positive numbers that sum to 1, and λ is a positive constant (75). (The λ of equation 5 is identical to the solution to equation 2.) Multiplying all the scores in a substitution matrix by a positive constant does not change their essence: an alignment that was optimal using the original scores remains optimal. Such multiplication alters the parameter λ but not the target frequencies q_{ij}. Thus, up to a constant scaling factor, every substitution matrix is uniquely determined by its target frequencies. These frequencies have a special significance (16, 75): a given class of alignments is best distinguished from chance by the substitution matrix whose target frequencies characterize the class.

To elaborate, one may characterize a set of alignments representing homologous protein regions by the frequency with which each possible pair of residues is aligned. If valine in the first sequence and leucine in the second appear in 1% of all alignment positions, the target frequency for (valine, leucine) is 0.01. The most direct way to construct appropriate substitution matrices for local sequence comparison is to estimate target and background frequencies, and calculate the corresponding log-odds scores of equation 5. These frequencies in general cannot be derived from first principles, and their estimation requires empirical input.

4.3.1 The PAM and BLOSUM substitution matrices

Although all substitution matrices are implicitly of log-odds form, the first explicit construction using Equation 5 was by Dayhoff and co-workers (10,

11). From a study of observed residue replacements in closely related proteins, they constructed the PAM (point accepted mutation) model of molecular evolution. One 'PAM' corresponds to an average change in 1% of all amino acid positions. After 100 PAMs of evolution, not every residue will have changed: some will have mutated several times, perhaps returning to their original state, and others not at all. Thus it is possible to recognize as homologous proteins separated by much more than 100 PAMs.

Using the PAM model, the target frequencies and the corresponding substitution matrix may be calculated for any given evolutionary distance. The best known of these matrices, the PAM-250, was the only one originally published by Dayhoff. However, when two sequences are compared, it is not generally known a priori what evolutionary distance will best characterize any similarity they may share. Therefore, it may be useful to employ multiple matrices, tailored to a variety of evolutionary distances (16, 22, 23). For database searching, at most three matrices (e.g. PAM-30, PAM-120 and PAM-250) should suffice (16, 23).

Dayhoff's formalism for calculating target frequencies has been criticized (13), and there have been several recent efforts to update her numbers using the vast quantities of derived protein sequence data generated since her work (18, 20). These newer PAM matrices do not differ greatly from the original ones (90).

An alternative approach to estimating target frequencies, and the corresponding log-odds matrices, has been advanced by Henikoff and Henikoff (19). They examine multiple alignments of distantly related protein regions directly, rather than extrapolate from closely related sequences. An advantage of this approach is that it cleaves closer to observation; a disadvantage is that it yields no evolutionary model. A number of tests have been advanced (90, 91) to show that the 'BLOSUM' matrices (*Figure 2*) produced by this method generally are superior to the PAM matrices for detecting biological relationships.

4.3.2 Gap scores

The theoretical development above concerning the optimality of matrices constructed using Equation 5 unfortunately is invalid as soon as gaps and associated gap scores are introduced, and no more general theory is available to take its place. However, if the gap scores employed are sufficiently large, one can expect that the optimal substitution scores for a given application will not change substantially. In practice, the same substitution scores have been applied fruitfully to local alignments both with and without gaps. There have been some efforts to put the selection of gap scores on a solid empirical footing (92, 93), but no theory exists to guide the process. Accordingly, appropriate gap scores have been selected over the years by trial and error, and most Smith–Waterman-like programs will have a default set of gap scores to go with a default set of substitution scores. If the user wishes to employ a different set of substitution scores, there is no guarantee that the same gap scores will remain appropriate. No clear guidance can be given except that affine gap scores as described above, with a large penalty for opening a gap and a much smaller one for extending it, have generally proved the most effective.

4.3.3 DNA substitution scores

Although substitution matrices have only been discussed in the context of protein sequence comparison, all the main issues carry over to DNA sequence comparison. One warning is that when the sequences of interest code for protein, it is almost always better to compare the protein translations than to compare the DNA sequences directly. The reason is that after only a small amount of evolutionary change, the DNA sequences, when compared using simple nucleotide substitution scores, contain less information with which to deduce homology than do the encoded protein sequences (17). Sometimes, however, one may wish to compare non-coding DNA sequences, at which point the same log-odds approach as before applies. An evolutionary model in which all nucleotides are equally common and all substitution mutations are equally likely yields different scores only for matches and mismatches (17). A more complex model, in which transitions are more likely than transversions, yields different 'mismatch' scores for transitions and transversions (17) (*Table 1*). The best scores to use will depend upon whether one is seeking relatively diverged or closely related sequences (17).

Table 1. Log-odds DNA substitution scores based on a biased mutation model in which a given transition mutation is threefold more likely than a given transversion (17)

PAM distance	Per cent conserved	Match score (bits)	Transition score (bits)	Transversion score (bits)	Average information per position (bits)
0	100.0	2.00	$-\infty$	$-\infty$	2.00
5	95.2	1.93	-3.13	-4.67	1.65
10	90.7	1.86	-2.19	-3.70	1.42
20	82.6	1.72	-1.32	-2.76	1.09
30	75.6	1.60	-0.86	-2.23	0.85
40	69.5	1.48	-0.57	-1.87	0.67
50	64.2	1.36	-0.37	-1.60	0.54
60	59.6	1.25	-0.23	-1.39	0.43
70	55.6	1.15	-0.12	-1.22	0.35
80	52.1	1.06	-0.04	-1.08	0.29
90	49.0	0.97	0.02	-0.96	0.23
100	46.3	0.89	0.06	-0.86	0.19
110	44.0	0.81	0.10	-0.77	0.16
120	41.9	0.74	0.12	-0.70	0.13
130	40.1	0.68	0.14	-0.63	0.11

PAM distances refer to nucleotide mutations. Scores are given as logarithms to the base 2 (bits). Average information is the expected score per position for an alignment of two homologous DNA sequences, diverged by the given PAM distance (16, 17).

4.3.4 Specialized scoring systems

Special purpose scores can be constructed to detect unusual mutation events such as frame shifts, sometimes attributable to experimental error (94). Also, the log-odds scores of Equation 5 are 'optimal' only within a very restricted framework. Much more complicated systems than those that assign scores only to aligned pairs of residues, or residues and gaps, are possible. One could, for example, assign scores to aligned pairs of dipeptides, or let scores depend on even greater context. Such more complex scores may well be able to detect biological relationships that simple substitution and gap costs would miss. Argos, for example, has described a 'sensitive' scoring system (95) that, although it eschews gaps, is claimed to be much more sensitive than standard local alignment scores (96). The drawback of this and many other complex scoring methods is that they may slow computation by several orders of magnitude (96).

5. Database search methods

When a stretch of coding DNA has been newly sequenced, potentially one of the most informative procedures is to compare its translation to all known protein sequences. A significant similarity to a well-characterized protein can immediately yield strong clues to structure or function. Most widely used programs for database similarity searching use local alignment methods that are variations on the Smith–Waterman algorithm. Due to the large and increasing sizes of sequence databases, rigorous implementations of this algorithm on standard machines are too slow for many applications, typically requiring several hours for a single search. Two broad strategies for increasing speed are the use of multiple processors, and the use of heuristic algorithms. An alternative, non-alignment based approach to database searching relies on comparing vectors of oligonucleotide or oligopeptide usage.

We review here a number of database search strategies, and discuss some of their intrinsic strengths and weaknesses. It should be noted, however, that the utility of a given system will depend not only on the algorithm it employs, but also on a number of factors that are sometimes thought peripheral (89). These include up-to-date and comprehensive sequence databases, suitable scoring systems, accurate statistical assessments for program output, and methods of filtering queries for low complexity regions. The practical user should consider these issues (89) as well as the underlying algorithm when selecting a particular database search system.

5.1 Parallel architectures

Parallel computers are well adapted to many sequence comparison algorithms, and many database search programs have been written to take advantage of these machines (96–100). Currently the best known such programs are *Blaze* (99) and *Blitz* (100), direct implementations of Smith–Waterman written for Maspar computers, and available respectively from IntelliGenetics and EMBL e-mail servers. The main choice for the user is the selection of substitution and gap scores.

In order to achieve the speed of parallel computation at lower costs, special purpose VLSI chips have been designed for local sequence alignment (101–103). Systems based on such chips have yet to gain widespread use, although a database search server employing the BioSCAN chip (103) is currently available from the University of North Carolina, Chapel Hill.

5.2 Heuristic algorithms

For database searching using standard computers, the best known heuristic methods are the *Fasta* (104, 105) and *BLAST* (71–73) algorithms. Both use the ability to find exactly matching words quickly to confine the search for interesting local alignments to a small fraction of the entire search space. Neither algorithm is guaranteed to find every locally optimal alignment that exceeds a cut-off score, but in practice neither is likely to miss any strong similarity. As with Smith–Waterman, the time required by either *Fasta* or *BLAST* is proportional to the product of the query and database lengths; the only difference is in the constant of proportionality.

A recent alternative approach, *FLASH* (106), indexes the entire database and uses generalized 'words' from the query to locate positions within the database with potential strong matches. Like *Fasta* and *BLAST*, *FLASH* is heuristic, but it can be shown to have a very low probability of missing matching regions that achieve the specified cut-off score (106). The indexing strategy used by *FLASH* achieves speed at the cost of very large memory requirements.

There is no space here to review the details of the *Fasta*, *BLAST*, and *FLASH* algorithms. Instead, we will discuss the strengths and weakness of each approach, and the effect of varying certain key parameters.

5.2.1 *Fasta*

With *Fasta*, in addition to the scoring system employed, there are two main choices for the user, both of which involve a trade-off of speed for sensitivity. The first choice is the value of the 'ktup' parameter (for proteins generally set either to 2 or 1) which affects an early heuristic screening step of the algorithm. A larger value of ktup causes the program to run more rapidly, but with a degradation of its ability to find weaker similarities. The second choice is whether to 'optimize' all alignments, or only those with the highest scores after an early stage. Optimizing all alignments costs time, but permits some biologically significant comparisons to jump greatly in score.

The principal advantage of *Fasta* is its ability, like Smith–Waterman, to find local alignments with gaps. Trials have shown that for protein sequence

comparison, with ktup equal to 1 and full optimization, *Fasta* finds biologically significant relationships as well as Smith–Waterman, but still executes in
a fraction of the time (107). One disadvantage of *Fasta*, and indeed of most
database search implementations of Smith–Waterman, is that it reports only
the single best local alignment of the query to each database sequence. Thus
a strong similarity may mask a weaker one of interest. As discussed above,
Smith–Waterman may be modified to permit the reporting of several distinct
alignments (63–66), but due to time considerations these refinements have
rarely been implemented in database search programs. Before reporting in
print any specific similarity found by *Fasta* or most other database search
programs, it is therefore advisable to compare the sequences in question
using an implementation of Smith–Waterman that reports all locally optimal
alignments over some cut-off score (63–66).

5.2.2 *BLAST*

The *BLAST* database search programs (71–73), like *Fasta* and Smith–Waterman, seek locally optimal alignments, but they do not allow the explicit introduction of gaps. A gapped alignment may be represented implicitly by
several distinct, ungapped segment pairs, for which a combined statistical
assessment is reported (108). In addition to the choice of scoring matrix, the
BLAST programs offer the user one key parameter affecting performance.
The '*T*' or threshold parameter controls a trade-off between speed and sensitivity. A high value of *T* decreases execution time but increases the probability that weak similarities will be overlooked (71). The default value of *T*
generally is set relatively low, rendering it quite unlikely for a statistically significant segment pair to be missed due to the heuristic nature of the algorithm. Other available parameters affect *BLAST* output in various ways; the
most important select score or significance level cut-offs for reporting an
alignment.

The principal disadvantages of the *BLAST* programs are their heuristic
nature, and the slight loss in sensitivity due to the indirect rather than explicit
treatment of gaps (91, 107). The main advantages are speed, the statistical
tractability of the similarity measure employed, and the ability of standard
implementations to report not just the best but multiple high-scoring regions
of similarity. For example, when comparing the yeast Ste6 protein (68) to a
protein database, *BLAST* finds similarities to both the rasGEF and SH3
domains of yeast Cdc25 (69) (*Figure 4*), whereas *Fasta*, *FLASH*, and most
other database search programs report only the stronger similarity.

5.2.3 *FLASH*

FLASH requires a computer with a very large amount of disk storage, on the
order of 100 times the size of the database to be searched (106). This is the
primary practical consideration for a site considering running the program
locally, but it is not a problem for the user of a network service. Excluding

parameters affecting the quantity of output, the only question for the user is which substitution and gap scores to employ.

FLASH is perhaps best compared to the *Blaze* and *Blitz* programs, because of its non-standard hardware requirements. *FLASH* has the advantage that massive storage generally is less expensive than a massively parallel architecture. However, the program remains heuristic and so can not claim rigorously to find every alignment that satisfies a given reporting threshold.

5.3 Vector-based comparison methods

A very different approach to database searching is based on representing a sequence by a vector describing the frequency with which it uses short words (109–113). A measure of the similarity of two vectors, such as their dot product, may then be used to measure the similarity of the sequences they represent. This does away almost completely with the need for alignment; sequences of any length are compared by means of the same small number of arithmetic operations.

The principal advantage of this approach is speed. The main disadvantages are that much of the order information embedded in each sequence is cast away, and that the method is essentially a global one. Many of the most exciting discoveries from sequence comparison, such as the homology between regions of the neurofibromatosis type 1 gene product and GAP proteins (114), have involved only small domains of much larger sequences. However vector methods, although able to cluster sequences that are similar across their entire lengths, fail to detect reliably these more subtle relationships. Represented by a point in a high-dimensional space, a multidomain protein will be marooned between the separate spatial regions corresponding to its various domains. Dividing such a sequence into arbitrary pieces may still leave two domains linked, or cut a single one in half. Furthermore, to the extent that sequences are divided into smaller pieces, the speed advantage of vector methods is sacrificed. Although some have seen great promise for these methods in screening databases (115), their broad utility has yet to be demonstrated convincingly.

6. Local multiple alignment

A researcher may be confronted with multiple sequences which are suspected to contain one or several similar regions, but not necessarily to be globally related. To locate and characterize these regions is the problem of local multiple alignment.

Much research has been devoted to this difficult problem, and a variety of formulations of the central question as well as algorithmic attacks upon it have been proposed (60, 116–130). An ideal solution would probably satisfy all of the desiderata listed in *Table 2*. Unfortunately, no such solution is

Table 2. Desiderata for an ideal local multiple alignment algorithm

Time required is linear in number of sequences
Employs appropriate measure of pattern quality
Measure can reflect known amino acid relationships
Measure sensitive to subtle sequence relationships
Pattern may be missing from a subset of sequences
Width of pattern not unduly constrained
Width need not be specified a priori
Algorithm can find multiple distinct patterns
Alignment of segments may contain gaps
Algorithm is rigorous optimization procedure
Output is independent of order of input sequences
Algorithm is deterministic

known, and each suggested attack has been forced to make compromises. It may be anticipated that yet other, perhaps fundamentally different approaches remain to be described.

A number of the basic algorithms that have been developed are reviewed here, and the strengths and weaknesses of each are discussed. The potential user may then make an informed choice of the approach most appropriate for the question at hand. To analyse time and space complexity, it will be assumed that the input consists of N sequences of average length L. Programs based on most of the methods described in this section are available from their respective authors upon request.

6.1 Consensus word methods

One definition of the problem is to locate a **consensus word** that is 'close' to some word in each, or at least a large number, of the input sequences (116–119). This may be thought of as finding an optimal star alignment, as illustrated in *Figure 3*. Confining attention to words of a fixed width, one may define for each word a neighbourhood of 'adjacent' words, each with an associated score (117). For example, the neighbourhood of a given eight-nucleotide DNA word w might consist of all eight-tuples with no more than two mismatches to w. The score for a word in this neighbourhood might be the number of identical nucleotides it shares with w. One may then seek that word which, when compared to the highest-scoring neighbour extant within each input sequence, has maximum aggregate score. If there are A possible words, and the neighbourhood of each word has size B, one algorithm for finding the 'best' (i.e. highest-scoring) consensus word requires time proportional to $N(LB + A)$ and space proportional to A (117).

Two principal advantages of this approach are that for fixed word and neighbourhood size, it requires time proportional only to the total length of the input data and that, given its definition of the problem, it is guaranteed to find the optimal solution. The main disadvantages are the constraints on word and neighbourhood size imposed by practicality. If all words of width W are possible consensus words, memory requirements generally limit W to about 13 for DNA sequences and to about six for protein sequences. This is not too severe for many DNA sequence comparison problems, but it is fairly restrictive for proteins. For the algorithm to execute in reasonable time on current workstations, the neighbourhood of a given word generally can not exceed about 100 000 words. For a DNA word of width 13, an acceptable neighbourhood thus will include words with no more than about four mismatches, and for a protein word of width six no more than three mismatches. The inability of the method to measure similarity beyond the range of a word's neighbourhood will limit to some degree its sensitivity. Gaps can in principle be allowed by this method, but the degree to which they increase the size of a word's neighbourhood generally makes it impractical to permit them. Some statistical results concerning the significance of 'best' words are available (117).

6.2 Template methods

A second approach begins by defining a set of template patterns to compare with each of the input sequences (120, 121). For protein sequence comparison an element of this set could be 'V*C**D', for example, where '*' is a wild card. If the size of the template set is A, it takes time proportional to NLA to compare all the templates with the input sequences, and record which are found within the most sequences.

This is essentially an inversion of the consensus word method, with each pattern considered in turn, rather than each position within the input sequences. As such, it has the same general advantages and disadvantages: a rigorous algorithm, time proportional to the size of the input, but a restriction of the size and type of pattern sought. If the set of templates is defined systematically, the method can avoid setting aside space for every possible template, and save just those that occur with sufficient frequency to be surprising. By modifying the published methods, one can imagine as well introducing scores in a manner analogous to the consensus word methods. A variation that gains considerable speed seeks only perfectly matching words, perhaps in some restricted alphabet (60, 122). Again, statistical analyses are available with which to assess results (121, 122).

6.3 Progressive alignment methods

An algorithm analogous to the progressive alignment approach to global multiple alignment has been described by Bacon and Anderson (123). Basically, all segments of fixed width W in the first input sequence are compared to all such segments in the second. With reference to a given scoring system, the B best segment pairs are retained, and each is compared with all segments of width W in the third sequence. The B best triples are then compared with segments from the fourth sequence, etc. The time required is proportional to $NLBW$. Proposed modifications to this approach include saving a 'best' alignment for each possible starting position within the first input sequence, and employing an alignment scoring system based on information theoretic considerations (124, 125).

The primary advantages of this method are that it requires time proportional to the number of input sequences, and that it can deal easily with patterns of arbitrary width and with complicated scoring systems. The main disadvantage is that the method is heuristic and therefore not guaranteed to optimize the measure of alignment quality it employs. Different ordering of the input sequences can yield different output.

6.4 Pairwise comparison methods

A fourth general approach first compares all input sequences with one another, and then combines results from these pairwise comparisons into consistent multiple alignments (126, 127). The method of Vingron and Argos (127) checks for consistency at the level of path graph 'cells', whereas the MACAW algorithm (126) checks at the level of path graph 'diagonals'. In brief, for each pair of sequences, MACAW marks all diagonals that contain a segment pair with score exceeding a set threshold. It then merges pairwise diagonals into consistent higher-dimensional ones, i.e. alignments all of whose implicit pairwise diagonals have been marked. Finally, each high-dimensional diagonal so constructed is parsed to find the best region it contains.

The advantages of this approach are that it can detect patterns of arbitrary width, it may use arbitrary scoring systems, it can find many distinct alignments in a single run, it naturally forms alignments from subsets of the input sequences, it has available a statistical theory (75), and it is rigorous given the imposed 'threshold' constraint on pairwise similarity. The primary disadvantage is that the algorithm requires for its first step time proportional to $(NL)^2$, limiting its practical application to collections of sequences whose aggregate length is not much greater than 10 000 residues. Also, the time needed for the second, 'merging' step of the procedure is strongly dependent on the threshold score set and the nature of the input data. For thresholds that are too low, the time required for the second step becomes dominant.

6.5 Statistically-based methods

Several methods based upon recent ideas from statistics have been proposed (128–130). In their simplest form, they seek a single, fixed length segment within each input sequence. Using Bayesian statistics, this collection of segments implies a position-dependent model of residue probability. The aim is to maximize the ratio of the probability that the segments found arise from the implied common model to the probability that they arise from background residue probabilities. This is closely related to the goal of one of the progressive methods described above (124, 125).

An expectation maximization (EM) attack on this optimization problem (128, 129) successively refines estimates of the 'common model', and estimates of the probability that it occurs at each possible position within the input sequences. Alternatively, iterative Gibbs sampling (130) uses the evolving common model to sample, with appropriate probabilities, positions for the segments within the input sequences. Convergence is detected when the common model does not improve for a number of iterations. Gibbs sampling has the advantage that it more easily avoids being trapped by local maxima, and that it has practical extensions to the simultaneous detection of multiple patterns (130). If the pattern sought has width W, and each sequence is sampled on average T times, the execution time required is proportional to $NLWT$. A method for choosing among results produced for different widths W has been described (130).

The advantages of this approach are that its time requirement grows linearly in the number of input sequences, it can handle large pattern widths, it is sensitive to quite subtle patterns, and although heuristic it appears rarely to be trapped by local maxima. The main disadvantages are that the optimality of the results produced can not be guaranteed, and that a specific range of pattern widths must be supplied.

6.6 General issues

Although most of the approaches described above can be generalized explicitly to allow gaps, it is usually too costly to do so. Instead gaps can be treated implicitly by seeking multiple local patterns within sequences; a gap will then appear as two nearly adjacent local alignments.

Many variations are possible for the interrelationships among a set of sequences, and it is difficult to specify a fully automatic procedure for uncovering all that might be found. An exploratory approach to the data is therefore frequently to be recommended, with the existence of one significant alignment suggesting a pattern search within restricted regions of the same sequences. Flexible alignment editors thus can form a useful adjunct to local multiple alignment search tools (126). Faced with extracting surprising patterns from the vast sea of potential local alignments, the more specific knowledge that can be used to restrict the search space the better.

7. Sequence motifs

When a multiple alignment is constructed, by either local or global methods, certain information may emerge that no individual sequence contains. Some sequence positions may be highly variable whereas others are almost completely conserved; insertions or deletions may be common in some sequence positions but rare or non-existent at others; the precise extent of sequence domains or motifs may become manifest. There is a hope that this information may be captured concisely and deployed to recognize more accurately sequences or sequence regions related to those in the multiple alignment. Many proposals for how this may be done have been advanced, and the field is still in ferment.

Two basic structures that have been used to represent the common sequence information in a multiple alignment are **consensus sequences** or **regular expressions** (131–133) and **weight matrices** or **profiles** (133–153). Very efficient algorithms exist to search for consensus sequences or regular expressions (154). Nevertheless, when available, weight matrices are almost always to be preferred because they are able in general to describe any pattern that a consensus sequence can, but in addition can capture much more subtle relationships. Furthermore, the weight matrix formalism springs naturally from statistical mechanical considerations applicable to the energetic interactions of molecules (140, 155).

7.1 Weight matrices

In its simplest form, a weight matrix describes a pattern of fixed width W; it is a rectangular array of numbers with W columns and one row for each possible residue (*Figure 6*). Such a matrix assigns a score to every sequence segment of width W, calculated simply as the sum of the relevant matrix elements. (More general weight matrices that allow variable length segments and gaps will be discussed below.) Weight matrices are constructed to assign high scores to patterns with specific biological significance, such as ribosome or integration host factor binding sites in DNA molecules (138, 147), or helix–turn–helix structural motifs in proteins (130, 146).

Many different methods for calculating the actual numbers to be used in weight matrices have been advanced (124, 125, 130, 134–153). The most frequent prescription, and the one with the best theoretical foundation, is that the numbers in column j should be of the form $\log (q_{ij}/p_i)$, where q_{ij} is the estimated probability that residue i is found at position j in instances of the pattern, and p_i is the background probability of finding residue i in non-pattern positions (124, 125, 130, 137, 138, 140, 141, 145, 146, 150–152). This leaves open the question of how best to estimate q_{ij} from a multiple alignment. Given a large number of independent observations of the pattern, it seems clear that the best estimate of q_{ij} will converge simply to the observed relative frequency of residue i at position j. Complicating considerations, however, include small sample size (138), correlations (particularly evolutionary ones) among the available observations (34), and a priori knowledge of relationships among the possible residues (10, 151, 152). These considerations generally are all more acute for protein than for DNA weight matrices.

Similarly to the substitution matrices discussed in Section 4.3, all the scores

(a)

SWISS-PROT code	Starting position	Segment	Ending position
IF41_HUMAN	352	NRENYIHRIGRGGR	365
RECQ_ECOLI	316	NIESYYQETGRAGR	329
UVRB_ECOLI	531	SERSLIQTIGRAAR	544
PRIA_ECOLI	574	FAQLYTQVAGRAGR	587
RAD3_YEAST	658	AMRHAAQCLGRVLR	671
VI08_VACCC	485	SKSMRDQRKGRVGR	498
UL09_HSV11	377	DMVSVYQSLGRVRT	390
POLG_TVMV	1457	SLGERIQRFGRVGR	1470
RECG_ECOLI	574	GLAQLHQLRGRVGR	587
NTP1_VACCC	466	NEASLRQIVGRAIR	479
ETF1_VACCC	444	TFSQYNQILGRSIR	457

(b)

Amino Acid	Position													
	1	2	3	4	5	6	7	8	9	10	11	12	13	14
A	0	0	1	-1	0	0	-4	-2	0	-4	-4	3	0	-4
C	-2	-1	-1	-1	-1	-1	-4	2	-1	-4	-4	-1	-1	-3
D	1	-1	0	0	-3	1	-4	-2	-2	-4	-4	-3	-3	-4
E	0	1	2	1	-2	0	-4	0	-1	-4	-4	-3	-2	-3
F	1	1	-2	-2	0	-2	-5	-1	1	-5	-5	-4	-3	-4
G	0	-3	0	-2	-3	-2	-4	-3	-3	5	-4	1	4	-4
H	0	0	0	2	-1	2	3	0	-1	-3	-3	-2	-2	-3
I	-3	1	-2	-2	-1	2	-5	2	2	-5	-5	-2	2	-4
K	0	1	0	0	-1	0	-4	0	1	-5	-4	-3	-2	-3
L	-4	1	-2	-1	2	-1	-5	0	1	-6	-6	-3	0	-5
M	-2	3	-1	1	-1	-1	-5	0	0	-5	-5	-2	-2	-4
N	3	-1	0	1	-2	1	-5	-1	-1	-5	-5	-3	-2	-4
P	-2	-3	-2	-2	-3	-3	-5	-3	-3	-5	-6	-3	-3	-5
Q	0	0	1	2	-2	0	6	-1	-1	-5	-5	-2	-2	-4
R	-2	0	1	0	2	1	-5	2	0	-5	6	-3	1	5
S	2	-1	1	2	-2	-1	-4	0	-1	-4	-4	1	-2	-4
T	1	-1	-1	-1	-2	1	-4	1	1	-4	-4	-2	-2	1
V	-3	0	0	-2	1	-1	-5	1	1	-5	-5	3	-2	-4
W	-3	-2	-2	-2	-1	-2	-5	-2	-2	-6	-6	-4	-4	-5
Y	-3	-1	-2	-1	4	2	-5	-1	-1	-5	-5	-3	-3	-4

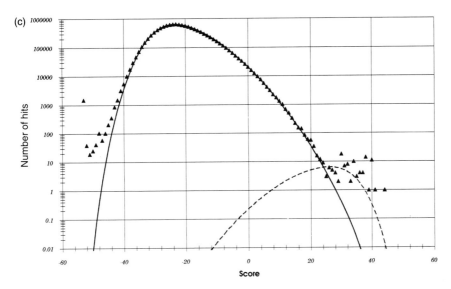

Figure 6. A large superfamily of helicases and nucleic acid-dependent ATPases may be characterized by seven conserved regions or motifs (164). Panel (a) shows a local alignment of regions comprising the C-terminal motif from 11 of these proteins. Using the method of Brown *et al.* (151), position-dependent amino acid frequencies q_{ij} were calculated from this alignment, and background frequencies p_i were estimated from the SWISS-PROT database, release 27 (165). A rounded log-odds weight matrix based on these frequencies is shown in panel (b). This matrix is used to evaluate every length-14 segment in the SWISS-PROT database, and the number of segments receiving each score is plotted in panel (c). Assuming background residue frequencies p_i, the probability for each score may be calculated explicitly (136, 152, 156). Given the size of the database, the solid curve represents the expected number of hits at each score; less than one is expected by chance at any score over 28. At low scores, the deviation of the observed number of hits from the solid curve is likely due to 'low-entropy' regions (88, 89); at high scores, the deviation is due to instances of the motif. Assuming the database contains approximately 125 such instances (164 and E. Koonin, personal communication), and that the derived q_{ij} correctly model the position-dependent motif residue frequencies, the dashed curve represents the expected number of true positive hits at each score.

in a weight matrix may be multiplied by an arbitrary positive constant without changing its essence. It is usually convenient, therefore, to scale and round so that all the elements of a matrix are integers. Given such an integral matrix and the set of background residue probabilities p_i, one may calculate explicitly the probability of obtaining any weight matrix score at a random position (136, 152, 156). To account for the multiple tests that arise from scanning one or many sequences with a weight matrix, it is approximately valid to assume the independence of scores at different positions (*Figure 6*).

7.2 Generalizations

Sometimes it is desirable to allow a weight matrix to permit gaps, or to permit a sequence segment to match only a portion rather that the whole of the matrix. Once again, no good theory of how to assign internal or terminal gap scores exists, and they have generally been chosen empirically (142). Also, no statistical theory exists for sequence matching to weight matrices with gaps. However, biologically related sequences do contain gaps, and they need therefore to be addressed either implicitly or explicitly. An indirect approach is to describe patterns that consist of several ungapped blocks (130, 149). Alternatively, the formalism of **hidden Markov models** has recently come into vogue for constructing weight matrices with appropriate gap costs (39, 151, 157–159). However defined, weight matrices may be employed in iterative sequence database searches, with the additional sequences found in each iteration used to construct a new weight matrix (152, 153). A final important generalization of weight matrices is the inclusion of scores for correlations between positions (155, 160, 161). The mushrooming quantity of sequence data, and the recognition that detectable cross-phylum sequence relationships can be described by a relatively small number of 'ancient conserved regions' (162), together insure that research into how best to represent common protein motifs will remain very active.

Acknowledgements

I thank Drs Mark Boguski, Eugene Koonin, and Roman Tatusov for assistance with several of the examples in this chapter.

References

1. Needleman, S. B. and Wunsch, C. D. (1970). *J. Mol. Biol.*, **48**, 443.
2. Sankoff, D. (1972). *Proc. Natl. Acad. Sci. USA*, **69**, 4.
3. Sellers, P. H. (1974). *SIAM J. Appl. Math.*, **26**, 787.
4. Sankoff, D. and Kruskal, J. B. (1983). *Time warps, string edits and macromolecules: the theory and practice of sequence comparison.* Addison-Wesley, Reading, MA.
5. Ukkonen, E. (1983). In *Proceedings of the International Conference on the Foundations of Computer Theory, Lecture Notes in Computer Science,* Vol. 158, p. 487. Springer-Verlag, Berlin.
6. Fickett, J. W. (1984). *Nucleic Acids Res.*, **12**, 175.
7. Spouge, J. L. (1989). *SIAM J. Appl. Math.*, **49**, 1552.
8. Myers, E. W. and Miller, W. (1988). *Comput. Appl. Biosci.*, **4**, 11.
9. McLachlan, A. D. (1971). *J. Mol. Biol.*, **61**, 409.
10. Dayhoff, M. O., Schwartz, R. M., and Orcutt, B. C. (1978). In *Atlas of protein sequence and structure,* Vol. 5, Suppl. 3 (ed. M. O. Dayhoff), p. 345. Natl. Biomed. Res., Found., Washington, DC.

11. Schwartz, R. M. and Dayhoff, M. O. (1978). In *Atlas of protein sequence and structure*, Vol. 5, Suppl. 3 (ed. M. O. Dayhoff), p. 353. Natl. Biomed. Res., Found., Washington, DC.
12. Feng, D. F., Johnson, M. S., and Doolittle, R. F. (1985). *J. Mol. Evol.*, **21**, 112.
13. Wilbur, W. J. (1985). *Mol. Biol., Evol.*, **2**, 434.
14. Taylor, W. R. (1986). *J. Theor. Biol.*, **119**, 205.
15. Risler, J. L., Delorme, M. O., Delacroix, H., and Henaut, A. (1988). *J. Mol. Biol.*, **204**, 1019.
16. Altschul, S. F. (1991). *J. Mol. Biol.*, **219**, 555.
17. States, D. J., Gish, W., and Altschul, S. F. (1991). *Methods*, **3**, 66.
18. Gonnet, G. H., Cohen, M. A., and Benner, S. A. (1992). *Science*, **256**, 1443.
19. Henikoff, S. and Henikoff, J. G. (1992). *Proc. Natl. Acad. Sci. USA*, **89**, 10915.
20. Jones, D. T., Taylor, W. R., and Thornton, J. M. (1992). *Comput. Appl. Biosci.*, **8**, 275.
21. Overington, J., Donnelly, D., Johnson M. S., Sali, A., and Blundell, T. L. (1992). *Protein Sci.*, **1**, 216.
22. Collins, J. F., Coulson, A. F. W., and Lyall, A. (1988). *Comput. Appl. Biosci.*, **4**, 67.
23. Altschul, S. F. (1993). *J. Mol. Evol.*, **36**, 290.
24. Fitch, W. M. and Smith, T. F. (1983). *Proc. Natl. Acad. Sci. USA*, **80**, 1382.
25. Gotoh, O. (1982). *J. Mol. Biol.*, **162**, 705.
26. Altschul, S. F. and Erickson, B. W. (1986). *Bull. Math. Biol.*, **48**, 603.
27. Miller, W. and Myers, E. W. (1988). *Bull. Math. Biol.*, **50**, 97.
28. Gusfield, D., Balasubramanian, K., and Naor, D. (1992). In *Proceedings of the Third Annual ACM-SIAM Symposium on Discrete Algorithms*, p. 432. ACM Press, New York, NY.
29. Waterman, M. S., Eggert, M., and Lander E. (1992). *Proc. Natl. Acad. Sci. USA*, **89**, 6090.
30. Deken, J. (1983). In *Time warps, string edits and macromolecules: the theory and practice of sequence comparison* (ed. D. Sankoff and J. B. Kruskal), p. 359. Addison-Wesley, Reading, MA.
31. Reich, J. G., Drabsch, H., and Däumler, A. (1984). *Nucleic Acids Res.*, **12**, 5529.
32. Fitch, W. M. (1983). *J. Mol. Biol.*, **163**, 171.
33. Altschul, S. F. and Erickson, B. W. (1985). *Mol. Biol. Evol.*, **2**, 526.
34. Altschul, S. F., Carroll, R. J., and Lipman, D. J. (1989). *J. Mol. Biol.*, **207**, 647.
35. Vingron, M. and Sibbald, P. R. (1993). *Proc. Natl. Acad. Sci. USA*, **90**, 8777.
36. Gerstein, M., Sonnhammer, E. L. L., and Chothia, C. (1994). *J. Mol. Biol.*, **236**, 1067.
37. Henikoff, S. and Henikoff, J. G. (1994). *J. Mol. Biol.*, **243**, 574.
38. Thompson, J. D., Higgins, D. G., and Gibson, T. J. (1994). *Comput. Appl. Biosci.*, **10**, 19.
39. Eddy, S. R., Mitchison, G., and Durbin, R. (1995). *J. Comp. Biol.*, **2**, 9.
40. Sankoff, D. (1975). *SIAM J. Appl. Math.*, **28**, 35.
41. Sankoff, D. and Cedergren, R. J. (1983). In *Time warps, string edits and macromolecules: the theory and practice of sequence comparison* (ed. D. Sankoff and J. B. Kruskal), p. 55. Addison-Wesley, Reading, MA.
42. Fredman, M. L. (1984). *Bull. Math. Biol.*, **46**, 553.
43. Murata, M., Richardson, J. S., and Sussman, J. L. (1985). *Proc. Natl. Acad. Sci. USA*, **82**, 3073.

44. Gotoh, O. (1986). *J. Theor. Biol.,* **121**, 327.
45. Carrillo, H. and Lipman, D. (1988). *SIAM J. Appl. Math.,* **48**, 1073.
46. Altschul, S. F. and Lipman, D. J. (1989). *SIAM J. Appl. Math.,* **49**, 197.
47. Lipman, D. J., Altschul, S. F., and Kececioglu, J. D. (1989). *Proc. Natl. Acad. Sci. USA,* **86**, 4412.
48. Altschul, S. F. (1989). *J. Theor. Biol.,* **138**, 297.
49. Waterman, M. S. and Perlwitz, M. D. (1984). *Bull. Math. Biol.,* **46**, 567.
50. Bains, W. (1986). *Nucleic Acids Res.,* **14**, 159.
51. Barton, G. J. and Sternberg, M. J. (1987). *J. Mol. Biol.,* **198**, 327.
52. Feng, D. and Doolittle, R. F. (1987). *J. Mol. Evol.,* **25**, 351.
53. Taylor, W. R. (1987). *Comput. Appl. Biosci.,* **3**, 81.
54. Corpet, F. (1988). *Nucleic Acids Res.,* **16**, 10881.
55. Hein, J. (1990). In *Methods in enzymology,* Vol. 183 (ed. R. F. Doolittle), p. 626. Academic Press, London.
56. Berger, M. P. and Munson, P. J. (1991). *Comput. Appl. Biosci.,* **7**, 479.
57. Higgins, D. G. and Sharp, P. M. (1992). *Comput. Appl. Biosci.,* **8**, 189.
58. Devereux, J., Haeberli, P., and Smithies, O. (1984). *Nucleic Acids Res.,* **12**, 387.
59. Johnson, M. S. and Doolittle, R. F. (1986). *J. Mol. Evol.,* **23**, 267.
60. Sobel, E. and Martinez, H. (1986). *Nucleic Acids Res.,* **14**, 363.
61. Vingron, M. and Argos, P. (1989). *Comput. Appl. Biosci.,* **5**, 115.
62. Smith, T. F. and Waterman, M. S. (1981). *J. Mol. Biol.,* **147**, 195.
63. Sellers, P. H. (1984). *Bull. Math. Biol.,* **46**, 501.
64. Altschul, S. F. and Erickson, B. W. (1986). *Bull. Math. Biol.,* **48**, 633.
65. Waterman, M. S. and Eggert, M. (1987). *J. Mol. Biol.,* **197**, 723.
66. Huang, X., Hardison, R. C., and Miller, W. (1990). *Comput. Appl. Biosci.,* **6**, 373.
67. States, D. J. and Boguski, M. S. (1991). In *Sequence analysis primer* (ed. M. Gribskov and J. Devereux), p. 89. Stockton Press, New York, NY.
68. Hughes, D. A., Fukui, Y., and Yamamoto, M. (1990). *Nature,* **344**, 355.
69. Broek, D., Toda, T., Michaeli, T., Levin, L., Birchmeier, C., Zoller, M., Powers, S., and Wigler, M. (1987). *Cell,* **48**, 789.
70. Boguski, M. S. and McCormick, F. (1993). *Nature,* **366**, 643.
71. Altschul, S. F., Gish, W., Miller, W., Myers, E. W., and Lipman, D. J. (1990). *J. Mol. Biol.,* **215**, 403.
72. Altschul, S. F. and Lipman, D. J. (1990). *Proc. Natl. Acad. Sci. USA,* **87**, 5509.
73. Gish, W. and States, D. J. (1993). *Nature Genet.,* **3**, 266.
74. Gumbel, E. J. (1958). *Statistics of extremes.* Columbia University Press, New York, NY.
75. Karlin, S. and Altschul, S. F. (1990). *Proc. Natl. Acad. Sci. USA,* **87**, 2264.
76. Dembo, A., Karlin, S., and Zeitouni, O. (1994). *Ann. Prob.,* **22**, 2022.
77. Dembo, A. and Karlin, S. (1991). *Ann. Prob.,* **19**, 1737.
78. Dembo, A. and Karlin, S. (1991). *Ann. Prob.,* **19**, 1756.
79. Smith, T. F., Waterman, M. S., and Burks, C. (1985). *Nucleic Acids Res.,* **13**, 645.
80. Mott, R. (1992). *Bull. Math. Biol.,* **54**, 59.
81. Waterman, M. S. and Vingron, M. (1994). *Proc. Natl. Acad. Sci. USA,* **91**, 4625.
82. Altschul, S. F. and Gish, W. (1996). In *Methods in enzymology,* Vol. 266 (ed. R. F. Doolittle), p. 460. Academic Press, London.
83. Altschul, S. F. and Erickson, B. W. (1986). *Bull. Math. Biol.,* **48**, 617.
84. Arratia, R., Gordon, L., and Waterman, M. S. (1986). *Ann. Stat.,* **14**, 971.

85. Arratia, R. and Waterman, M. S. (1989). *Ann. Prob.,* **17**, 1152.
86. Arratia, R. and Waterman, M. S. (1994). *Ann. Appl. Prob.,* **4**, 200.
87. Claverie, J.-M. and States, D. J. (1993). *Comput. Chem.,* **17**, 191.
88. Wootton, J. C. and Federhen, S. (1993). *Comput. Chem.,* **17**, 149.
89. Altschul, S. F., Boguski, M. S., Gish, W., and Wootton, J. C. (1994). *Nature Genet.,* **6**, 119.
90. Henikoff, S. and Henikoff, J. G. (1993). *Proteins,* **17**, 49.
91. Pearson, W. R. (1995). *Protein Sci.,* **4**, 1145.
92. Pascarella, S. and Argos, P. (1992). *J. Mol. Biol.,* **224**, 461.
93. Benner, S. A., Cohen, M. A., and Gonnet, G. H. (1993). *J. Mol. Biol.,* **229**, 1065.
94. Claverie, J.-M. (1993). *J. Mol. Biol.,* **234**, 1140.
95. Argos, P. (1987). *J. Mol. Biol.,* **193**, 385.
96. Vogt, G. and Argos, P. (1992). *Comput. Appl. Biosci.,* **8**, 49.
97. Coulson, A. F. W., Collins, J. F., and Lyall, A. (1987). *Comput. J.,* **30**, 420.
98. Jones, R. (1992). *Comput. Appl. Biosci.,* **8**, 377.
99. Brutlag, D. L., Dautricourt, J.-P., Diaz, R., Fier, J., Moxon, B., and Stamm, R. (1993). *Comput. Chem.,* **17**, 203.
100. Sturrock, S. S. and Collins, J. F. (1993). *MPsrch version 1.3.* Biocomputing Research Unit, University of Edinburgh, UK.
101. Chow, E. T., Hunkapiller, T., Peterson, J. C., Zimmerman, B. A., and Waterman, M. S. (1991). In *Proceedings of the 1991 International Conference on Supercomputing*, p. 216. ACM Press, New York, NY.
102. Hughey R. P. (1991). Ph.D. Thesis, Brown University, Providence, RI.
103. White, C. T., Singh, R. K., Reintjes, P. B., Lampe, J., Erickson, B. W., Dettloff, W. D., *et al.* (1991). In *Proceedings of the 1991 IEEE International Conference on Computer Design: VLSI in Computers and Processors*, p. 504. IEEE Comp. Soc. Press, Los Alamitos, CA.
104. Lipman, D. J. and Pearson, W. R. (1985). *Science,* **227**, 1435.
105. Pearson, W. R. and Lipman, D. J. (1988). *Proc. Natl. Acad. Sci. USA,* **85**, 2444.
106. Califano, A. and Rigoutsos, I. (1993). In *Proceedings of the First International Conference on Intelligent Systems for Molecular Biology* (ed. L. Hunter, D. Searls, and J. Shavlik), p. 56. AAAI Press, Menlo Park, CA.
107. Pearson, W. R. (1991). *Genomics,* **11**, 635.
108. Karlin, S. and Altschul, S. F. (1993). *Proc. Natl. Acad. Sci. USA,* **90**, 5873.
109. Blaisdell, B. E. (1986). *Proc. Natl. Acad. Sci. USA,* **83**, 5155.
110. Blaisdell, B. E. (1989). *J. Mol. Evol.,* **29**, 526.
111. Torney, D. C., Burks, C., Davison, D., and Sirotkin, K. M. (1990). In *Computers and DNA* (ed. G. I. Bell and T. G. Marr), p. 109. Addison-Wesley, New York, NY.
112. van Heel, M. (1991). *J. Mol. Biol.,* **220**, 877.
113. Pizzi, E., Attimonelli, M., Liuni, S., Frontali, C., and Saccone, C. (1992). *Nucleic Acids Res.,* **20**, 131.
114. Xu, G. F., O'Connell, P., Viskochil, D., Cawthon. R., Robertson, M., Culver, M., *et al.* (1990). *Cell,* **62**, 599.
115. Taylor, W. (1991). *Nature,* **353**, 388.
116. Queen, C. M., Wegman, N., and Korn, L. J. (1982). *Nucleic Acids Res.,* **10**, 449.
117. Waterman, M. S., Arratia, R., and Galas, D. J. (1984). *Bull. Math. Biol.,* **46**, 515.
118. Galas, D. J., Eggert, M., and Waterman, M. S. (1985). *J. Mol. Biol.,* **186**, 117.
119. Staden, R. (1989). *Comput. Appl. Biosci.,* **5**, 293.

120. Posfai, J., Bhagwat, A. S., Posfai, G., and Roberts, R. J. (1989). *Nucleic Acids Res.*, **17**, 2421.
121. Smith, H. O., Annau, T. M., and Chandrasegaran, S. (1990). *Proc. Natl. Acad. Sci. USA*, **87**, 826.
122. Leung, M. Y., Blaisdell, B. E., Burge, C., and Karlin, S. (1991). *J. Mol. Biol.*, **221**, 1367.
123. Bacon, D. J. and Anderson, W. F. (1986). *J. Mol. Biol.*, **191**, 153.
124. Stormo, G. D. and Hartzell, G. W. III (1989). *Proc. Natl. Acad. Sci. USA*, **86**, 1183.
125. Hertz, G. Z., Hartzell, G. W. III, and Stormo, G. D. (1990). *Comput. Appl. Biosci.*, **6**, 81.
126. Schuler, G. D., Altschul, S. F., and Lipman, D. J. (1991). *Proteins*, **9**, 180.
127. Vingron, M. and Argos, P. (1991). *J. Mol. Biol.*, **218**, 33.
128. Lawrence, C. E. and Reilly, A. A. (1990). *Proteins*, **7**, 41.
129. Cardon, L. R. and Stormo, G. D. (1992). *J. Mol. Biol.*, **223**, 159.
130. Lawrence, C. E., Altschul, S. F., Boguski, M. S., Liu, J. S., Neuwald, A. F., and Wootton, J. C. (1993). *Science*, **262**, 208.
131. Smith, R. F. and Smith, T. F. (1990). *Proc. Natl. Acad. Sci. USA*, **87**, 118.
132. Bairoch, A. (1993). *Nucleic Acids Res.*, **21**, 3097.
133. Staden, R. (1988). *Comput. Appl. Biosci.*, **4**, 53.
134. McLachlan, A. D. (1977). *Biopolymers*, **16**, 1271.
135. Stormo, G. D., Schneider, T. S., Gold, L., and Ehrenfeucht, A. (1982). *Nucleic Acids Res.*, **10**, 2997.
136. McLachlan, A. D. (1983). *J. Mol. Biol.*, **169**, 15.
137. Staden, R. (1984). *Nucleic Acids Res.*, **12**, 505.
138. Schneider, T. S., Stormo, G. D., Gold, L., and Ehrenfeucht, A. (1986). *J. Mol. Biol.*, **188**, 415.
139. Taylor, W. R. (1986). *J. Mol. Biol.*, **188**, 233.
140. Berg, O. G. and von Hippel, P. H. (1987). *J. Mol. Biol.*, **193**, 723.
141. Dodd, I. B. and Egan, J. B. (1987). *J. Mol. Biol.*, **194**, 557.
142. Gribskov, M., McLachlan, A. D., and Eisenberg, D. (1987). *Proc. Natl. Acad. Sci. USA*, **84**, 4355.
143. Patthy, L. (1987). *J. Mol. Biol.*, **198**, 567.
144. Stormo, G. D. (1988). *Annu. Rev. Biophys. Biophys. Chem.*, **17**, 241.
145. Bucher, P. (1990). *J. Mol. Biol.*, **212**, 563.
146. Dodd, I. B. and Egan, J. B. (1990). *Nucleic Acids Res.*, **18**, 5019.
147. Goodrich, J. A., Schwartz, M. L., and McClure, W. R. (1990). *Nucleic Acids Res.*, **18**, 4993.
148. Gribskov, M., Lüthy, R., and Eisenberg, D. (1990). In *Methods in enzymology*, Vol. 183 (ed. R. F. Doolittle), p. 146. Academic Press, London.
149. Henikoff, S. and Henikoff, J. G. (1991). *Nucleic Acids Res.*, **19**, 6565.
150. Bowie, J. U., Lüthy, R., and Eisenberg, D. (1991). *Science*, **253**, 164.
151. Brown, M., Hughey, R., Krogh, A., Mian, I. S., Sjölander, K., and Haussler, D. (1993). In *Proceedings of the First International Conference on Intelligent Systems for Molecular Biology* (ed. L. Hunter, D. Searls, and J. Shavlik), p. 47. AAAI Press, Menlo Park, CA.
152. Tatusov, R. L., Altschul, S. F., and Koonin, E. V. (1994). *Proc. Natl. Acad. Sci. USA*, **91**, 12091.

153. Yi, T.-M. and Lander, E. S. (1994). *Protein Sci.,* **3**, 1315.
154. Kolakowski, L. F. Jr., Leunissen, J. A., and Smith, J. E. (1992). *Biotechniques,* **13**, 919.
155. Bryant, S. H. and Lawrence, C. E. (1993). *Proteins,* **16**, 92.
156. Staden, R. (1989). *Comput. Appl. Biosci.,* **5**, 89.
157. Haussler, D., Krogh, A., Mian, S., and Sjölander, K. (1992). *C.I.S. Tech. Report UCSC-CRL-92-23,* University of California, Santa Cruz, CA.
158. Tanaka, H., Ishikawa, M., Asai, K., and Konagaya, A. (1993). In *Proceedings of the First International Conference on Intelligent Systems for Molecular Biology* (ed. L. Hunter, D. Searls, and J. Shavlik), p. 395. AAAI Press, Menlo Park, CA.
159. Baldi, P., Chauvin, Y., Hunkapiller, T., and McClure, M. A. (1994). *Proc. Natl. Acad. Sci. USA,* **91**, 1059.
160. Gibrat, J.-F., Garnier, J., and Robson, B. (1987). *J. Mol. Biol.,* **198**, 425.
161. Bryant, S. B. and Lawrence, C. E. (1991). *Proteins,* **9**, 108.
162. Green, P., Lipman, D., Hillier, L., Waterston, R., States, D., and Claverie, J.-M. (1993). *Science,* **259**, 1711.
163. Boguski, M. S., Hardison, R. C., Schwartz, S., and Miller, W. (1992). *New Biol.,* **4**, 247.
164. Gorbalenya, A. E. and Koonin, E. V. (1993). *Curr. Opin. Struct. Biol.,* **3**, 419.
165. Bairoch, A. and Boeckmann, B. (1993). *Nucleic Acids Res.,* **21**, 3093.

8

Simple sequences of protein and DNA

JOHN C. WOOTTON

1. Introduction

Protein and nucleic acid sequences are very different from random strings of letters. Many regions, segments, or patches of sequence, over a wide range of length scales, are biased in residue composition. The term 'simple' conveys only part of the diversity and richness of variegation in natural sequences, which may be very subtle. It is rather as if blocks or domains of sequence, which may even overlap, are written in a mosaic of different languages, and alphabetic simplicity is just one of the characteristics of some of these languages (1, 2). This has profound implications for understanding molecular structure, function, variation, and evolution.

Simple DNA sequences commonly show intricate subpatterns arising from mutational processes such as unequal crossing over, replication slippage, biased substitution, and transposition. Examples include microsatellites, VNTRs, long tandemly repeated satellite DNAs, telomeric sequences, irregular low-complexity sequences in intergenic regions, CpG islands, and recombinational hotspots. Also, major segments of DNA sequence show more subtle compositional biases, for example the trinucleotide quasi-periodicity in coding sequences arising from the different biases in the three codon positions.

From recent analyses of protein sequence databases (3, 4), approximately 25% of the residues occur in compositionally biased segments and more than half of proteins contain at least one such segment. Biased sequences include numerous interspersed near-homopolymeric regions and also many non-globular regions of fibrous and structural proteins. They are abundant in large, multidomain polypeptides crucial in morphogenesis, embryonic development, transcriptional control, signal transduction, and both intracellular and extracellular structure and integrity. Many have molecular interactions with important biological consequences, shown by protein engineering experiments, and variations in human disease and locations of autoimmune epitopes.

Recently, theoretical advances have been made in understanding mathematically defined sequence attributes of complexity, pattern, and periodicity (Section 2). A few useful, robust programs that analyse simple sequences have become generally available via Internet (Section 3). Some of the basic concepts and practical applications of these methods are described in this chapter, with emphasis on informational complexity measures. Sequence analysis algorithms in general must be resistant to the effects of simple sequences and statistical heterogeneity, in order to avoid serious pitfalls. For example, attempts to infer long-range correlations in genomic sequences using random walks may be misled by non-random patchiness (1), and sequence comparison and alignment methods may be confounded by low-complexity regions (5, Chapter 7, and Section 4). Indeed, it rarely makes sense from either structural or evolutionary viewpoints to attempt to align simple sequences position by position.

2. Some practical guidelines to a complex body of theory

The analysis of simple sequences and sequence mosaicism has many fascinating mathematical ramifications, most of which are omitted from this chapter (for key references see Section 2.6). At present, software implementations lag well behind these theoretical developments. The following points provide a simplified perspective.

2.1 Complexity, pattern, and periodicity are distinct properties of simple sequences

Complexity, pattern, and periodicity are three different abstract attributes of sequences which can be clearly distinguished. This distinction is necessary to be able to understand the properties of different algorithms that analyse simple sequences. A very general definition of 'complexity' is required (Sections 2.2, 2.3) based only on residue composition. For example, the following three simple sequences have identical (low) complexity because of their identical compositions (G_8, A_8):

- G A A G G A A A G G G A G A G A has neither significant pattern nor periodicity

- G G A G G A A A A G G A A G G A has notable k-gram patterns (GGA and AGGA) but these are irregularly spaced and do not show periodicity

- G A G A G A G A G A G A G A G A has periodicity, modulo-2, and hence significant k-gram patterns as a consequence

See Sections 2.2 and 2.3 for terminology. Examples of how different algorithms are sensitive to these different attributes of simple sequences are given in Section 3.2.

2.2 Terminology

(a) **Complexity** (of which simplicity is one facet) may be defined in many different ways following fundamental concepts in information theory, coding theory, complexity theory or data compression. Only one very general sense is used in this chapter, namely *compositional complexity*, which may be a local or a global property of sequences. This property is computable from *complexity state vectors* (Section 2.3), and provides a general representation of compositional bias that is independent of pattern and periodicity.

(b) **Patterns** in simple sequences (including irregular repeats) are usually analysed by their content and spacing of residues and k-grams, although several other formalisms are possible (references in Section 2.6 and Chapter 7). k-grams are k-letter-words, thus ATTG is a 4-gram.

(c) **Periodicity** is repetition of residue types or k-grams at a constant interval (period, modulo, or distance). For DNA sequences, it is customary to distinguish **true periodicity**, i.e. tandem repetition, exact or with variations, of a sequence pattern of constant length, and **quasi-periodicity**, where repetition arises as a secondary consequence of different compositional biases in different phases, for example modulo-3 in coding sequences. Also, non-integral periods arise from helical secondary structures in polynucleotide or polypeptide chains.

Classical reassociation kinetics (6, 7) has used terms such as 'simple sequence DNA' and 'sequence complexity', but in a very different sense from this chapter. Hybridization experiments are sensitive, within the length ranges tractable to these methods, to sequence repetitiveness, and other properties such as GC content, rather than compositional bias as discussed here.

'Complexity' (and hence 'low complexity' for simple sequences), is a strongly preferable term to 'entropy' or 'information content', even though the formal definitions for complexity given in Sections 5.1 and 5.2 resemble the familiar Boltzmann and Shannon equations for entropy. The term 'entropy' has suffered from inconsistent usage in the biological literature: strictly, it is not a local property and refers to system or population distributions as a whole.

2.3 Local compositional complexity

Local compositional complexity is a property of the **complexity state** of each window, as represented by an ordered list of numbers, a **complexity state**

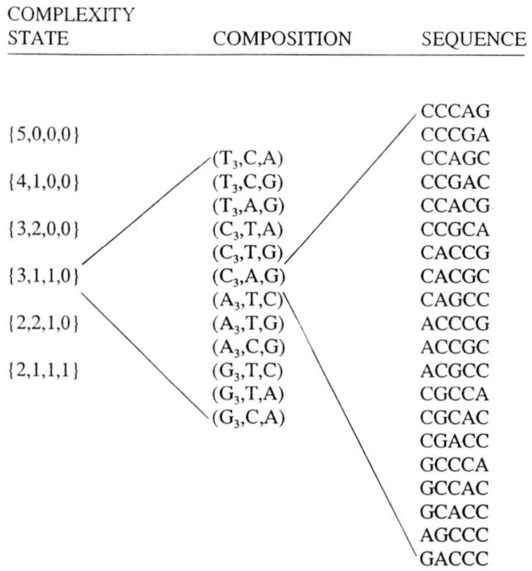

COMPLEXITY
STATE   COMPOSITION   SEQUENCE

COMPLEXITY STATE	COMPOSITION	SEQUENCE
{5,0,0,0}		CCCAG
		CCCGA
	(T_3,C,A)	CCAGC
{4,1,0,0}	(T_3,C,G)	CCGAC
	(T_3,A,G)	CCACG
{3,2,0,0}	(C_3,T,A)	CCGCA
	(C_3,T,G)	CACCG
{3,1,1,0}	(C_3,A,G)	CACGC
	(A_3,T,C)	CAGCC
{2,2,1,0}	(A_3,T,G)	ACCCG
	(A_3,C,G)	ACGCG
{2,1,1,1}	(G_3,T,C)	ACGCC
	(G_3,T,A)	CGCCA
	(G_3,C,A)	CGCAC
		CGACC
		GCCCA
		GCCAC
		GCACC
		AGCCC
		GACCC

Figure 1. Complexity state vectors, compositions and sequences. The four-letter DNA alphabet at window length $L = 5$ generates six complexity states, 56 compositions, and $4^5 = 1024$ sequences. For any one complexity state, all compositions have the same number of possible sequences, and this number provides the basis of complexity measures (Sections 2.3 and 5).

vector. *Figure 1*, left column, shows the six possible complexity state vectors of a five-nucleotide window. One of these complexity states, {3,1,1,0}, has 12 different compositions, with different nucleotides assigned to the four numbers in this vector, and each of these compositions has 20 possible sequences (*Figure 1*). The possibility of 20 sequences per composition makes {3,1,1,0} more complex than, for example {3,2,0,0} which has only 10 possible sequences per composition. Complexity measures are computed from complexity state vectors using the equations in Sections 5.1 and 5.2.

The theoretical number of complexity state vectors, which is computable from well-established principles of number theory (3), becomes very large at longer windows. For example, an amino acid window of length 40 generates 35 251 complexity states and a (rounded) total of 1.1×10^{52} sequences. The concept of complexity state vectors is also readily generalized to non-contiguous positions in a sequence: *periodic* compositional complexity may be calculated (using the definitions in Section 5) from complexity state vectors whose numbers are counts of residues spaced at equal intervals, and *pattern* compositional complexity may be measured for residues in any arbitrary pattern.

2.4 Low complexity is more clear-cut for proteins than DNA

For proteins, high compositional complexity is the norm. More than half the residues in current sequence databases are evidently in globular domains, which resemble (with some interesting statistical differences) random sequences in their sequence complexity (4, 8). Simple amino acid sequences may then be thought of as deviating from a random model to different extents (3, 4). In contrast, DNA is so strongly variegated that we do not know what 'random' should mean. Very few functional classes of DNA are well represented by simple random or Markov models. It is rarely clear what magnitudes of deviation of (say) k-gram frequencies should be considered 'surprising'.

2.5 Unbiased inference

Faced with mosaicism in both protein and DNA sequences, we may treat a sequence or database initially as a heterogeneous mixture with unknown statistical properties, and then attempt to infer these properties. An initial assumption of equal uniform probabilities for the appearance of residues, as with the SEG algorithm (3, Section 3.2), places all possible low complexity segments on an equal footing. For example, regions rich in generally common amino acids such as Leu, Ala, and Ser are treated as no more or no less surprising than segments rich in His, Met, or Trp. As a separate stage of investigation, the biases of individual classes of segments may be evaluated against more specific statistical models, as in the SAPS methodology (9, 10, Section 3.1).

2.6 Sources for mathematical background

Most of these key papers are published in relatively unfamiliar sources, such as biomathematical journals and special issues of *Computers and Chemistry*.

- analyses based on complexity measures (3, 4, 11–13)
- 'linguistic' analyses based on k-grams (2, 14–20)
- treatments of repetition that account for overlaps and equivalences (18–21)
- statistical analyses of patterns, clusters and distance relationships (1, 9, 10)
- principles of optimal segmentation of sequences into statistically contrasting regions, including use of non-homogeneous and hidden Markov models (3, 22–25)
- application of concepts from data compression and fractal representation (13, 26–28)

2.7 Visual inspection is complementary to mathematical analysis

For any sequence suspected of containing simple sequences, visual examination of a self–self dot matrix plot is recommended (many implementations available). An unfiltered plot is maximally informative, preferably using a suitable colour-coded scale of pixels to show the scores for all diagonals (see *Figure 2* for a grey-scale example). Many regions of local repetitiveness and low-complexity show as medium-scoring blocks (mid-grey in *Figure 2*).

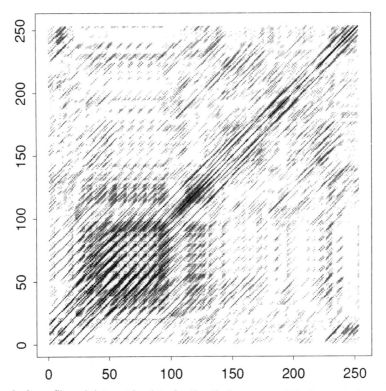

Figure 2. An unfiltered dot matrix plot of self-self alignments of the human prion protein precursor. The sequence is from SWISS-PROT, accession number P04156, name PRIO_HUMAN. Each grey-scale-coded point represents the maximum-scoring Smith–Waterman diagonal that goes through that residue pair (see *Figure 3* legend of Chapter 7 for more details of this method). Low-complexity features are conspicuous as variously-patterned mid-grey blocks: five tandem repeats of an eight-residue glycine-rich segment (55–94); other glycine-rich segments at 41–54 and 114–131 (the latter also hydrophobic); threonine-rich segment (188–201); leucine-rich hydrophobic segment (237–253). The N-terminal signal sequence is also hydrophobic, shown by weak matches to the C-terminal region. A more precise, objective definition of the low-complexity segments is given by the *SEG* algorithm (see *Figure 3a*, from which these residue numbers are taken).

Although these scores are not rigorous complexity measures (3), graphic display of them may provide an intuitive preliminary to more thorough analyses outlined in Section 3.

3. Software and examples of applications

3.1 Available software

SEG, *SAPS*, *XNU*, *PYTHIA*, and *SIMPLE34* are programs that use different principles to analyse and report simple sequences. *SEG* and *XNU* also function as filters to mask segments of sequences before database searches (Section 4). These five programs are relatively portable, easy to use, and available over Internet with documentation. For typical query sequences, results are returned in seconds or less on common Unix workstations.

- *SEG* (3)—anonymous FTP from `ncbi.nlm.nih.gov` in subdirectory `/pub/seg`

- *SAPS* (9)—anonymous FTP from `gnomic.stanford.edu`—see SAPS_README and subdirectory `/pub`

- *XNU* (29)—anonymous FTP from `ncbi.nlm.nih.gov` in subdirectory `/pub/jmc/xnu`

- *PYTHIA* (13)—e-mail server: send e-mail with the word 'help' in the Subject field to `pythia@anl.gov`

- *SIMPLE34* (30, 31)—anonymous FTP from `life.anu.edu.au` in subdirectory `/pub/molecular-biology/software/simple`

Also notable is program *TRACTS* (32) which reports repetitive tracts such as dinucleotide repeats in DNA sequences. Another approach applies the Lempel–Ziv data compression algorithm to analyse the 'complexity' of genome segments (26). 'Complexity' here means algorithmic compressibility, distinct from complexity as defined in this chapter. In practice, compressibility of a sequence may easily be estimated approximately using the $-v$ option of the Unix 'compress' function, which employs the Lempel–Ziv method. The *GENMARK* system (25) is also interesting because it uses the potentially relevant methodology of non-homogeneous Markov chain models, although the present implementation (e-mail server: send e-mail with the word 'help' in the Subject field to `genmark@ford.gatech.edu`) is designed to find coding sequences rather than low-complexity regions.

3.2 Comparison of different algorithms and programs

Table 1 gives a detailed comparison of the properties of *SEG*, *SAPS*, *XNU*, and *SIMPLE34*. For protein sequences, *SEG*, together with associated statistical programs (3), and *SAPS* are generally complementary in approach. As a rough guide, *SEG* rigorously explores sequence complexity at different strin-

Table 1. Comparison of programs for analysing simple sequences

	SEG	*SAPS*	*XNU*	*SIMPLE34*
Sequences analysed				
Protein	•	•	•	
Nucleic acid	•			•
Single sequences	•	•	•	•
Entire databases	•		•	
Recommended or built-in length limits (residues)				
Minimum	3	200	5	64
Maximum	None	10 000	None	50 000
Mathematical basis of algorithms				
Rigorous	•	•		
Heuristic			•	•
Statistical strategy for searches				
Unbiased inference (Section 2.4)	•			
Empirical frequency models		•	•	•
User-defined stringency	•		•	
Statistical evaluation (models used)				
Random	•	•		•
Markov		•		•
Periodic		•		•
Sensitive to				
Complexity as general property	•			
Residue or k-gram clustering		•		•
Repetition/periodicity		•	•	
Residue spacing		•		
Outputs (default or options)				
Sequence masking (Section 4)	•		•	
Rigorously optimized segments	•			
Distinct segment boundaries	•	•	•	
Structured for automated parsing	•		•	
Summarized statistics	•	•		•
'Profiles' as graphics files				•
'Profiles' as data files	•			

gencies in the absence of specific prior models (Section 2.5 and *Figure 3*), and *SAPS* rigorously evaluates specific sequence attributes such as residue clustering, spacing, and periodicity against various statistical models. *XNU* tests for self–self similarity of sequence windows offset at different intervals using a scoring matrix such as PAM-120, and is therefore primarily sensitive to local tandem repetitiveness.

For DNA sequences, *SEG*, *PYTHIA*, and *SIMPLE34* locate low-complexity regions in different ways. (While this chapter was in press, *SEG* was superseded by *NSEG* (35) for the analysis of nucleotide sequences—also available by anonymous *FTP* from ncbi.nlm.nih.gov, subdirectory /pub/seg.) *PYTHIA* uses a mathematically rigorous method based on minimal encoding lengths and their probabilities (13). *SIMPLE34* estimates the significance of

(a)

LOW-COMPLEXITY SEGMENTS		HIGH-COMPLEXITY SEGMENTS
ppqgggggwgqphgggwgqphgggwgqphgggwgqggg	1-49	MANLGCWMLVLFVATWSDLGLCKKRPKPGGWNTGGSRYPGQGSPGGNRY
	50-94	THSQWNKPSKPKTNMKHM
agaaagavvgglggymlgsams	95-112	
	113-135	RPIIHFGSDYEDRYYRENMHRYPNQVYRPMDEYSNQNNFVHDCVNITIKQH
tvtttkgenftet	136-187	
	188-201	DVKMMERVVEQMCITQYERESQAYYQRGSSMVLFS
sppvillisfliflivg	202-236	
	237-253	

(b)

```
>PRIO_HUMAN
MANLGCWMLVLFVATWSDLGLCKKRPKPGGWNTGGSRYPGQGSPGGNRYxxxxxxxxxxxxxxxxxxxxxxxxxxxxxxxxxxxxxxxxxTHSQWN
KPSKPKTNMKHMxxxxxxxxxxxxxxxxxxxxxxxxxxxRPIIHFGSDYEDRYYRENMHRYPNQVYRPMDEYSNQNNFVHDCVNITIKQHxxxxxxxxxxx
xDVKMMERVVEQMCITQYERESQAYYQRGSSMVLFSxxxxxxxxxxxxxxxxx
```

Figure 3. Automated segmentation of the human prion protein by local compositional complexity using the *SEG* algorithm. (a) A readable display of the low- and high-complexity segments (lower and upper case, respectively). The sequence segments all read from left to right and their order in the polypeptide is top to bottom, as shown by the central column of residue numbers. (b) The masked sequence ready for use as a database search query sequence. This is in *FASTA* format with the low-complexity sequences replaced with 'x' characters. These results were obtained using the default parameters of the *SEG* program, which are (as defined in Section 5.4): initial window length *L* = 12; trigger complexity *K21* = 2.2 bits; extension complexity *K22* = 2.5 bits. (a) Given by the command line seg prion.aa (equivalent to seg prion.aa 12 2.2 2.5) where prion.aa is a *FASTA*-formatted file of the prion amino acid sequence. (b) Given by seg prion.aa -x. It is valuable to use *SEG* as an exploratory tool at various different stringencies (see Section 2.5), for example seg myprotein.aa 45 3.4 3.75 is diagnostic of many long non-globular regions of proteins (4).

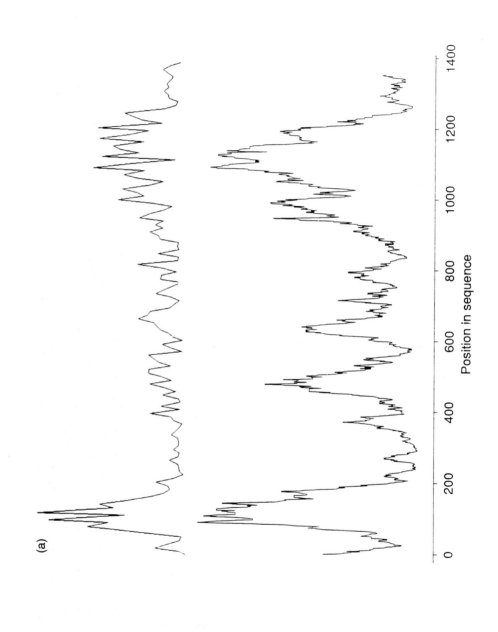

(a)

Position in sequence

(b)

LOW-COMPLEXITY SEGMENTS		HIGH-COMPLEXITY SEGMENTS
	1-87	AATATTGCAAATATGTTTATATTTGCAAATGAATGCGATTGAGGGAAAATGAAAAATTTG TTTTCGTGGTCTCACATGGTACGTCGC
gaaaaaaaaattttcagaaaaaaaaaatttttcagaaaaaa aaatttttcagaaaaaaaaatttttcaaaaaaaaatgaaggaattgataatgaattaa	88-206	
	207-217	CGCTATTTTAG
gcatttaaaagtaaaatataaacctgt	218-245	
	246-474	CGTGGTTTCGCTTGCCGACCACGAGAATATGAAAAATGTCAATATTGCGGAAGTGCACATT ACGGTTGTCAGCTTCAAAAAGTACTATAGGAGGTATGACCAGCTCGCGGCCATGTAACATGT GTGAGCAATATTATGGACCGTAGAAAAATGAATTTTTTTCTACAGATGGTCAAGATGGTA ATATAAAAATGGTCATGTGGGCAACCTCTCGCATATATAAGGATAAGC
tcaaatgatgaaaagtaaaattgtataatttttactcaaatatcatccatatt	475-529 530-639	GTAAAGAAAAACCACCTTATATATGGAAAATGAAAGATGGTGAGATGAGCGGCCACACAT ACAAATTTGCGATGAAGATGAGCGGACCTCATCGCAAATATTTTATATC
tagttttgaatgaagatgaaaaaatagtagcaatataacaa	640-683 684-814	GAGCAAATAATCATAATTTGACGGTCGGTCTTTAAAAGAGACCAATCGCATGATAATAG AAAAATTTGTCATTTGTAATGACTAAAACAAGAACGCGTGTCAGTTTGCTTTTTGTGCT TCTGAATTGAC
agtgtttatatttattataaaatgtgaag	815-843 844-941	CGCAAGAGAATTATTTCAACGGCACAAGTGCTCGTGAACGTATATGGTTTGTTCTTTTG TAATGCAATTGAATTTGCATATGCGAGCGAATGCATCC
atactttattgatatattttatatattagataagagagttctagagaagagtaa aaaaaaaaccccctaaatgaaaaaagagggaaagaaaaatgaagcgaatttcgaaact tattaattcaattcaattgaatgcataaaataaaaattataatgtttggaga tggaaaaataaactgaaaataaaatatattatatgaatgtgattaatattaatg aaattgaagaaaatgcaaagaaatgaaaaaattgttcgttcttgtaaaatatatat aatataacaacaatcctg	942-1261	
	1262-1416	GTTGATCCTGCCAGTAGTTATATGCTTGTCTCCAAAGATTAAGCCATGCATGTCTAAGTAC ACAGAATTAAAAGTGAAACCGCAAAAGGCTCATTATATCAGTTATGGTTCCTTAGATCG TTAACAGTTACTTGGATAACTGTGGTAATTCTAGA

Figure 4. Comparison of *SIMPLE34* and *SEG* for identification of simple DNA sequences. The sequence is a part of tsetse fly ribosomal RNA gene cluster that includes an intergenic spacer and the 5'-end of an 18S rRNA gene (GenBank accession number X05007, locus GMRDNA1). (a) 'Simplicity profiles' as produced by *SIMPLE34* (upper line) and the complexity measure used in the first stage of *SEG* (lower line; Section 5.4; the function plotted is $[2 - K2]$ rather than $K2$ in order to represent 'simplicity' instead of 'complexity'). The window is 64 residues for both plots. (b) Automated segmentation of this sequence using *SEG*. While this chapter was in press, *SEG* was superseded by *NSEG* for analyses of nucleotide sequences (35). Clearly, from the profiles in (a), both *SIMPLE34* and *SEG* find the two major regions of simple DNA, and *SEG* reports additional shorter regions of AT-rich low-complexity sequence (b), corresponding to narrower peaks present in (a) in the *SEG* profile but not the *SIMPLE34* profile.

3-gram and 4-gram clusters (31). *Figure 4a* shows an example for which the *SEG* and *SIMPLE34* 'profiles' are compared. Both of these programs identify the two major regions of simple DNA, and *SEG*, with appropriate parameters, reports two additional shorter AT-rich regions (*Figure 4b*). The e-mail server implementation of *PYTHIA* identifies one region corresponding to the largest (first) peak of this profile.

3.3 Future software developments

While this chapter was in press, extensions to *SEG* have been made that enable rigorous, optimized segmentation of periodic and quasi-periodic repetitive sequences using periodic compositional complexity (ref. 35 and Section 2.3). These are incorporated in programs *PSEG* for amino acid sequences and *NSEG* for nucleotide sequences (35). A longer term challenge is to locate and classify more comprehensively all types of simple regions in long genomic sequences and their translated protein products. Theoretical advances, as listed in Section 2.6, will provide a crucial part of meeting this challenge. Also, integrated software packages that include multilevel browsing facilities for sequence and genome analysis will increasingly incorporate algorithms such as those outlined in this chapter.

4. Masking of low-complexity sequences for searching databases

4.1 The problem

Ideally, database search methods should compare all classes of sequences, whether simple or complex, but most current search algorithms base their statistics on random models and are commonly confounded by low-complexity regions (ref. 5 and Chapter 7). These may interfere with the discovery of new sequence similarities unless special precautions are taken. Problems include ambiguities in the phase of sequence alignments, misleading, spuriously high scores, and overwhelmingly large output lists in which interesting similarities may be inconspicuously buried. Low-complexity segments of proteins generally evolve rapidly and do not give meaningful alignments position by position in ways that reflect protein structure and mutational history. Permutations, shuffles, or reversals of most low-complexity sequences give high alignment scores that are similar to those of the original sequence. These scores primarily reflect compositional biases rather than significant positional similarity.

4.2 Masking methods

This problem is essentially solved at present by automatic masking, using *SEG* and/or *XNU*, of the low-complexity segments in database search query sequences, replacing their residues with 'x' characters. *Figure 3b* shows the prion protein filtered in this way using *SEG*. This sequence had 34 true homologues in *BLASTP* searches of the NCBI non-redundant database of 15 January 1994. Run unmasked, it gave an overwhelming output list of 945 database sequences above the score threshold. Almost all of these 'hits' were spurious low-complexity matches. *SEG* masking reduced the *BLASTP* output list to a manageable 59 sequences and revealed some previously buried true matches to weakly homologous partial scrapie sequences in addition to the strong prion homologues. As an alternative to *SEG*, *XNU* required non-default parameters to force masking of the modulo-8 repeat, and this reduced the output list to 86 sequences.

The prion protein is a relatively small, uncomplicated example. For sequences such as multidomain transcription factors, for example, many of which have numerous interspersed low-complexity regions, masking may result in a very dramatic simplification of the output list. An important application of masking with *SEG* and *XNU* has been in the production of pre-computed sets of all amino acid sequence homologues ('neighbours') released with the NCBI databases and incorporated into the *Entrez* retrieval system (33). Since, September 1992, the entire sequence databases and all updates have been masked prior to searches, in order to avoid spurious matches in the neighbour lists.

5. Complexity definitions and segmentation algorithm

5.1 Definition 1

Based on the complexity state vector of a sequence window (Section 2.3), local compositional complexity, K_1, is defined (3, 12) as the information needed per position, given the window's composition, to specify a particular residue sequence. For an N-residue alphabet (usually, $N = 4$ or 20) and a window of length L:

$$K_1 = \frac{1}{L} \log_N \left(\frac{L!}{\prod_{i=1}^{N} n_i!} \right) \tag{1}$$

where the n_i are the N numbers in the complexity state vector. Here, the logarithm is taken to base N to place this measure in the range 0 to 1. To express complexity in frequently used information units, logarithms may be taken to base 2, giving bits, or to base e, giving nats.

5.2 Definition 2

Another informational measure of local compositional complexity, K_2, usually expressed in bits, is sometimes used (11, 34) instead of K_1.

$$K_2 = - \sum_{i=1}^{N} \frac{n_i}{L} \left(\log_2 \frac{n_i}{L} \right) \tag{2}$$

K_2 is an approximation that converges towards K_1 at large window lengths.

5.3 Probabilities of complexity states

Assuming equal (uniform) probabilities for the appearance of the four nucleotides or 20 amino acids (see Section 2.5), the probability P_0 for the occurrence of any complexity state is:

$$P_0 = \frac{1}{N^L} \left(\frac{L!}{\prod_{i=1}^{N} n_i!} \right) \left(\frac{N!}{\prod_{k=0}^{L} r_k!} \right) \tag{3}$$

Here, r_k is the count of the number of times the number k occurs among the n_i of the complexity state vector.

5.4 Segmentation algorithm based on compositional complexity

SEG is a two-stage algorithm (3). The first stage identifies approximate 'raw segments' of low-complexity and the second stage is a local optimization. The optimized segments produced correspond well to intuitive concepts of simple sequences.

The length-range and stringency are determined at the first stage by three user-defined parameters. In the order of the *SEG* program command-line (*Figure 3* legend), these are the initial window length L_1, the 'trigger' complexity $K_2 1$ and the 'extension' complexity $K_2 2$. These complexity parameters use Definition 2 (Section 5.2) for computational efficiency and units of *bits*. At the first stage, 'trigger windows' (length L_1 and complexity less than or equal to $K_2 1$) are extended into contigs in both directions by merging with 'extension windows' (overlapping windows of length L_1 and complexity less than or equal to $K_2 2$). Each contig is a raw segment.

At the second stage, each raw segment is reduced to a single optimal low-complexity segment, which may be the entire raw segment but is usually a subsequence. This is found by exhaustive search for the subsequence of the raw segment with the minimum value of P_0 (Equation 3), using precomputed look-up tables for efficiency. P_0 is a particularly suitable function for this optimization because it gives closely similar expected values at all window lengths (3).

References

1. Karlin, S. and Brendel, V. (1993). *Science,* **259**, 677.
2. Pevsner, P. A., Borodovsky, M. Y., and Mironov, A. A. (1989). *J. Biomol. Struct. Dyn.,* **6**, 1013.

3. Wootton, J. C. and Federhen, S. (1993). *Comput.Chem.,* **17**, 149.
4. Wootton, J. C. (1994). *Comput. Chem.,* **18,** 269.
5. Altschul, S. F., Boguski, M., Gish, W., and Wootton, J. C. (1994). *Nature Genet.,* **6**, 119.
6. Britten, R. J. and Kohne, D. E. (1968). *Science,* **161**, 529.
7. Wetmur, J. G. and Davidson, N. (1968). *J. Mol. Biol.,* **31**, 349.
8. Ptitsyn, O. B. and Volkenstein, M. V. (1986). *J. Biomol. Struct. Dyn.,* **4**, 137.
9. Brendel, V., Bucher, P., Nourbakhsh, I. R., Blaisdell, B. E., and Karlin, S. (1992). *Proc. Natl. Acad. Sci. USA,* **89**, 2002.
10. Karlin, S. and Brendel, V. (1992). *Science,* **257**, 39.
11. Konopka, A. K. and Owens, J. (1990). *Genet. Anal. Tech. Appl.,* **7**, 35.
12. Salamon, P. and Konopka, A. K. (1992). *Comput. Chem.,* **16**, 117.
13. Milosavljevic, A. and Jurka, J. (1993). *Comput. Appl. Biosci.,* **9**, 407.
14. Trifonov, E. N. (1989). *Bull. Math. Biol.,* **51**, 417.
15. Pietrokovski, S., Hirshon, J., and Trifonov, E. N. (1990). *J. Biomol. Struct. Dyn.,* **7**, 1251.
16. Fickett, J. W. and Tung, C. S. (1992). *Nucleic Acids Res.,* **20**, 6441.
17. Claverie, J. M., Sauvaget, I., and Bougueleret, L. (1990). In *Methods in enzymology,* Vol. 183 (ed. R. F. Doolittle), p. 237. Academic Press, London.
18. Konopka, A. K. and Smythers, G. W. (1987). *Comput. Appl. Biosci.,* **3**, 193.
19. Konopka, A. K. and Owens, J. (1990). In *Computers and DNA* (ed. G. I. Bell and T. G. Marr), p. 147. Addison-Wesley, Redwood City.
20. Pevsner, P. A. (1992). *Comput. Chem.,* **16**, 103.
21. Bell, G. I. and Torney, D. C. (1993). *Comput. Chem.,* **17**, 185.
22. Auger, I. E. and Lawrence, C. E. (1989). *Bull. Math. Biol.,* **51**, 39.
23. Chappey, C. and Hazout, S. (1992). *Comput. Appl. Biosci.,* **8**, 255.
24. Churchill, G. A. (1989). *Bull. Math. Biol.,* **51**, 79.
25. Borodovsky, M. and McIninch, J. (1993). *Comput. Chem.,* **17**, 123.
26. Gusev, V. D., Kulichkov, V. A., and Chupakhina, O. M. (1993). *Biosystems,* **30**, 183.
27. Jeffrey, H. J. (1990). *Nucleic Acids Res.,* **18**, 2163.
28. Solovyev, V. V. (1993). *Biosystems,* **30**, 137.
29. Claverie, J. M. and States, D. J. (1993). *Comput. Chem.,* **17**, 191.
30. Tautz, D., Trick, M., and Dover, G. A. (1986). *Nature,* **322**, 652.
31. Hancock, J. M. and Armstrong, J. S. (1994). *Comput. Appl. Biosci.,* **10**, 67.
32. Yagil, G. (1993). *J. Mol. Evol.,* **37**, 123.
33. Benson, D., Lipman, D. J., and Ostell, J. (1993). *Nucleic Acids Res.,* **21**, 2963.
34. Staden, R. (1984). *Nucleic Acids Res.,* **12**, 521.
35. Wootton, J. C. and Federhen, S. (1996) In *Methods in enzymology,* Vol. 266 (ed. R. F. Doolittle), p. 554. Academic Press, London.

9

Repetitive sequences in DNA

JÖRG T. EPPLEN and OLAF RIESS

1. Introduction

A characteristic feature of eukaryotic genomes is the presence of various amounts of different repetitive DNA sequences (1). So far this generalized statement lacks any exceptions, but not a single particular repetitive element can be found in all of the different eukaryotes. Apparently the whole evolution of eukaryotes has been accompanied by repetitive DNA (2). Nevertheless the true meaning of genomic redundancy is far from being perceived and understood. According to an extreme point of view, repetitive DNA sequences have long been evaluated within the framework of prokaryotic dogma without sufficient consideration of the higher-order phenomena that characterize the molecular biology of eukaryotic organisms (3). An increasing number of prokaryote species have recently been identified which host some repetitive DNA. The amounts, abundance, and the interspersion patterns of this sequence class with single copy elements, however, is still to be clarified.

From the very beginning of the computer-aided analysis of nucleic acids the existence of repetitivity in the sequences was hampering rather than helping to unravel the secrets of the genomic information. This chapter gives a condensed overview on the repetitive elements in order to increase the awareness of possibly confounding results during sequence analysis and the design of experimental strategies (e.g. primer development for polymerase chain reaction, PCR). Whenever possible direct examples are included, particularly from the human genome, because of the overwhelming amount of work so far already invested into this single species. The chapter is written for the non-expert user of workstations and mainframe computers which are used to search through currently existing data banks harbouring more than 5 $\times$ 10^8 nucleotides (e.g. EMBL/GenBank database as of 28 April, 1996, *Table 1*) or the respective amounts of protein sequence information. Efficient analysis/search strategies need to be applied especially by the less experienced clients in order to keep the cpu hours and thus costs in a manageable range.

Table 1. Representation of simple tandem repetitive DNA sequences in the EMBL/GenBank data bank

Simple repeat motif (monomer)	6.1.'96 EMPRI	28.4.'96 GEALL
$(g)_{40}$	0	7
$(a)_{40}$	84	917
$(g)_{20}(a)_{20}$	0	0
$(g)_{20}(t)_{20}$	0	2
$(g)_{20}(c)_{20}$	0	0
$(a)_{20}(g)_{20}$	0	2
$(a)_{20}(t)_{20}$	0	0
$(t)_{20}(a)_{20}$	0	0
$(t)_{20}(g)_{20}$	0	0
$(c)_{20}(g)_{20}$	0	0
$(ga)_{20}$	34	624
$(gt)_{20}$	1102	3994
$(gc)_{20}$	0	0
$(at)_{20}$	30	526
$(gt)_{10}(ga)_{10}$	41	239
$(ga)_{10}(gt)_{10}$	30	62
$(ag)_{10}(gt)_{10}$	0	1
$(ga)_{10}(tg)_{10}$	0	0
$(gt)_{10}(ag)_{10}$	0	2
$(tg)_{10}(ga)_{10}$	1	1
$(gt)_{10}(at)_{10}$	18	59
$(at)_{10}(gt)_{10}$	10	43
$(gt)_{10}(ta)_{10}$	0	0
$(tg)_{10}(at)_{10}$	0	2
$(ta)_{10}(gt)_{10}$	1	3
$(at)_{10}(tg)_{10}$	0	0
$(ga)_{10}(at)_{10}$	0	0
$(at)_{10}(ga)_{10}$	0	1
$(at)_{10}(ag)_{10}$	3	11
$(ga)_{10}(ta)_{10}$	0	1
$(ag)_{10}(ta)_{10}$	0	0
$(ta)_{10}(ag)_{10}$	0	0
$(gc)_{10}(at)_{10}$	0	0
$(gc)_{10}(ta)_{10}$	0	0
$(at)_{10}(cg)_{10}$	0	0
$(ta)_{10}(cg)_{10}$	0	0
$(gga)_{13.3}$	1	30
$(ggt)_{13.3}$	6	19
$(ggc)_{13.3}$	3	16
$(gaa)_{13.3}$	2	75
$(gat)_{13.3}$	27	26
$(gac)_{13.3}$	0	3
$(gta)_{13.3}$	14	23
$(gtt)_{13.3}$	4	41
$(gca)_{13.3}$	35	86
$(aat)_{13.3}$	70	238

Simple repeat motif (monomer)	6.1.'96 EMPRI	28.4.'96 GEALL
$(ggga)_{10}$	3	20
$(gggt)_{10}$	0	1
$(gggc)_{10}$	0	0
$(ggaa)_{10}$	27	272
$(ggat)_{10}$	4	76
$(ggta)_{10}$	0	1
$(ggtt)_{10}$	0	20
$(ggca)_{10}$	1	10
$(gagt)_{10}$	0-	2
$(gaaa)_{10}$	32	399
$(gaat)_{10}$	14	20
$(gaac)_{10}$	0	0
$(gata)_{10}$	34	1685
$(gatt)_{10}$	0	17
$(gaca)_{10}$	2	28
$(gact)_{10}$	0	5
$(gtaa)_{10}$	0	4
$(gtat)_{10}$	32	65
$(gtta)_{10}$	0	0
$(gttt)_{10}$	0	10
$(gcaa)_{10}$	1	7
$(gcat)_{10}$	0	0
$(gcta)_{10}$	0	0
$(gcca)_{10}$	0	2
$(aaat)_{10}$	16	131
$(aatt)_{10}$	0	6
$(ga)_{10}(gaca)_5$	1	4
$(ga)_{10}(gata)_5$	2	2
$(gt)_{10}(gaca)_5$	0	0
$(gt)_{10}(gata)_5$	0	0
$(gaca)_5(ga)_{10}$	0	2
$(gaca)_5(gt)_{10}$	0	0
$(gata)_5(ga)_{10}$	1	1
$(gata)_5(gt)_{10}$	0	0
$(gata)_5(aat)_{6.7}$	0	0
$(ata)_{6.7}(gata)_5$	0	0
$(gaaa)_5(gata)_5$	0	0
$(gata)_5(gaaa)_5$	1	1

2. Types of repetitive sequences

Due to the many different entities (and research approaches), the field of repetitive DNA is flooded with descriptive but unsystematic and partly contradictory nomenclature. In order to clarify the subjects of our deliberations, it is necessary to define a few terms concerning different forms of repetitive DNA sequences. We cover tandemly organized repetitive elements (Sections 2.1, 2.2, and 2.4) in addition to those lacking obvious tandem arrangement. The best defined characteristic for the latter category concerns intimate interspersion with single copy (or other repetitive) sequences (Section 2.3). Rapidly reassociating hairpin or fold-back repeats are not mentioned in detail since they are dealt with in Chapters 12 and 14. This outline follows the extensive discussions held at the 2nd international meeting on DNA fingerprinting (4). The order of the repeat classes described below reflects their historical detection and description rather than their present-day meaning.

2.1 Satellite DNA

DNA satellites were found and named before their exact sequence contents could even be analysed. In density gradients run in ultracentrifuges, genomic DNA forms bands of uniform specific density depending on the G-C contents of the respective part of the genome. These smaller 'satellite' peaks (in relation to the bulk of genomic DNA) were known to consist of highly repetitive DNA sequences, usually millions of tandem repeats of short motifs (1). More rarely the motif lengths may amount to a few hundred (thousand) base pairs (bp). On the chromosomes, satellites are mostly located in the heterochromatic parts (e.g. in and around the centromeres). In general only few satellite loci are present per genome, but never more than one or two per chromosome. Different loci may harbour locus- (chromosome-) specific repeat variants (5). Other satellite DNAs appear quite stable during longer evolutionary time spans. The so-called alpha satellite DNA is localized at the centromeric region of primate chromosomes (6). Human alphoid DNA is organized into chromosome-specific subfamilies after amplification of segments composed of tandemly arranged copies of 171 bp monomeric or 340 bp dimeric repeat units (7). It is believed that the different alphoid satellite families arose prior to the emergence of the human, ape, and monkey species after which the families have remained more or less unaltered (8). The chromosome specificity of the alphoid subfamilies implies that transfer of sequences between non-homologous human chromosomes occurs quite rarely.

With increasing pace of molecular genetic techniques, especially molecular cloning in prokaryotic hosts, satellites have lost a great deal of their (methodologically based) attraction, having represented more or less the first tools to unravel the secrets of large eukaryote genomes. Ironically, in the foreseeable future these long stretches of monotonous repetitivity will remain the only true white spots in the detailed maps of the 'totally sequenced' genomes.

Nevertheless, their early characterization has influenced the naming of many other classes of repeats (see below). This knowledge should help to unravel some of the mysteries of the erratic nomenclature concerning genomic redundancy.

2.2 Simple repetitive DNA sequences

The lengths of motifs in simple repeats should by definition be limited to one to six bases because otherwise partial overlaps (and confusion) with other sequence classes cannot be avoided. Simple sequence motifs had been detected early as long polypyrimidinic stretches (9) or as components of classical satellites (5). Today mainly the shorter versions of five to several hundred tandem motif repetitions are meant by this term. These elements are ubiquitously interspersed throughout eukaryotic genomes, show a high degree of polymorphism (informativity), and are amenable to PCR amplification techniques. Therefore, simple sequences are currently the most widely applied markers for genomic mapping (10) and indirect gene diagnoses in humans. One particular form of these dinucleotide stretches, $(CA)_n/(GT)_n$, has been estimated to occur with more than 50 000 copies per human haploid genome (11, 12). In mammals sequences with fewer than 10 repeat units were found to be not sufficiently informative or even non-polymorphic (13). On the basis of the famous 'minisatellite sequences' (14) the term 'microsatellite' was coined for simple repeats, since the basic repetition units are shorter. Furthermore, the latter designation means that these loci are used for DNA profiling (in the widest sense differentiating individuals of a given species) employing PCR systems for efficient typing (15, 16).

2.3 Short and long interspersed nucleotide elements (SINEs and LINEs)

Ultimately the designation of interspersed repeated DNA resulted from extensive DNA denaturation and reassociation studies defining the overall arrangements of single copy and repetitive sequence arrays (for reviews see 2, 17). Quite simply, these entities were categorized according to their size into short (up to 500 bp) and long interspersed nucleotide elements (SINEs and LINEs respectively) (18). With respect to their general biological relevance, phenomena like (retro-) transposition, recombination, evolution, and protein coding capabilities have always been worthy of discussion, albeit that many controversies still exist. While both SINEs and LINEs may occupy similar proportions of the human genome (5–10%), the former are of more immediate importance in modern molecular genetic techniques, because of their intimate interspersion with genes and the possibility to differentiate human genomic counterparts from those of other species, e.g. in somatic hybrid cell lines (19). The most abundant subclass of the primate SINEs may encompass 5×10^5 members or more, and less than

300 bp of length (see e.g. ref. 20). It is designated as the Alu (element) family, because of a restriction site present in most members. This Alu family as well as the rodent equivalent B1 are probably derived from 7SL RNA (20), whereas the rodent B2 and rabbit C repeats could well stem from tRNA sequences. Thus both of these abundant superfamilies are ancestrally related to sequences with functions in translation (see also Section 3.2. on expression).

By 1994, Alu family members were represented more than 3500 times in the EMBL/GenBank databases as nucleotide sequences (2513 in the EMPRI sublibrary (*BLASTN* program, random match probability $p < 0.01$); 952 in the EMNEW sublibrary with $p < 0.01$; 17 April 1994). Two tandem monomer units in each of the Alu repeats harbour short poly(A)$^+$ tails of varying lengths. Whereas the 5' ends commonly start with pure runs of As, (A)$^+$-rich 3' ends are often highly structured generating tandemly repeated units. Interestingly, these poly(A)$^+$ tails often give rise to the presence of dinucleotide $(CA)_n/(GT)_n$ microsatellites. The interrelationship of the different repeat sequence classes is exemplified here. As a practical consequence of this relationship, the design of primer pairs for PCR and microsatellite analysis should be controlled by databank searching in order to avoid (cross-priming) pitfalls. Fortunately Claverie (21) has already designed an Alu-specific entry into the main sequence databases which acts as a warning against the presence of unrecognized Alu-derived amino acid sequences.

2.4 Minisatellites

This illustrious expression was ingeniously coined by Dr Alec J. Jeffreys (14) who was the first to exploit fully the power of these interspersed elements. The basic units are 10–100 bp long organized in up to several thousand loci. Each locus appears to exhibit a distinctive repeat unit. Many minisatellite loci appear to be located (clustered) at the telomeres of human chromosomes (22). Collectively, many minisatellite loci were the cross-hybridization targets for multilocus DNA fingerprinting (23)—a technique which allowed for the first time a demonstration of genetic individuality, e.g. in man. At times these same or similar elements have also been termed **variable number of tandem repeat** (VNTR; 24) loci. Since this expression lacks discriminatory potential, it was used to cover all loci with tandem repeat substructures. Owing again to the legacy of Jeffreys, some of the same authors (25) could not refrain from constructing yet another entity—'midisatellites'. Fortunately this additional expression is currently no longer employed in meaningful literature citations.

3. Repeats in genomic DNA (and protein) databanks

Jurka *et al.* (26) have compiled consensus sequences of different sorts of repetitive DNAs. Such collections help, at least in part, in searching databanks.

3.1 Evolutionary aspects

In general, repetitive DNA sequences are characterized by short time-histories in evolutionary time spans: satellites differ in closely related species; various versions of SINEs like Alu repeats characterize the genomes of primate species; population differences exist. The mutation rates for individual human simple tandem repeats can reach 8×10^{-3} (27). In highly inbred mice a probe containing $(GGGCA)_n$ motifs allows us to differentiate individuals because of significantly elevated rates of mutations (28). In an extremely exceptional case, a minisatellite system in chicken surpassed these mutation frequencies and came close to the theoretical maximum of 100% (29). On the contrary, perhaps merely as an exception that verifies the rule, particular microsatellites may be preserved across species barriers (30) or even persist in the same genomic location for millions of years for completely unknown reasons (31).

Obviously the different classes of repeated sequences exhibit quite different characteristic modes of evolutionary behaviour (large-scale amplifications and deletions for classical satellites; slippage mutations for simple repeats; polar mutations for minisatellites; transposition for SINEs and LINEs). Yet this does not mean that simple repeats cannot be contained in large-scale amplification (32) and/or transposition units (jumps of certain Alu repeat family members that contain microsatellite units). Recently Gaillard and Strauss (33) produced evidence that $(CA)_n/(GT)_n$ stretches can be associated into four-stranded complexes bound by HMG 1 and 2 proteins. These novel DNA/protein structures could well be envisioned to take part in DNA wrapping and chromosomal looping, recombination, and repair mechanisms.

3.2 Expression of repeats

In humans Alu sequences constitute about 10% of heterogeneous nuclear RNA (hnRNA) sequences, some of which even persist in mature mRNA (34). The presence of 100–1000 Alu repeat copies per cell has been taken as evidence that these transcripts bear some functional meaning. In this context immediately obvious and attractive—though by no means proven—were roles in the regulation of gene expression or translation. Whereas $(CA)_n/(GT)_n$ simple repeats appear frequently in the 5'- and 3'-untranslated regions of mRNAs (35), particular simple repeat elements, for example $(CAC)_n/(GTG)_n$ are clearly underrepresented (36). For example, using $(GAA)_6$ as an oligonucleotide probe only 0.1% of a fetal brain cDNA library was positive (37). When different tissues are compared, mRNAs containing $(CGG)_n$ and $(CAG)_n$ polymorphic trinucleotide repeats are especially abundant in the fetal brain (38). The reasons for this differential representation, besides their varying a priori abundance, may be due to their interspersion in different chromosomal compartments (rich or poor in genes), the distances from transcription units and the sequence composition influencing the secondary structure of the transcription unit. Increased interest in simple (trinucleotide) repeats because of their expression on the mature mRNA level (38) and even sometimes in simple proteins (39; see also Section 5).

3.3 Repeats as tools

In eukaryotes (with average genome sizes with respect to their organismal complexity) one can expect to encounter at least one simple repeat stretch per 10 kb of DNA. This frequent interspersion mode has been exploited also for contig mapping of human chromosomes (40). The frequent occurrence, the simplified techniques available nowadays for the amplification of these length differences, and the high degree of variability rendered them as highly informative DNA markers for genome mapping and linkage studies. Linkage maps of the human genome have been constructed based on the segregation analysis of currently more than 6000 polymorphic loci containing short tracts of $(CA)_n/(GT)_n$ repeats (41–43; also see respective databank sublibraries). Genetic linkage mapping has become an increasingly important technology for studying human biology and, in particular, for isolating disease genes. Landmark discoveries over the past few years on the basis of informative linked markers include the isolation of genes responsible for cystic fibrosis (44), FRA X-linked mental retardation (45), myotonic dystrophy (46), and Huntington disease (47).

The cumulative degree of high allelic variability allows the use of DNA profiling technology to identify and to distinguish individuals in a variety of different applications in forensics, paternity testing, transplantation analysis (48), and tumour screening (32, 49) as well as ecological and evolutionary genetics (50, 51). Similar applications with practically unlimited discrimination potential can be achieved by using multilocus DNA fingerprinting employing either minisatellite probes (14) or oligonucleotides containing simple repeat motifs (29, 50).

4. Short consensus motifs for the identification of functional sequences in DNA which appear repetitively in and around genes

At face value those short consensus sequences which constitute the hallmarks of genes could be called repetitive (*Figure 1*; 52). Yet, due to the haploid human genome size of 3×10^9 bp, all sequences shorter than 17 bases may occur repetitively by pure chance. Nevertheless, the tags are mentioned here since elaborated search strategies via short oligonucleotide probes allow us to identify potential gene sequences in the barren stretches of the genomic desert. Accordingly this approach has also been applied successfully to contigs from various genomic regions and the human Y chromosome (53).

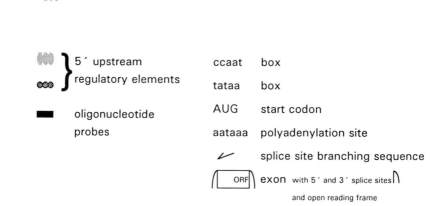

Figure 1. Schematic representation of recurrent DNA sequence consensus elements which can be used for the identification of genes in eukaryotes via short oligonucleotide probes.

5. Diseases caused by expansion of simple trinucleotide repeats

Increased interest in microsatellite mutations has been stimulated by the discoveries that fragile X (FRAXA, -E) syndrome, myotonic dystrophy, spinocerebellar ataxia 1, dentatorubropallidoluysian atrophy, Kennedy's, Machado Joseph, Huntington's disease, and Friedreich's ataxia (45–47, 54–58) are caused by elongated simple trinucleotide repeats. This disease category was divided into subgroups (57). Type 1 is caused by the expansion of CAG trinucleotides in the open reading frame of genes encoding poly-glutamines. Transcription/translation of these genes are not affected. The repeats blocks rarely exceed 100 trinucleotide units. Type 2 disorders are caused by elongated CTG, CCG, or GAA repeats outside the open reading frame, alter transcription/translation or cause mRNA instability. These repeats may be expanded to >1000 units and they show somatic instability. Hypermethylation has been shown as a consequence of extended reiterations of CCG repeats. It remains an open question why repeat expansions occur most frequently via maternal transmission in FRAX syndrome or in myotonic dystrophy (type 2) yet through paternal transmission in type 1 disorders. With the exception that instability of short tandem repeats is par-ticularly prevalent in certain types of colon tumours (59) disease causing mutations based on di-, tetra-, penta-, or hexanucleotide repeats have not yet been identified.

6. Conclusions

During the last 30 years repetitive DNA has received only marginal attention from the general research community of biologists and human geneticists. This overall neglect sometimes vanished in waves when the redundancy principle could be used as an efficient tool (satellites as probes, minisatellites, and simple repeats to reveal individualization and genetic relationships, microsatellites for disease mapping, forensics, etc.). Without sufficient awareness of their intricacies, it is no surprise that, in the context of modern computing and databanking, repetitive sequences represent annoying disturbances rather than a means to effect and accelerate the standard analyses. Efficient methods to circumvent these obstacles include searching in phases involving initially small databases containing domains representative of large sequence families as proposed by Altschul *et al.* (60). In order to understand some of the further implications of repetitive sequences on DNA in general, it appears mandatory to acquire via novel algorithms more sophisticated knowledge with respect to the structure and hence their influences: e.g. on gene expression.

References

1. Britten, R. J. and Kohne, D. E. (1968). *Science,* **161**, 529.
2. Britten, R. J. and Davidson, N. (1973). *Q. Rev. Biol.* **48**, 565.
3. Epplen, J. T. and Ohno, S. (1984). In *Research perspectives in cytogenetics* (ed. R. S. Sparkes and F. F. de la Cruz), p. 17. University Park Press, Baltimore.
4. Tautz, D. (1993). In *DNA fingerprinting: state of the science* (ed. S. D. J. Pena, R. Chakraborty, J. T. Epplen, and A. J. Jeffreys), p. 21. Birkhäuser, Basel.
5. Skinner, D. M. (1977). *BioScience,* **27**, 790.
6. Choo, K. H., Vissel, B., Nagy, A., Earl, E., and Kalitsis, P. (1991). *Nucleic Acids Res.* **19**, 1179.
7. Willard, H. F. and Waye, J. S. (1987). *Trends Genet.* **3**, 192.
8. Gillespie, D., Donhower, L., and Strayer, D. (1982). In *Genome evolution* (ed. G. A. Dover and R. B. Flavell), p. 113. Academic Press, London.
9. Birnboim, H. C. and Straus, N. A. (1975). *Can. J. Biochem.,* **53**, 640.
10. Weber, J. L. (1990). *Genomics,* **7**, 524.
11. Hamada, H. and Kakunaga, T. (1982). *Nature,* **298**, 396.
12. Stallings, R. L., Torney, D. C., Hildebrand, C. E., Longmire, J. L., Deaven, L. L., Jett, J. H., *et al.* (1990). *Proc. Natl. Acad. Sci. USA,* **87**, 6218.
13. Georges, M. and Massey, J. (1992). International patent application published under the cooperation treaty (No. WO 92/13102. WIPO, Geneva, Switzerland).
14. Jeffreys, A. J., Wilson, V., and Thein, S. L. (1985). *Nature,* **316**, 67.
15. Weber, J. L. and May, P. E. (1989). *Am. J. Hum. Genet.,* **44**, 388.
16. Litt, M. and Luty, J. A. (1989). *Am. J. Hum. Genet.,* **44**, 397.
17. Schmidtke, J. and Epplen, J. T. (1980). *Hum. Genet.,* **55**, 1.
18. Singer, M. F. (1982). *Cell,* **28**, 433.

19. Brooks-Wilson, A. R., Smailus, D. E., Weier, H.-U. G., and Goodfellow, P. J. (1992). *Genomics,* **13**, 409.
20. Deininger, P. (1989). In *Mobile DNA* (ed. D. Howe and P. Berg), p. 619. American Society for Microbiology.
21. Claverie, J.-M. (1992). *Genomics,* **12**, 838.
22. Royle, N. J., Clarkson, R. E, Wong, Z., and Jeffreys, A. J. (1988). *Genomics,* **3**, 352.
23. Jeffreys, A. J., Wilson, V., and Thein, S. L. (1985). *Nature,* **316**, 76.
24. Nakamura, Y., Leppert, M., O'Connell, P., Wolff, R., Holm, T., Culver, M., *et al.* (1987). *Science,* **235**, 1616.
25. Nakamura, Y., Julier, C., Wolff, R., Holm, T., O'Connell, P., Leppert, M., *et al.* (1987). *Nucleic Acids Res.,* **15**, 2537.
26. Jurka, J., Walichiewicz, J., and Milosavljevic, A. (1992). *J. Mol. Evol.,* **35**, 286.
27. Weber, J. L. and Wong, C. (1993). *Hum. Mol. Genet.,* **2**, 1123.
28. Kelly, R., Bulfield, G., Collick, A., Gibbs, M., and Jeffreys, A.J. (1989). *Genomics,* **5**, 844.
29. Epplen, J. T., Ammer, H., Kammerbauer, C., Schwaiger, W., Schmid, M., and Nanda, I. (1991). *Adv. Mol. Genet.,* **4**, 301.
30. Schlötterer, C., Amos, B., and Tautz, D. (1991). *Nature,* **354**, 63.
31. Schwaiger, F. -W., and Epplen, J. T. (1995). *Immunol. Rev.,* **1**, **43**, 199.
32. Nürnberg, P., Zischler, H., Fuhrmann, E., Thiel, G., Losanova, T., Kinzel, D., *et al.* (1992). *Genes Chromosomes Cancer,* **3**, 79.
33. Gaillard, C. and Strauss, F. (1994). *Science,* **264**, 433.
34. Liu, W.-M., Maraia, R. J., Rubin, C. M., and Schmid, C. W. (1994). *Nucleic Acids Res.,* **22**, 1087.
35. Boguski, M. S., Lowe, T. M. J., and Tolstoshev, C. M. (1993). *Nature Genet.,* **4**, 332.
36. Epplen, C. and Epplen, J. T. (1994). *Hum. Genet.,* **93**, 35.
37. Siedlaczck, I., Epplen, C., Riess, O., and Epplen, J. T. (1994). *Electrophoresis,* **14**, 973.
38. Riggins, G. J., Lokey, L. K., Chastain, J. L., Leiner, H. A., Sherman, S. L., Wilkinson, K. D., *et al.* (1992). *Nature Genet.,* **2**, 186.
39. Stallings, R. L., Ford, A. F., Nelson, D., Torney, D. C., Hildebrand, C. E., and Moyzis, R. K. (1991). *Genomics,* **10**, 807.
40. Wootton, J. C. and Federhen, S. (1993). *Comput. Chem.,* **17**, 149.
41. Weissenbach, J., Gyapay, G., Dib, C., Vignal, A., Morisette, J., Millasseau, P., *et al.* (1992). *Nature,* **359**, 794.
42. Cox Matise, T., Perlin, M., and Chakravarti, A. (1994). *Nature Genet.,* **6**, 384.
43. Buetow, K. H., Weber, J. L., Ludwigsen, S., Scherpbier-Heddema, T., Duyk, G. M., Sheffield, V. C., *et al.* (1994). *Nature Genet.,* **6**, 391.
44. Rommens, J. M., Iannuzzi, M. C., Kerem, B.-S., Drumm, M. L., Melmer, G., Dean, M., *et al.* (1989). *Science,* **245**, 1059.
45. Fu, Y.-H., Kuhl, D. P. A., Pizzuti, A., Pieretti, M., Sutcliffe, J. S., Richards, S., *et al.* (1991). *Cell,* **67**, 1047.
46. Brook, J. D., McCurrach, M. E., Harley, H. G., Buckler, A. J., Church, D., Aburatani, H., *et al.* (1992). *Cell,* **68**, 799.
47. Huntington's Disease Collaborative Research Group (1993). *Cell,* **72**, 971.
48. Kremens, B., Gomolka, M., Ottinger, H., Grosse-Wilde, H., Schäfer, U. W., and Epplen, J. T. (1994). *Bone Marrow Transplant.,* **12**, 661.

49. Bock, S., Epplen, J. T., Noll-Puchta, H., Rotter, M., Höfler, H., Block, T., *et al.* (1993). *Genes Chromosomes Cancer,* **6**, 113.
50. Epplen, C., Melmer, G., Siedlaczck, I., Schwaiger, F-W., Mäueler, W., and Epplen, J. T. (1993). In *DNA fingerprinting: state of the science* (ed. S. D. J. Pena, R. Chakraborty, J. T. Epplen, and A. J. Jeffreys), p. 29. Birkhäuser, Basel.
51. Buitkamp, J., Schwaiger, W., Epplen, C., Gomolka, M., Weyers, E., and Epplen, J. T. (1993). In *DNA fingerprinting: state of the science* (ed. S. D. J. Pena, R. Chakraborty, J. T. Epplen, and A. J. Jeffreys), p. 87. Birkhäuser, Basel.
52. Melmer, G. and Epplen, J. T. (1993). *Adv. Electrophoresis,* **6**, 179.
53. Träger, T., Schmidt, P., and Epplen, J. T. (1994). *Electrophoresis,* **15**, 871.
54. Orr, H. T., Chung, M., Banfi, S., Kwiatkowski, T. J., Servadio, A., Beaudet, A. L., *et al.* (1993). *Nature Genet.,* **4**, 221.
55. Campuzano, V., Montermini, L., Moltò, M. D., Pianese, L., Cossée, M., Cavalcanti, F., Monros, E., *et al.* (1996). *Science,* **271**, 1423.
56. Ashley, C. T. and Warren, S. T. (1995). *Annu. Rev. Genet.,* **29**, 703.
57. Ross, C. A. (1995). *Neuron,* **15**, 493.
58. Richards, R. I. and Sutherland, G. R. (1992). *Nature Genet.,* **1**, 7.
59. Ionov, Y., Peinado, M. A., Malkhosyan, S., Shibata, D., and Perucho, M. (1993). *Nature,* **363**, 558.
60. Altschul, S. F., Boguski, M. F., Gish, W., and Wootton, J. C. (1994). *Nature Genet.,* **6**, 119.

10

Isochores and synonymous substitutions in mammalian genes

GIORGIO BERNARDI, DOMINIQUE MOUCHIROUD,
and CHRISTIAN GAUTIER

1. Introduction

The mammalian genome is a mosaic of isochores, long DNA segments (>300 kb on average) that are remarkably homogeneous in base composition and that can be subdivided into a small number of families characterized by different GC levels (1–4) (GC is the molar ratio of guanine + cytosine). In the human genome, which is representative of the majority of mammalian genomes (5, 6), isochores cover a broad GC range, 30–60% (4, 7). The third codon positions of human genes are compositionally correlated with the isochores in which the corresponding genes are located (1, 8), but they cover a much broader GC range, 25–97.5% (6).

The generation and maintenance of the large compositional heterogeneity of the isochores forming the mammalian genomes, and the genomes of warm-blooded vertebrates in general, have been the subject of contrasting views. In the traditional view of molecular evolution, the rate of point mutation is uniform over the genome of an organism and variation in the rate of nucleotide substitution among DNA regions reflects differential selective constraints (9–11).

In 1989, it was claimed (11) that the mutation rate at synonymous positions varied significantly among regions in the mammalian genome and was correlated with the base composition of genes and their flanking DNA. It was further proposed (11) that the differences arise because mutation patterns vary with the timing of replication of different chromosomal regions in the germline and that this hypothesis could account for both the origin of isochores in the mammalian genomes (1) and the observation (12) that synonymous nucleotide substitutions in different mammalian genes do not have the same molecular clock. More recently, it was claimed (13) that patterns of either damage or excision repair rates along the genome will produce patterns of mutation rates and that such patterns cause the DNA sequence patterns (i.e. the isochore patterns) which exist along the genome.

The specific point which will be discussed here, essentially following (14), is the evidence that synonymous substitution rates vary over different isochore families and over the genes contained in them. Obviously, this question is of crucial importance for the models (11, 13) just mentioned in connection with the generation and maintenance of isochores.

Recent developments in this area are described in refs 40–43.

2. Methods

The analyses concerning pairwise comparisons of homologous sequences carried out in (14) were performed on sequences retrieved from GenBank Release 76 (March 1993) using the database management system ACNUC (15). The procedure used to select the homologous pairs was described by Mouchiroud and Gautier (16). Comparisons comprised the mammalian genome pairs for which a large enough number of homologous sequences was available, namely, the human/other primates, human/artiodactyls, human/rabbit, mouse/rat, and human/rat pairs (6). The features exhibited by human/mouse homologous gene pairs were similar to those of the human/rat pairs.

To quantify dissimilarity between homologous sequences, the synonymous difference frequency (SDF) was used. SDF is the percentage divergence in third codon positions of synonymous codons. SDF does not rely on any hypothesis concerning the nature of the substitution process, in contrast with K_s, the substitution rate per synonymous site, as calculated according to Li *et al.* (17). However, when relatively small ranges are considered, K_s (or K_4, the substitution rate per fourfold degenerate site) and SDF show a strong linear correlation (14). This means that comparisons between our SDF data with K_s or K_4 data of other authors, are valid. In every case, the average SDF values were estimated not only for all points, but also for those characterized by low (< 47%), intermediate (47–74%), and high (> 74%) GC levels, in order to detect possible differences. These GC values roughly correspond to the borders between genes that are located in human isochore families L1, L2–H2, and H3, respectively (7).

3. Results

Plots of frequencies of substitution in third codon positions of synonymous codons (SDF) against GC levels of those positions show no significant correlation in the comparisons human/other primates, human/sheep, human/rabbit (14), rat/mouse (*Figure 1*), and human/rat (*Figure 2*). Such lack of correlation holds, therefore, both in the cases characterized by a conserved base compo-

RAT / MOUSE

(a)

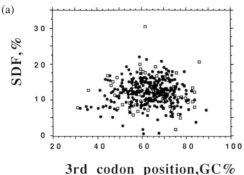

3rd codon position, GC%

(b)

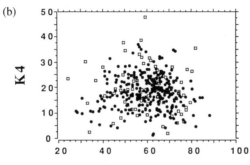

Fourfold degenerate sites, GC%

Figure 1. (a) SDF between homologous genes of mouse and rat is plotted against the average GC level of third codon positions. Filled circles correspond to coding sequences longer than 180 codons. For correlation coefficients and average SDF values for the L (<47% GC), M (47–74% GC), and H (>74% GC) sections see *Table 1*. (b) K4, the frequency of substitutions per fourfold degenerate sites, is plotted against the average GC level of those sites. Filled circles correspond to coding sequences longer than 180 codons.

sition in third codon positions (rat/mouse) and in those characterized by a different base composition (human/rat). A correlation which is weak, yet significant, was found in the comparisons man/calf (*Figure 3*) and human/pig (*Table 1*), in which SDF showed a slight increase with increasing GC. However, this increase appears to be due to slightly lower SDF values for low GC third codon positions (*Table 1*). Since this phenomenon is not found in any other case, in particular in the two comparisons (mouse/rat and human/rat) involving the largest number of genes, an explanation which can be offered at present is that it is due to the particular, small sample of genes present in the low GC section of the plot. An alternative explanation, not exclusive of the

MAN / RAT

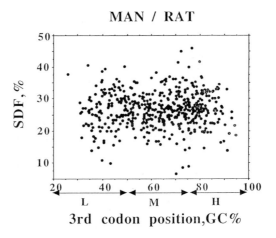

Figure 2. SDF between homologous genes of man and rat is plotted against GC of the third codon position of human genes. Other indications as in *Figure 1* (14).

MAN / CALF

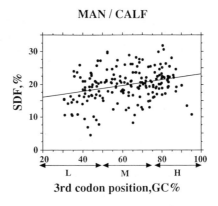

Figure 3. SDF between homologous genes of man and calf is plotted against GC of third codon positions of human genes. Other indications as in *Figure 1* (14).

former one, is that some changes become very frequent at the two ends of the spectrum of synonymous GC levels and they are underestimated. In the case of the man/calf comparison, A↔T and A→G are very frequent at the low GC end, G↔C are very frequent at the high GC end (*Table 2* and Figure 4). Under these circumstances, it is possible that back mutations become very frequent leading to an underestimate of the frequency of changes. A similar, yet weaker, phenomenon is probably responsible for the higher SDF value of the middle GC section of the human/rabbit comparison (*Table 1* and ref. 14).

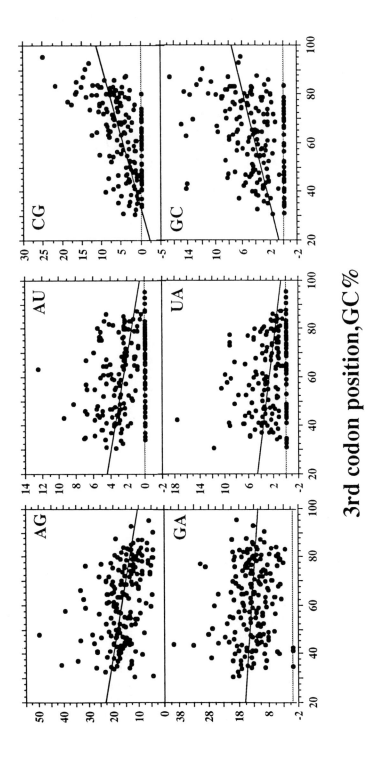

3rd codon position, GC%

Figure 4. Individual changes between bovine and human genes (G → C, A → T, etc.) as percentages of all changes in synonymous third codon positions are plotted against GC levels of human third codon positions.

Table 1. Statistical analyses of synonymous difference frequencies in mammalian genes

Plots	Human/primate	Human/calf	Human/sheep	Human/pig	Human/rabbit	Rat/mouse	Human/rat
Number of genes	44	194	39	79	97	386	523
SDF vs GC$_3$ (human)[a]	$R = 0.07$	$R = 0.32$**[b]	$R = 0.16$	$R = 0.29$**	$R = 0.16$	$R = 0.045$	$R = 0.03$
Average SDF							
All sequences	5.2 (2.3)[c]	19.7 (4.9)	21.4 (4.8)	19.9 (4.3)	20.8 (5.3)	12.5 (3.7)	26.8 (5.4)
L sequences (<47% GC)	4.2 (2.1)	17.5 (5.3)	19.7 (6.3)	16.5 (5.1)	18.2 (5.8)		26.7 (6.1)
M sequences (47–74% GC)	5.5 (2.5)	21.6 (4.5)	21.4 (4.8)	20.7 (3.6)	22.7 (4.6)		26.7 (5.3)
H sequences (>74% GC)	5.2 (2.3)	21.3 (4.1)	22.6 (3.7)	20.6 (4.4)	17.8 (4.1)		27.1 (4.9)

[a] GC$_3$ is the third codon position GC of human sequences. In the mouse/rat comparison the average between the third codon position GC of the sequence pairs was used.
[b] Asterisks refer to statistical significancy (*, 5%; **, 1%).
[c] Values in parentheses are standard deviations.

Table 2. Frequencies of individual synonymous substitutions in homologous genes from human/calf and their correlations with GC levels of synonymous positions

Changes[a]	Human/calf	
	%	R
A→C	4.1	−0.19
A→G	16.5	−0.37
A→U	2.4	−0.35
C→A	3.2	+0.03
C→G	4.8	+0.57
C→U	19.2	+0.22
G→A	13.8	−0.14
G→C	4.3	+0.39
G→U	2.3	+0.10
U→A	2.7	−0.12
U→C	23.8	−0.12
U→G	2.9	+0.16

[a] Changes are indicated in the human → calf direction. Bold-type changes correspond to p-values equal to 10^{-4}.

4. Discussion

This discussion will be divided in three parts. First, we discuss the previous reports in the light of our present findings. Then we analyse the interpretations claimed to account for previous reports. Finally, we discuss some general issues.

4.1 The frequencies of synonymous substitutions do not exhibit differences related to regions of the mammalian genome

The results of *Figures 1–3* unambiguously show that, although the frequencies of synonymous differences exhibit relatively large fluctuations in different genes (covering up to a 20-fold range in the mouse/rat case), they do not show any significant trend over the very extended range of synonymous position GC under consideration (explanations for the trend of *Figure 3* were provided in the Section 3). The above findings raise the question why discrepant results on the dependence of synonymous position GC level were previously reported.

1. The data by Miyata *et al.* (18), showing a lack of differences in synonymous substitution rates, are in apparent agreement with the results by Bernardi *et al.* (14). They fail, however, to prove the point made by the latter authors because of the small size of the samples studied. Indeed, the data only concern a total of 17 comparisons of genes from human, rat, rabbit, and monkey. Although some of the coding sequences investigated did show differences in synonymous position GC levels, the lack of differences in synonymous mutation rates may well be due to the small size of the sample studied,

the largest pairwise comparison, human/rat, comprising only nine genes. No definite conclusion can, therefore, be drawn from those data (and none was drawn by the authors). Indeed, if a systematic variation existed, it would have been missed. The same conclusion applies to the 12 comparisons of primate genes made by Filipski (19), in which no correlation was found, and also to the 13 human/Old World monkeys comparisons made by Wolfe *et al.* (11).

2. The second set of results concerns the rat/mouse comparison in which variations of mutation rates were reported, even if the results were different in two series of data. Indeed, in one case a strong increase with decreasing GC was found (19), whereas in another case a peak at 50% GC was reported (11). It is clear that these discrepant results were both due to the fact that the sizes of the gene samples used (30 genes in (19), 23 large genes in (11)) were still too small. Indeed, the variation in synonymous divergence being relatively large even for genes having exactly the same GC levels in synonymous codon positions (see *Figures 1–3*), any correlation may be found when using a small sequence sample.

3. Ticher and Graur (20) reported a correlation between silent substitutions rate and the percentage of different nucleotides at silent positions. This correlation was positive for A and T, negative for C, and non-significant for G. It concerned 42 homologous genes from human and rat having GC levels in third codon positions higher than 45%. This correlation could not be confirmed in the present work.

4. The last set of data (21) indicated no significant rate difference for 17 human/artiodactyl gene pairs that showed no GC differences in synonymous positions. This is, however, again a small sample from which no general rule can be drawn (see paragraph 1 above). Human or artiodactyl/rodent comparisons showed that some nine pairs of genes, with a small or no difference in silent position GC, exhibited lower rates than another nine or so pairs of genes, which showed large differences. The significance of these differences in rates is, however, doubtful in view of the present results on the human versus rat (or mouse) comparisons. Indeed, since in such a case differences in silent position GC exist for both GC-poor and GC-rich genes, the so-called 'minor shift' (5, 6, 16), one should notice higher numbers of substitutions for those 'extreme' genes compared with genes having a more balanced composition, which is not the case.

4.2 Differences in repair efficiency do not cause differences in the rates of synonymous substitutions of genes located in different isochore families

The higher rate of accumulation of mutations in GC-poor sequences in rodents compared with primates (and their compositional bias) was explained (19) as being due to less-efficient DNA repair in the GC-poor regions of the rodent genome. Obviously, the non-existence of such a higher rate does not question the existence of a less efficient repair of DNA lesions in rodent cells than in human cells (22), nor the evidence for between-gene differences in efficiency of DNA repair (23). In fact, the latter do exist among genes located in the same or in different isochore families. The lack of rate differences over genes located in different families of isochore indicates, therefore, that such less efficient or differential repair does not influence, on average, the rates of silent substitutions, even if between-gene differences do exist. This is an important conclusion, because DNA repair has been repeatedly considered to be a cause for differences in rates (and biases) in the mutation process (24–27). Indeed, this conclusion implies that there is no 'organization of mutation along the genome' which would be a 'prime determinant of genome evolution' as recently claimed (13) and that it is plainly wrong to consider (13) that R-bands and G-bands are characterized by a slow and by a fast molecular clock, respectively. Incidentally, these claims ignore the strong compositional heterogeneity of R-bands (see also Section 4.3) and the fact that repair efficiency is very different in transcribed versus nontranscribed sequences.

4.3 Differences in the process of mutation associated with replication timing do not affect the rates nor the biases of synonymous substitutions of genes located in different isochore families

The main conclusion drawn by Wolfe *et al.* (11), was 'that much of the intragenomic variation in silent substitution rate and base composition in mammals results from variation in the process of mutation, rather than from natural selection (28, 29). According to Wolfe *et al.* (11) 'the variation in both silent substitution rate and base composition is due to systemic differences in the rate and pattern of mutation over regions of the genome, the differences arising because mutation patterns vary with the timing of replication of different chromosomal regions in the genome', the proposal being that 'isochores arise as a result of the synchronous replication of megabase stretches of DNA under varying dNTP pool conditions'.

The fact that the conclusion of Wolfe *et al.* (11) 'that the substitution rate and the base composition of silent sites vary together in a systematic way' is wrong has two important consequences.

First, if changes in the nucleotide pools in the germline do exist (as assumed on the basis of what happens in somatic cells), the fact that mutation patterns do not vary with the timing of replication (because mutation rates are the same on average for genes which replicate early or late) means that changes in nucleotide pools do not cause biases in mutation patterns. In fact, it had already been pointed out that in the somatic tissues of mammals, late replicating DNA, like satellites and the inactive X chromosome, may be either GC-rich or GC-poor (30), and it has also been shown that early and late replicating genes may be either GC-poor or GC-rich (31). This finding can be understood (32–35) because of the abundance of GC-poor isochores in R-bands (which replicate early) and, to a much lesser extent, of GC-rich isochores in G-bands (which replicate late). Finally, early and late replication patterns also exist in cold-blooded vertebrates (2, and papers quoted therein), which never developed strong compositional differences in their genomes (36–38).

Secondly, if mutation patterns do not vary with the timing of replication of different chromosomal regions in the germline, the explanation of Wolfe *et al.* (11) for the origin of isochores in mammalian genomes no longer holds.

5. Conclusions

In conclusion, differences in average mutation rates (and in mutational biases) of synonymous codon positions of genes located in different isochore families of mammalian genomes have been claimed by several authors (11,19–21), but they could not be confirmed (14). The differences under consideration appear to be due to three reasons:

(a) The existence of relatively large individual fluctuations from gene to gene.

(b) The use of small, non-representative gene samples.

(c) The underestimate of A↔T and G↔C transversions which become quite frequent at the two ends of the compositional spectrum of synonymous positions.

A lack of correlation between synonymous site divergence and GC levels in the mouse/rat comparison (as well as between K_4 and GC_4 (the GC levels at fourfold degenerate sites)) was also independently reported by Wolfe and Sharp (39). However, these authors found a variation in K_4 (more specifically, a peak of K_4 values at 60% GC), but only when K_4 was averaged over genes within each 1% interval of GC_4. This effect might be simply due to sampling effects. Alternatively, it might be real and due to an underestimate of the rate associated with extreme compositions, as already discussed. In any case, this effect is admittedly small. In fact, small enough not to support anymore the claim 'that the substitution rate and the base composition of silent sites vary together in a systematic way' and the ensuing speculations on the maintenance and origin of isochores (11).

Under these circumstances, explanations other than differences in mutation rates and in mutational biases have to be taken into consideration in order to account for the generation and maintenance of the large differences in GC levels of third codon positions of genes located in different isochore families from mammalian genomes (3, 28, 30).

References

1. Bernardi, G., Olofsson, B., Filipski, J., Zerial, M., Salinas, J., Cuny, G., *et al.* (1985). *Science,* **228**, 953.
2. Bernardi, G. (1989). *Annu. Rev. Genet.,* **23**, 637.
3. Bernardi, G. (1993). *Mol. Biol.Evol.,* **10**, 186.
4. Bernardi, G. (1993). *Gene,* **135**, 57.
5. Sabeur, G., Macaya, G., Kadi, F., and Bernardi, G. (1993). *J. Mol. Evol.,* **37**, 93.
6. Mouchiroud, D. and Bernardi, G. (1993). *J. Mol. Evol.,* **37**, 109.
7. Mouchiroud, D., D'Onofrio, G., Aïssani, B., Macaya, G., Gautier, C., and Bernardi, G. (1991). *Gene,* **100**, 181.
8. Aïssani, B., D'Onofrio, G., Mouchiroud, D., Gardiner, K., Gautier, C., and Bernardi, G. (1991). *J. Mol. Evol.,* **32**, 493.
9. Kimura, M. (1985). *The neutral theory of evolution.* Cambridge University Press.
10. Sharp, P. M. and Li, W.-H. (1987). *Mol. Biol. Evol.,* **4**, 222.
11. Wolfe, K. H., Sharp, P. M., and Li, W.-H. (1989). *Nature,* **337**, 283.
12. Li W.-H., Tanimura, M., and Sharp, P. M. (1987). *J. Mol. Evol.,* **25**, 330.
13. Holmquist, G. P. and Filipski, J. (1994). *Trends Ecol. Evol.,* **9**, 65.
14. Bernardi, G., Mouchiroud, D., and Gautier, C., (1993). *J. Mol. Evol.,* **37**, 583.
15. Gouy, M., Gautier, C., Attimonelli, M., Lanave, C., and di Paola, G. (1985). *Comput. Appl. Biosci.,* **1**, 167.
16. Mouchiroud, J. D., and Gautier, C. (1990). *Mol. Evol.,* **31**, 81.
17. Li, W. -H., Wu, C. I., and Luo, C. C. (1985). *Mol. Biol. Evol.,* **2**, 150.
18. Miyata, T., Hayashida, H., Kikuno, R., Hasegawa, M., Kobayashi, M., and Koike, K. (1982). *J. Mol. Evol.,* **19**, 28.
19. Filipski, J. (1988). *J. Theor. Biol.,* **134**, 159.
20. Ticher, A. and Graur, D. (1989). *J. Mol. Evol.,* **28**, 286.
21. Saccone, C., Pesole, G., and Preparata, G. (1989). *J. Mol. Evol.,* **29**, 407.
22. Hart, R. W. and Setlow, R. B. (1974). *Science,* **71**, 2169.
23. Bohr, V. A., Philips, D. H., and Hanawalt, P. C. (1987). *Cancer Res.,* **47**, 6426.
24. Filipski, J. (1987). *FEBS Lett.,* **217**, 184.
25. Sueoka, N. (1988). *Proc. Natl. Acad. Sci. USA,* **85**, 2653.
26. Sueoka, N. (1992). *J. Mol. Evol.,* **34**, 95.
27. Sueoka, N. (1993). *J. Mol. Evol.,* **37**, 137.
28. Bernardi, G. and Bernardi, G. (1986). *J. Mol. Evol.,* **24**, 1.
29. Gillespie, J. H. (1986). *Genetics,* **113**, 1077.
30. Bernardi, G., Mouchiroud, D., Gautier, C., and Bernardi, G. (1988). *J. Mol. Evol.,* **28**, 7.
31. Eyre-Walker, A. (1992). *Nucleic Acids Res.,* **20**, 1497.
32. Gardiner, K., Aïssani, B., and Bernardi, G. (1990). *EMBO J.,* **9**, 1853.
33. Pilia, G., Little, R. D., Aïssani, B., Bernardi, G., and Schlessinger, D. (1993). *Genomics,* **17**, 456.
34. Saccone, S., De Sario, A., Della Valle, G., and Bernardi, G. (1992). *Proc. Natl. Acad. Sci. USA,* **89**, 4913.
35. Saccone, S., De Sario, A., Wiegant, J., Raap, A. K., Della Valle, G., and Bernardi, G. (1993). *Proc. Natl. Acad. Sci. USA,* **90**, 11929.
36. Bernardi, G. and Bernardi, G. (1990). *J. Mol. Evol.,* **31**, 265.
37. Bernardi, G. and Bernardi, G. (1990). *J. Mol. Evol.,* **31**, 282.

38. Bernardi, G. and Bernardi, G. (1991). *J. Mol. Evol.,* **33**, 57.
39. Wolfe, K. H. and Sharp, P. M. (1993). *J. Mol. Evol.,* **37**, 441.
40. Mouchiroud, D., Gautier, C., and Bernardi, G. (1995). *J. Mol. Evol.,* **40**, 107.
41. Cacciò, S., Zoubak, S., D'Onofrio, G., and Bernardi, G. (1995). *J. Mol. Evol.*, **40**, 280.
42. Zoubak, S., D'Onofrio, G., Cacciò, S., Bernardi, G., and Bernardi, G. (1995). *J. Mol. Evol.*, **40**, 293.
43. Bernardi, G. (1995). *Annu. Rev. Genet.*, **29**, 445.

11

Identifying genes in genomic DNA sequences

ERIC E. SNYDER and GARY D. STORMO

1. Introduction

This chapter describes methods for identifying genes in eukaryotic genomic DNA sequences and determining their fine structure. Because all current methods rely on an ensemble of sequence classification methods, we briefly review the types of evidence used to classify sequences as coding or non-coding. Next, we discuss the methods used by four currently available programs and discuss their strengths and weaknesses. Finally, we test each program on a database of genomic sequences and compare the performance of each program.

We focus on the problem of identifying coding regions in eukaryotes. In contrast to prokaryotic coding regions which are almost universally contiguous and delineated by relatively well-defined signal sequences, eukaryotic coding regions are more often than not interrupted by non-coding regions (introns) and embedded in DNA sequence of unknown function. This presents a challenging problem to the biologist trying to deduce the translated protein from the genomic sequence.

It is not the intent of the computational methods of gene identification that are discussed in this chapter to mimic the biochemistry of transcription, mRNA processing, and translation. Rather, these methods make use of all available information in an effort to best identify the majority of coding regions. For example, all current methods use some measure of codon bias to identify putative exons. These exons are further defined by searching for likely splice sites. After assembling the exons into a complete coding region, the program may look for likely promoter elements 5′ to the first exon. This is almost surely exactly the reverse of what happens *in vivo*. However, this approach is universally applied because, for example, it allows one to escape the vagaries of promoter identification until other information about the putative coding region is known. Users of these programs should keep in mind the implications of this approach when considering unusual cases, such as pseudogenes and alternative splicing. Software developers should think about what can be learned by attempting to model the underlying biology more closely.

1.1 Low-level motif identification

All gene assembly programs available today make use of multiple lines of evidence from which to draw conclusions about the positions of coding exons in a genomic sequence. This information can take two forms: search by content and search by signal (1). Content statistics measure the bulk properties of a length of sequence whereas signal-based methods look for short conserved sequence elements.

1.1.1 Content statistics

The simplest content measure for identification of coding sequences is to look for open reading frames (ORFs). The presence of an ORF is a necessary but insufficient condition for a coding exon. Since internal exons can be as short as seven nucleotides (nt) (exon 3, quail troponin I) and the coding regions of terminal exons can be even shorter (2), other measures are required.

Most organisms have a biased usage of synonymous codons and most proteins have a biased usage of amino acids. These facts lead to the **codon usage** statistic developed by Staden (3). A contextual effect on codon usage has also been observed (4, 5). This propensity can be measured by the **in-frame hexamer** or **dicodon** statistic, which measures the frequency of pairs of adjacent codons. It is also possible to measure compositional bias between codon positions (6, 7). This property reflects the pattern of nucleotide usage in the most frequently used codons.

There are other methods which are not related to codon usage. Periodicities in nucleotide usage occur which are characteristic of sequence type. Exons tend to display a three-base repeat interval between identical bases (8). In contrast, introns display a two-base repeat interval (9). These properties can be measured by **autocorrelation** (8) or **Fourier** (10) methods. Exons and introns can also be distinguished on the basis of their 'simple sequence' DNA content. Introns show a higher concentration of 'simple sequence' DNA, characterized by low **local compositional complexity** (11) than do exons. Fickett and Tung (12) have reviewed and studied the performance of these and a number of other coding statistics.

A third type of content information (i.e. comparative) has been applied to sequence classification. As the amount of information in the biological sequence databases increases, there is a significant (and increasing) chance that any newly sequenced gene will have a homologous sequence already characterized in the database. Thus finding a protein sequence with similarity to the translated product of a query sequence can be used as evidence that the new sequence itself codes for a protein (13). This idea can also be used to identify non-coding regions. For example, intron sequences will contain many more repetitive sequences since these sequences rarely code for proteins (14).

1.1.2 Site identification

The traditional approach to identifying functional sites is to search for similarity to a consensus sequence. This consensus represents the common features of a group of sequences with similar function but fails to distinguish quantitatively between highly conserved sites and sites which may tolerate some level of variation. A more sophisticated approach is to represent the consensus as a weight matrix in which each position on a sequence is represented (in the case of nucleic acids) by four numbers based on the number of occurrences of each base at that particular position (15). Typically, each matrix element will represent the frequency with which a base occurs at a position. Other possibilities exist. For example, the discriminatory power of a matrix can be optimized using neural networks (16, 17) or multiple linear regression (18). Information theory has also been applied, measuring the non-randomness of the signal and calculating the number of bits required to define it (19). Approaches to finding appropriate matrices for any signal have been reviewed (20, 21).

In the context of gene structure determination, weight matrices are used to locate donor and acceptor splice sites. It is often useful to search for a translational initiation codon, ATG, with a weight matrix since additional information is present 5' of the start site (22). Stop codons do not appear to exist in specific contexts (Snyder and Stormo, unpublished observations) and are usually searched for explicitly.

1.2 Assembling complete genes using multiple pieces of evidence

Given the large amount of evidence that can be brought to bear on the problem, how can it be applied to maximize predictive utility? First we will discuss how different statistics can be used together to evaluate likely coding regions. Second, we will examine different methods used to assemble putative exons into the coding sequence of a complete gene.

1.2.1 Combining the evidence

There are two basic approaches to applying evidence to classify sequences, rule-based (*GeneID* and *GeneModeler*) and neural network methods (*GeneParser, GRAIL*).

In the rule-based approach, criteria are applied serially to identify possible exons and then rank them or eliminate them from consideration. For example, *GeneModeler* identifies all ORFs above a minimum length in a sequence, then finds splice sites which exceed a threshold. The ORFs and splice sites are then merged to identify putative exons. These exons are evaluated using a codon usage measure and only intervals which exceed the codon usage threshold are passed on for further analysis. *GeneID* uses a similar approach. Various content measures are applied, further filtering the intervals prior to intron–exon assembly.

GRAIL and *GeneParser* use a different approach to applying statistics. Instead of applying the statistics and cut-offs sequentially, they are applied in parallel and weighted according to their importance. That is, for each interval, an ensemble of statistics is calculated and weighted to determine a composite score which can be used to rank intervals. In *GRAIL2* (newest version), a cut-off is applied to the weighted score and these intervals are passed to an assembly algorithm.

GeneModeler, *GeneID*, and *GRAIL* all attempt to prune search space by applying a cut-off to potential exon intervals. *GeneModeler* and *GeneID* apply a cut-off to the interval for each statistic. Thus, to be considered a viable exon, the interval must exceed the cutoff for all classification statistics. In contrast, *GRAIL* and *GeneParser* weight the classification statistics with a neural network, obtaining a single score for the putative exon. In *GRAIL*, a cut-off is applied to this composite score to determine if an interval is an exon. In contrast, *GeneParser* allows all $\sim \frac{1}{2}N^2$ intervals (for a sequence of length N) to potentially enter the solution. This is advantageous because low-scoring exons may enter the solution if they allow a high-scoring gene model to be formed. However, this is at the expense of significantly increased computational complexity.

1.2.2 Interval calculations

Compromises must also be made when calculating content statistics over an interval. *GRAIL* simplifies this task by using a scanning window approach. Although this approach is very fast, it makes the information more difficult to interpret. For example, a long exon may be represented by several (overlapping or non-overlapping) windows. Clearly, some of the statistical information is lost when the window values are averaged over the length of the exon. In contrast, the other programs examine each putative exon over its entire length. *GeneModeler* uses a measure of codon asymmetry and several compositional measures but also uses a sliding window measure of G + C content to find internal exons. *GeneID* relies on four measures of correlation between codon position measured by the chi-square test calculated over the length of the exon. *GeneParser* uses a log-likelihood ratio to evaluate deviation from random usage for hexamers and in-frame hexamers (dicodons).

1.2.3 Exon assembly

It should be clear that the precise identification of coding regions in higher eukaryotes is a very difficult problem. Given a sequence of N nucleotides consisting of only two classes of sequence, intron and exon, there are 2^N ways to divide the sequence into those components. It is computationally intractable to evaluate all possible solutions for a sequence of any biologically relevant length. All methods must then make some simplifying assumptions about the factors determining the position of exons within a gene.

2. Programs

2.1 *GeneModeler*

Until recently, *GeneModeler* (23) was the most complete gene assembly package publicly available. The package includes exon (5′-, internal and 3′-) and intron identification as well as identification of TATA box and polyadenylation signals and splice sites. The X-Windows graphical user interface allows the display of many possible solutions alongside the nucleic acid and putative amino acid sequence. The user can point to features on the pictorial gene display and the program will highlight these features on the text sequence.

One strength of *GeneModeler* is the ability to easily change the parameters used to identify gene components. For example, one can easily lower thresholds for splice site detection to find sites which are not close matches to the consensus. This allows the program to suggest possible products of alternative splicing. Codon usage requirements can also be lowered to aid in the detection of genes that are not highly expressed (such genes often have lower scores for codon bias).

GeneModeler was originally developed for use on the *Caenorhabditis elegans* genome. The authors tested the program on a set of five multi-exon genes from the nematode and found that it predicted the majority of the coding region of these genes.

Currently, *GeneModeler* supports the analysis of sequences from *Homo sapiens*, *Drosophila melanogaster*, *C. elegans*, and dicotyledonous and monocotyledonous plants. *GeneModeler* is available by anonymous FTP from `atlas.lanl.gov` as `pub/gm/gm2.tar.Z`.

2.2 *GeneID*

GeneID (24) is one of the more accurate gene structure prediction programs as well as one of the easiest to use. The program identifies candidate first, internal, and last exons as well as translational initiation codons and splice sites. The program uses the concept of 'exon equivalence' to build and report solutions. This speeds execution and allows the program to show suboptimal solutions ranked by equivalence class, eliminating solutions which differ trivially from one another, and emphasizing the variety of possible gene structures.

GeneID has been well evaluated by the authors. On a training set of 169 vertebrate genes, 28% of the optimum solutions contained the actual gene structure. However, since a solution may contain multiple gene structures (by virtue of exon equivalence), this figure may be artificially large. The authors also performed a nucleotide-by-nucleotide analysis of the top predicted structure. They found 80% of the coding region is predicted as coding while 88% of that which is predicted to be coding is actually coding. This corresponds to a correlation coefficient of 0.79. On a set of 28% genes not involved in the optimization of the program, *GeneID* predicted only 69% of the coding region as coding but the specificity remained high, 84%. This corresponds to a correlation coefficient of 0.70.

GeneID was trained on genes from a variety of vertebrates, including human, chicken, and *Xenopus*. Given the diversity of its training set, one can expect the program to do well on most vertebrate genes. However, the training set is highly biased towards mammals (humans in particular) so this should be kept in mind.

GeneID provides a number of useful options. The user can request that a *BLAST* search (25) be run with the sequence prior to gene assembly. High similarity to protein coding regions in the database can be used as evidence that a region in the query sequence is an exon. The authors claim a correlation coefficient of 0.87 for exon prediction when *GeneID* is run using the additional *BLAST* information (13). When available, users can also provide additional a priori evidence for the positions of exons to force a solution to contain known information about the gene's structure.

GeneID is available for use through an electronic mail server. For more information, send the e-mail message 'help' to `geneid@bir.cedb.uwf.edu`.

2.3 *GRAIL*

GRAIL is probably the most widely used of the coding sequence identification programs. It is available in three forms. The original version of *GRAIL* (14), used via an e-mail server, provides a table of neural network output scores as a function of sequence position along with associated ORFs. These data are further analysed to give likely coding windows with ORF, strand, reading frame, 'quality', and coding probability annotated. In this version, start and stop codons and splice sites are not found. The newly available *GRAIL2* uses the same basic *GRAIL* algorithm but further attempts to identify complete coding exons by finding these signals. *GRAIL2* reports likely exons on the forward and reverse strand and makes a final table of results similar to the original *GRAIL* now with precise exon boundaries.

The newest program from the Oak Ridge group, *XGRAIL*, provides a graphical user interface as well as additional analysis functionalities. The program is implemented as a client-server. The user must first acquire the *XGRAIL* client software for use on a local Unix workstation running X-Windows. This client allows the user to work interactively on a local machine while running the CPU intensive functions on supercomputers at Oak Ridge National Laboratory. This process is completely transparent to the user. However, a direct connection to the Internet is required.

In addition to allowing graphic display of *GRAIL*- and *GRAIL2*-style output, *XGRAIL* provides an assembly function to link compatible exons to form complete coding regions. Individual exons can be pointed to and their DNA sequence and translation can be displayed. The exons can even be selected for homology search on protein databases allowing the researcher to rapidly determine if the sequence has homology to known sequences. The interface also provides graphic display of G + C content, potential polyadenylation sites, and repetitive sequence elements.

Information on the *GRAIL* e-mail server is available by sending the message 'help' to `grail@ornl.gov`. The *XGRAIL* client software is available by anonymous FTP from `arthur.epm.ornl.gov`. Currently, Sun and DEC workstations are supported.

2.4 *GeneParser*

GeneParser, developed in our group, employs a novel algorithm for the sequence parsing problem. Instead of determining candidate intervals using filters, *GeneParser* scores all subintervals in a sequence. Combining the data from many different statistics and weighting them with a neural network, the program approximates a log-likelihood that each interval belongs to each of the possible sequence types: first, internal, or last coding exon or intron. Dynamic programming is used to find the maximum likelihood parsing of the sequence. This permits an efficient and exhaustive search of all possible sequence parsings.

GeneParser provides graphic display of the classification statistics and log-likelihood scores for all subintervals as well as diagrams of solutions. The dynamic programming method allows rapid calculation of ranked suboptimal solutions; these solutions can also be displayed graphically. In addition, a pointing device (mouse) can be used to identify a potential intron–exon junction and generate the optimal solution containing it.

GeneParser binaries are available for Sun, DEC, and Silicon Graphics workstations by anonymous FTP from `beagle.colorado.edu`.

3. Performance statistics

This section analyses the performance of the four programs discussed above. We will concentrate our attention on the problem of identifying genes in human sequences since many of the available programs have been optimized for performance on this organism. It is desirable to study the performance of these methods in the context of a single organism to avoid problems associated with generalization of classification statistics between organisms. It should be understood that most methods discussed in this chapter are applicable to other organisms without fundamental modifications to the algorithm. However, all methods require organism-specific information such as codon usage tables and all require some optimization to tune the performance for different organisms.

To evaluate performance, we will focus on the number of nucleotides predicted to be coding and the number of exons correctly predicted. To facilitate the comparison of methods, we will use three measures of performance: sensitivity (Sn), specificity (Sp), and the correlation coefficient (CC) (17). *Table 1* illustrates how each of these scores is calculated. We also report two additional numbers to evaluate performance at the level of complete exons. 'Exons correct' refers to the number of exons predicted exactly (donor/start site and acceptor/stop site correct). 'Exons overlapped' refers to the number of actual exons which overlap predicted exons.

3.1 Test data

A test set of human genes was compiled based on the human genes used in the testing of *GRAIL* and *GeneID*. These genes were not used in the development of these programs and were not present in the training set of *GeneParser*. This is important because it is often possible to obtain arbitrary levels of performance when a program is trained on a small set of genes at the expense of performance on sequences not in the training set. These genes are listed in *Table 2*.

As we observe here and others have reported (26), the performance of many gene identification programs is sensitive to G + C content. This is probably due in part to the mosaic structure of warm-blooded vertebrate genomes. There appear to be important biases in the distribution of genes between isochores of different G + C content with constitutively expressed (housekeeping) genes strongly favouring the G + C-rich isochores (27). Bernardi adds that these genes tend to be the most biased in codon usage. This observation has important implications for programs that attempt to identify genes based on 'universal' statistical properties such as hexamer usage.

In order to quantitate this effect, the genes in the test set were classified on the basis of G + C content by comparison with a distribution of all full-length

Table 1. Calculation of performance measures

	Predicted Positives (*PP*)	Predicted negatives (*PN*)
Actual positives (*AP*)	True positives (*TP*)	False negatives (*FN*)
Actual negatives (*AN*)	False positives (*FP*)	True negatives (*TN*)

Sensitivity (*Sn*)[a] $= \dfrac{TP}{TP + FN}$

Specificity (*Sp*)[b] $= \dfrac{TP}{TP + FP}$

Correlation coefficient (*CC*)[c] $= \dfrac{(TP)(TN) + (FP)(FN)}{\sqrt{(PP)(PN)(AP)(AN)}}$

[a] Sensitivity measures the fraction of the true set that is correctly predicted.
[b] Specificity measures the fraction of the predicted set that is correct.
[c] The correlation coefficient combines both measures in a single number which can range from −1 (predictions are always incorrect), through 0 (any random prediction), to +1 (predictions are exactly correct).

Table 2. Test set I[a]

Locus name	Accession	Length (bp)	CDS (bp)	G + C (%)
High G + C				
HUMALPHA	J03252	4556	1599	62.116
HUMAPRT	Y00486	3016	543	65.782
HUMCYP2DG	M33189	5503	1503	61.566
HUMGAPDHG	J04038	5378	1008	60.785
HUMPNMTA	J03280	4174	849	61.811
HUMRASH	J00277	6453	570	68.185
HUMTRPY1B	M33494	2609	828	65.811
Medium G + C				
HUMEMBPA	M34462	3608	669	51.414
HUMFOS	K00650	6210	1143	51.369
HUMIBP3	M35878	10884	876	48.833
HUMLD78A	D90144	3176	279	47.009
HUMLD78B	D90145	3112	282	49.968
HUMMETIA	K01383	2941	186	52.737
HUMP45C17	M19489	8549	1527	50.684
HUMPAIA	J03764	17509	1209	50.214
HUMPDHBET	D90086	8872	1080	45.525
HUMPOMC	K02406	8658	804	51.444
HUMPP14B	M34046	8076	543	54.842
HUMPRCA	M11228	11725	1386	56.913
HUMSAA	J03474	3460	369	49.653
HUMTBB5	X00734	8874	1335	56.198
HUMTCRAC	X02883	5089	426	50.147
HUMTHB	M17262	20801	1869	50.589
HUMTKRA	M15205	13500	705	53.274
Low G + C				
HUMHBBAAZ	M36640	2149	444	40.624
HUMHNRNPA	X12671	5368	963	43.256
HUMPRPH1	M13057	4946	501	42.398
HUMREGB	J05412	4251	501	42.249

[a] Genes taken from *GeneID* and *GRAIL* test sets. Loci are grouped according to G + C content. Loci HUMPLP-SPC and HUMNMYCA were removed from the original set. HUMPLPSPC has alternative splicing annotated in the 3'-exon and HUMNMYCA contains a second ORF. Loci names are from GenBank 64 for consistency with the original papers.

human gene loci in GenBank (data not shown). Genes within one standard deviation of the mean were considered 'medium G + C content'. Genes with a G + C content greater than one standard deviation above or below the mean were classified as 'high' and 'low' G + C content, respectively.

A second test set was also developed to test the effect that specific genes may have on performance. All human loci from GenBank release 77 containing complete protein coding regions and containing at least two exons were considered. Pseudogenes and loci with multiple CDS (coding sequence) fields, alternative splicing or CDS fields annotated as 'putative' were removed from the collection. Finally immunoglobulin genes and genes of the major histocompatibility complex were removed and the genes were ranked by G + C content. From this list, every third locus was selected. From the selected genes, loci that were present the training or test sets of *GeneID* or

GRAIL were deleted, as were loci longer than 30 kb in length. This resulted in the 34 loci in test set II shown in *Table 3*. We do not claim that this collection is representative of human genes in general or even of human genes in GenBank. However, it does provide a test set largely independent of previous test and training data sets.

We have assumed up to this point that the GenBank loci used in this study are error free. Although this is most likely not the case in general, it is very likely true that the sequences are free from the frame-shift errors in the coding region since in most cases, some independent evidence has been used to establish that the CDS as annotated is the CDS of the gene in question. To give an indication of how the different programs handle data with a known error frequency, we randomly introduced substitution and frame-shift errors into the medium G + C content sequences of test set I at a rate of 0.5% for each error type.

3.2 Comparison of currently available programs

Although it is desirable to compare programs with standardized test data to evaluate relative performance, it should be remembered that not all programs were designed with the same intent and to some extent any comparison is bound to be unfair. However, such a comparison does serve to illustrate the strengths and weaknesses of the programs in this specific context.

The conditions used for testing each program were as follows:

(a) *GeneModeler.* GeneModeler2 was run on the test sets using the default parameters and a set of parameters empirically derived (by us) chosen to optimize performance. Since *GeneModeler* generates a list of solutions consistent with the user-supplied parameters and does not attempt to rank solutions, the first solution listed in the output was chosen (this corresponds to the longest model).

(b) *GeneID.* GeneID was run using the e-mail server using default parameters plus '-small output' to display only exon maps and '-noexonblast' to disable the use of homology data for exon prediction (since the test data contain genes already in the database, this would give an over-optimistic estimate of performance). The top ranked gene was chosen for analysis.

(c) *GRAIL.* GRAIL was run in two ways. First, the test sequences were submitted to the e-mail server using the '-2' option to invoke exon boundary identification. Potential exons on the forward strand (with a 'quality' of 'marginal', 'good', or 'excellent') were extracted from the 'Final Exon Prediction'. These exons, while they are bounded by splice sites (except for first and last exons) cannot necessarily be spliced together in frame. Secondly, *XGRAIL* was run manually using the client server software and genes were constructed using the 'assemble' function. The exons from the assembled gene were recorded and used for analysis.

Table 3. Test set II[a]

Locus name	Accession	Length (bp)	CDS (bp)	G + C (%)
High G + C				
HUMAZCDI	M96326	5002	756	60.756
HUMMKXX	M94250	3308	432	65.478
HUMTRPY1B	M33494	2609	828	65.811
Medium G + C				
HUMAGAL	M59199	13662	1125	53.045
HUMAPEXN	D13370	3730	957	48.418
HUMCACY	J02763	3671	273	57.123
HUMCHYMB	M69137	3279	744	51.113
HUMCOLA	M95529	3401	333	53.661
HUMCOX5B	M59250	2593	390	49.826
HUMCRPGA	M11725	2480	675	48.145
HUMGOS24B	M92844	3135	981	60.223
HUMHAP	M92444	3046	957	48.194
HUMHEPGFB	M74179	6100	2136	59.852
HUMHLL4G	M57678	4428	408	56.843
HUMHPARS1	M10935	11551	1221	46.152
HUMI309	M57506	3709	291	48.665
HUMKAL2	M18157	6139	786	56.524
HUMMHCP42	M12792	5141	1485	58.880
HUMNUCLEO	M60858	10942	2124	45.403
HUMPEM	M61170	4243	1428	58.944
HUMPP14B	M34046	8076	543	54.842
HUMPROT1B	M60331	1306	156	51.991
HUMPROT2	M60332	1861	309	57.174
HUMRPS14	M13934	5985	456	48.805
HUMSHBGA	M31651	6087	1209	53.754
HUMTNFBA	M55913	2140	618	57.477
HUMTNFX	M26331	3103	702	53.529
HUMTNP2SS	L03378	1782	417	48.316
HUMTRHYAL	L09190	9551	5697	50.717
Low G + C				
HUMGFP40H	M30135	4379	435	41.676
HUMHIAPPA	M26650	7160	270	33.268
HUMIL8A	M28130	5191	300	33.288
HUMMGPA	M55270	7734	312	39.436
HUMPALD	M11844	7616	444	41.360

[a]Selected complete genomic sequences from GenBank release 77.

(d) *GeneParser*. A beta-version of *GeneParser2* was run locally and the optimum solution recorded. This version first scans the sequence for base composition and picks tables and network weights optimized for the G + C content of the sequence.

3.3 Results

Table 4 summarizes the results from the *GeneID*, *GRAIL*, and *GeneParser* programs when tested on test sets I and II.

In our hands, we found it difficult to find satisfactory parameters for *Gene-Modeler*. Using the default parameters we obtained a CC on test set I of

0.13. By tuning the parameters we could increase performance to CC = 0.28. Interestingly, the low score is due in large part to the program failing to generate models in many cases. Five of the genes were predicted with a CC of greater than 0.70, indicating the program is capable of doing well, at least occasionally.

GeneID, the *GRAIL* programs, and *GeneParser* all performed well on the test sets with *GRAIL2*+Assembly coming out on top in terms of the number of exon nucleotides correctly predicted. *GeneParser* and *GRAIL* both perform best on sequences with high G + C content. The reason for this is probably related to the observation that highly expressed genes tend to be found in G + C rich isochores (27) and that these genes have a more biased codon usage. Interestingly, *GeneID* displays the reverse tendency on test set I.

The differences between performance on test sets I and II is important to note because it shows how sensitive the different programs are to different data. *GeneID*, for example did significantly less well on set II, whereas *GRAIL* decreased somewhat less, and *GeneParser* increased slightly.

The comparison of the data from *GRAIL* with and without the assembly routine confirms an observation in our laboratory that checking reading frame compatibility between exons does not necessarily lead to an increase in predictive performance. Indeed, such methods often require considerably more computational time and can even decrease the total number of exonic nucleotides predicted. However, in cases where interesting features of the protein lie on exon junctions (for example, three of the four Ca^{2+} EF-hand sites in troponin C), the assembly routine may help by increasing the number of amino acids correctly predicted at exon junctions.

The performance of *GeneID*, *GRAIL*, and *GeneParser* on sequences with introduced errors is shown in *Table 5*. The performance of all programs suffered on these data. Not unexpectedly, *GeneID* and *GRAIL2*+Assembly suffered the most, very likely because frame-shift errors make it impossible to assemble the true solution correctly whilst maintaining reading frame compatibility. Since *GRAIL2* (without the assembly routine) and *GeneParser* do not try to enforce compatibility of exons, these methods suffer less from frame-shift errors.

4. Recommendations for users

Clearly no single program will suit the needs of every researcher for every gene; each method uses different heuristics and has different strengths and weaknesses. *GeneID* and *GeneParser* provide ranked suboptimal solutions in addition to the 'best' solution. Similarly, *GeneModeler* gives an unranked list of possible solutions in an undoubtedly realistic effort to discourage the researcher from putting too much confidence in the relative scores of the solutions. Multiple solutions are valuable since even the best program only

Table 4. Performance of the programs of test sets I and II[a]

Program	GeneModeler	GeneID	GRAIL2	GRAIL2'	GeneParser
Set I: total					
CC	0.13	0.69	0.80	0.83	0.78
Sn	0.07	0.69	0.86	0.83	0.87
Sp	0.42	0.77	0.80	0.87	0.76
Exons correct	0.02	0.42	0.34	0.52	0.47
Exons overlapped	0.09	0.73	0.88	0.81	0.87
Set I: high G+C					
CC		0.65	0.81	0.88	0.89
Sn		0.72	0.86	0.87	0.90
Sp		0.73	0.84	0.95	0.93
Exons correct		0.38	0.27	0.67	0.64
Exons overlapped		0.80	0.96	0.89	0.96
Set I: med G+C					
CC		0.67	0.80	0.83	0.75
Sn		0.65	0.88	0.86	0.86
Sp		0.77	0.77	0.84	0.70
Exons correct		0.37	0.37	0.51	0.41
Exons overlapped		0.67	0.85	0.83	0.84
Set I: low G+C					
CC		0.81	0.75	0.62	0.72
Sn		0.82	0.74	0.51	0.79
Sp		0.85	0.83	0.87	0.75
Exons correct		0.80	0.35	0.25	0.40
Exons overlapped		0.85	0.80	0.55	0.85
Set II: total					
CC		0.55	0.79	0.75	0.80
Sn		0.50	0.74	0.68	0.82
Sp		0.75	0.91	0.91	0.86
Exons correct		0.33	0.33	0.31	0.46
Exons overlapped		0.64	0.66	0.58	0.76
Set II: high					
CC		0.73	0.79	0.80	0.71
Sn		0.85	0.83	0.80	0.65
Sp		0.73	0.82	0.88	0.87
Exons correct		0.43	0.43	0.50	0.57
Exons overlapped		0.86	0.86	0.79	0.79
Set II: med					
CC		0.52	0.79	0.75	0.82
Sn		0.47	0.75	0.68	0.84
Sp		0.76	0.91	0.91	0.87
Exons correct		0.29	0.32	0.32	0.46
Exons overlapped		0.62	0.68	0.58	0.79
Set II: low					
CC		0.62	0.64	0.62	0.67
Sn		0.56	0.49	0.45	0.71
Sp		0.71	0.88	0.89	0.67
Exons correct		0.47	0.37	0.16	0.37
Exons overlapped		0.63	0.42	0.42	0.58

[a] *GeneModeler* was run with default parameters, *GeneID* using the '-no_blast' option, and *GRAIL* using the '-2' option to invoke exon termini identification. *GRAIL2'* indicates results obtained running *GRAIL2* and 'assembly' using the interactive *XGRAIL* package. *GeneParser* was run using hexamer tables and network weights optimized for G + C content.

occasionally gets a gene exactly correct. The exact structure can often be found by looking at the suboptimal solutions. Furthermore, by examining a group of predicted structures, the user can get an intuitive feel for what parts of the structure are most likely to be correct by noting what exons are found in the majority of structures.

GeneID and *GeneParser* also allow the user to provide additional experimental evidence in the form of constraints which the predicted structures must satisfy. The program can thus be guided by this evidence. This may prove useful in situations where the complete genomic sequence is known together with information from a partial cDNA or EST. *XGRAIL* allows the user to interactively define which areas of the sequence should be used in the assembly procedure. This allows regions known not to be protein coding to be excluded from the final solution (even if it appears to be coding by the program prior to assembly).

GeneModeler emphasizes user input in all aspects of the program allowing great flexibility in how the program interprets the information in the sequence. Interestingly, it is the only one of the programs considered here which allows the user control over its basic search parameters. Since all other programs are optimized for performance on the 'average' gene, this flexibility may be advantageous for predicting the structure of genes in difficult or very atypical cases.

GeneID and *GeneParser* have introduced the use of homology data from *BLAST* searches and integrated the information for use as evidence in the gene identification procedure. Although we have not tested the performance of these programs with homology data included, results from the authors (13, 28) suggest that substantial increases in accuracy can be obtained.

GRAIL has the advantage of giving a single very accurate prediction quickly and easily. The mail server is remarkably fast and the graphical user interface is very well designed and user friendly. Using the *XGRAIL* client, a researcher can identify potential coding regions in a sequence and automatically search for homologous sequences in the protein databases

Table 5. Performance of data with 0.5 frame-shift and 0.5 substitutions errors introduced to sequences

Program	GeneID	GRAIL2[a]	GRAIL2 with Assembly[a]	GeneParser
Set I: med G + C				
CC	0.42	0.63	0.52	0.69
Sn	0.38	0.64	0.42	0.72
Sp	0.59	0.70	0.76	0.72
Exons correct	0.17	0.26	0.17	0.27
Exon overlap	0.53	0.87	0.63	0.77

[a] *GRAIL2* was run on sequences less than 10 kb in length due to a length restriction introduced prior to running this test.

in a matter of minutes. *XGRAIL* is well suited for use in high volume sequencing laboratories which seek to identify potential genes for further investigation.

We have attempted to quantitate the effect of sequencing errors on prediction. Although not a rigorous test of error tolerance, our data illustrate that the programs tested are sensitive enough to sequencing errors that results from these programs should be interpreted with caution when run on sequences with unknown or high error rates. Although a reasonably high rate ($\sim 1\%$) of substitution sequencing errors can probably be tolerated by most programs (Snyder and Stormo, unpublished observations), frame-shift errors are much more problematic and probably nearly as frequent in real data (29). Since frame-dependent statistics are so important in exon identification, any sequencing error that disrupts the reading frame of an exon greatly decreases the strength of the signal. Depending on how the score is calculated, the exon could be interpreted as two separate exons in different reading frames or it could be missed altogether. This problem is more acute when the program attempts to assemble a complete gene with a continuous reading frame spanning many exons. Here, a single frame shift could dramatically alter the predicted solution. Care should be taken when interpreting results obtained on sequences with unusually high error frequency. Interestingly, *GeneParser* performs slightly better than the others on such data, probably because of its emphasis on non-frame-dependent statistics.

5. Conclusions

A great deal of progress has been made in coding sequence identification since the early work of Fickett and Staden. Although no program can claim a significant probability of exactly predicting the coding region of a gene in its entirety (with 100% specificity), there are several programs currently available which routinely find the majority of the coding sequence with a not unreasonable level of false positives. Perfect prediction of coding regions remains the holy grail of sequence analysis. However, as more is learned about the prevalence of alternative splicing and developmental and tissue-specific regulation of gene expression, the problem may well become more complex as we ask more and more difficult questions of sequence analysis software. Can promoter identification methods eventually predict expression patterns? Can this information coupled with current or novel methods of gene parsing predict developmentally regulated alternative splicing? These questions remain to be explored.

References

1. Stormo, G. D. (1987). In *Nucleic acid and protein sequence analysis, a practical approach* (ed. M. J. Bishop and C. J. Rawlings), pp. 231–58. IRL Press, Oxford.

2. Hawkins, J. D. (1988). *Nucleic Acids Res.,* **16**, 9893–908.
3. Staden, R. and McLachlan, A. D. (1982). *Nucleic Acids Res.,* **10**, 141–56
4. Lipman, D. J. and Wilbur, W. J. (1983). *J. Mol. Biol.,* **163**, 377–94.
5. Yarus, M. and Foley, L. S. (1985). *J. Mol. Biol.,* **182**, 529–40.
6. Fickett, J. W. (1982). *Nucleic Acids Res.,* **10**, 5303–18.
7. Smith, T. F., Waterman, M. S., and Sadler, J. R. (1983). *Nucleic Acids Res.,* **11**, 2205–20.
8. Michel, C. J. (1986). *J. Theor. Biol.,* **120**, 223–36.
9. Arques, D. G. and Michel, C. J. (1987). *Nucleic Acids Res.,* **15**, 7581–92.
10. Silverman, B. D. and Linsker, R. (1986). *J. Theor. Biol.,* **118**, 295–300.
11. Konopka, A. K. (1990). In *Structure and methods,* Vol. 1: *Human genome initiative and DNA recombination* (ed. R. H. Sarma and M. H. Sarma), pp. 113–25. Adenine Press, Schenectady, NY.
12. Fickett, J. W. and Tung, C.-S. (1992). *Nucleic Acids Res.,* **20**, 6441–6450.
13. Guigó, R., Knudsen, S., Drake, N., and Smith, T. (1993). GeneID online system for prediction of gene structure, version 2.0 (GeneID documentation).
14. Uberbacher, E. C. and Mural, R. J. (1991). *Proc. Natl. Acad. Sci. USA,* **88**, 11261–5.
15. Hawley, D. K. and McClure, W. R. (1983). *Nucleic Acids Res.,* **11**, 2237–55.
16. Stormo, G. D., Schneider, T. D., Gold, L., and Ehrenfeucht, A. (1982). *Nucleic Acids Res.,* **10**, 2997–3010.
17. Brunak, S., Engelbrecht, J., and Knudsen, S. (1991). *J. Mol. Biol.* **220**, 49–65.
18. Stormo, G. D., Schneider, T. D., and Gold, L. (1986). *Nucleic Acids Res.,* **14**, 6661–79.
19. Schneider, T. D., Stormo, G. D., Gold, L., and Ehrenfeucht, A. (1986). *J. Mol. Biol.,* **188**, 415–31.
20. Stormo, G. D. (1988). *Annu. Rev. Biophys. Chem.,* **17**, 241–63.
21. Stormo, G. D. (1990). *Methods in enzymology,* Vol. 183 (ed. R. F. Doolittle), p. 211–21. Academic Press, London.
22. Penotti, F. E. (1990). *J. Mol. Biol.,* **213**, 37–52.
23. Soderlund, C., Schanmugam, P., White, O., and Fields, C. (1992). In *Proceedings of the twenty-fifth Hawaii International Conference on System Sciences, Biotechnology Computing Minitrack, Vol. 1*, pp. 653–62. IEEE Computer Society Press, Los Alamitos, CA.
24. Guigó, R., Knudsen, S., Drake, N., and Smith, T. (1992). *J. Mol. Biol.,* **226**, 141–57.
25. Altschul, S. F., Gish, W., Myers, E. W., and Lipman, D. J. (1990). *J. Mol. Biol.,* **215**, 403–10.
26. Uberbacher, E. C. and Mural, R. J. (1993). GRAIL documentation.
27. Bernardi, G. (1989). *Annu. Rev. Genet.,* **23**, 637–61.
28. Snyder, E. E. and Stormo, G. D. (1993). GeneParser documentation.
29. Krawetz, S. A. (1989). *Nucleic Acids Res.,* **17**, 3951–7.

12

Prediction of mRNA sequence function

KEITH VASS

1. Introduction

In some mRNAs structural elements determine translational control, mRNA stability, subcellular location, and timing of expression in embryogenesis. I will attempt to suggest ways to help identify such elements in your RNA sequences.

A well-defined hairpin loop, the iron-response element, controls the translation or degradation of a number of mRNAs for proteins involved in iron metabolism or which utilize iron as a cofactor. The existence of this hairpin loop was suggested by some laboratory and cloning artefacts, not from a deliberate attempt to probe the structure of the mRNA. Several rat ferritin pseudogenes were found to have a common 5' sequence, not present in cDNA clones or detected by primer-extension experiments with the mRNA. S1-protection analysis normally also gave the same result but, within a narrow annealing temperature-window, a fragment of 70 nucleotides longer was protected, suggesting some secondary structure feature at the 5' end of the mRNA. This region can be folded into a compelling hairpin loop and has subsequently been found in all eukaryotic ferritin genes.

This rough guide to mRNA functional regions is not concerned with verifying structures, but with recognizing good candidates for control elements from sequences. Before searching for similar elements, have you any evidence for post-transcriptional control or any unexplained experimental results, cloning problems, or PCR difficulties? Many functional regions are located in features that are probably recognizable by intelligent observers if they avoid being overwhelmed by data.

- carry out as many analyses as possible
- get to know your molecules
- print out a widely spaced sequence and annotate it with features as you find them.

Suggestions of the sort of analytical tools to use are:

(a) Dinucleotide analysis: 'fingerprint' the mRNA; look for conserved   features in different species.

(b) Sequence alignment revealing conserved regions in the same RNA from different species is often crucial.

(c) Look for gross similarities of base composition.

(d) Folding—do the same structures exist in all the mRNAs?

(e) Look for similarities between different mRNAs which are co-ordinately regulated.

In the worked examples in this chapter the Unix version of the Wisconsin GCG programs (1) are used. They are easily available and designed to do simple tasks, producing output which can be used as input to other programs. The ability easily to modify output before passing it on to other programs is essential for 'hacking' data without doing any programming! (No suggestion is made that GCG is the best package.) The examples use c-*jun* mRNA sequences.

2. Analysis of sequence data

2.1 Short sequence patterns

Protocol 1. Identification of primary sequence patterns

1. Use a text editor to make a data file of simple patterns, ranging from simple dinucleotides to patterns such as AUUUA or ACCU with known recognition specificity (*Table 1*).

2. Make a file specifying the whole mRNA and the coding region. The file m57467.tick (*Table 2*) tells *mapplot* where to draw boxes.

3. Run *mapplot*: % mapplot -dat=rna.pat -in=gb_ov:m57467 -mark=m57467.tick -default

   The pattern obtained (*Figure 1*) is far from uniform, particularly for CG- and AU-containing dinucleotides.

4. Compare the patterns for the same mRNA from different species.

2.2 Repeated sequences

Examination of RNA for direct and inverted repeats is most simply done by dot-matrix analysis as the output is concise and easy to view. Use the GCG programs *compare* and *dotplot*. If the length of functional elements is short reduce the length of the 'window' used. The use of relatively short perfect 'words' is often useful. Inverted repeats may indicate hairpin loops, as is done more thoroughly and slowly by RNA folding programs such as *mfold*.

Table 1. rna.pat

My_RNA_Patterns

..

aa	0	aa	0	!
ac	0	ac	0	!
ag	0	ag	0	!
at	0	at	0	!
ca	0	ca	0	!
!	etc			
AUUUA	0	auuua	0	!examples of some
PolyU	0	uuuu	0	!suggested functional
XIhbox	0	accu	0	!elements

Table 2. m57467.tick
Ranges for chick c-*jun* sequence m57467

..

mRNA	500	2216	.	Black	|	|	Block
Coding	813	1745	.	Black	|	|	Block

2.3 Conserved sequences

Protocol 2. Multiple alignment of sequences

Separately align single regions of the mRNAs, say the 3′ UTR, rather than the whole sequences as the output is more easily handled. I use *pileup* for this purpose. Use a file to specify the sequences for consistency and to preserve your sanity!

1. Produce a file ('jun.fil', *Table 3*) specifying the sequences you want to align (an output file from *stringsearch* is a good place to start).

2. Check the database sequence header to make sure that you are aligning the same area for each mRNA.

3. Run *pileup*: %pileup -in=@jun.fil.
Vary the gap and gap length penalties, The default values may not suit your sequence or purpose. For example the strongest block of conserved sequence in the 3′ UTR of c-*jun* is shown in *Figure 2*.

If you find conserved sequences, what next?

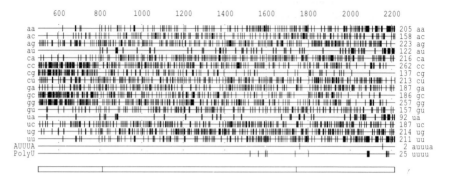

Figure 1. 'Fingerprint' of dinucleotide and other simple sequence patterns in chicken c-*jun* mRNA sequence (accession no. M57467, nucleotides 500–2216.) Each tick represents an occurrence of that pattern in the sequence.

2.4 Database searching

Protocol 3. Database searching

Use the program *pretty* to construct a consensus sequence from the output of *pileup* (*Figure 2*). At this stage it is a good idea to spend some time looking at the multiple alignment. In the case of the c-*jun* 3′ UTR, blocks of similar sequences are present, broken by a G-rich region in a subset of the sequences.

1. Extract the consensus sequence. Use a text editor, or in Unix: % cat pretty.pretty|grep Consens > con.seq. This takes the output from *pretty* and extracts the lines containing the consensus sequence into a file called 'con.seq'.

2. Edit 'con.seq'. Remove unwanted sequences and the 'Consensus' labels. I substitute 'N' for all dashes.

3. Reformat to make the sequence usable (% reformat con.seq).

4. Search the sequence databases with consensus sequence. Use programs such as *wordsearch*, *fasta*, or *blast* (2) (non-GCG). Do not rely on the programs to tell you what the biologically important matches are—many will be inappropriate such as non-transcribed regions.

A *fasta* search with the consensus from *Figure 2* found complementary homology with mouse LINE sequences. Is it possible that c-*jun* mRNA and LINE RNA can base-pair *in vivo*?

A more rigorous alternative is *profilesearch*, in which a weighted consensus

Table 3. jun.fil
'stringsearch file specifying sequences'
..

```
gb_ov:m57467    -begin:1746  -end:2216
gb_ov:x15547    -begin:1462  -end:1938
b_ro:mmjun      -begin:1359  -end:2565
gb_ro:mmjunc    -begin:1619  -end:2335
gb_ro:rsjunap1  -begin:1357  -end:2020
gb_pr:humjuna   -begin:1910  -end:2115
!
```

is used. This profile can be made directly from the output of *pileup* by using the program *profilemake*.

2.5 Secondary structure

Use *fold* and *mfold* (3) to look for clues to secondary structure, crucial in many recognition operations.

There are several things to be aware of:

- the programs look for energetically optimized structures
- they take a fairly long time to run
- they only fold sequences up to 1200 or 850 nucleotides long, respectively
- the pattern of folding often varies with the subfragments folded

There may also be kinetically favoured structures which are not stable enough to be shown by this method. The folding data should be examined together with the sequence alignment. For further hints read the excellent article by Robin Gutell (4).

```
Plurality: 5.00  Threshold: 1.00  AveWeight 1.00  AveMatch 1.00  AvMisMatch 0.00

PRETTY of:  {*}  February 18, 1994  12:20  ..
              291                                                                  360
  {M57467}  cagatcttcgtttaacattgaccaagacctgcatggacctaacattcgatgatcattcagtattaaagg.
  {X15547}  cagatcttcgtttaacattgaccaagacctgcatggacctaacattcgatgatcattcagtattaaagg.
  {MMJUN}   tggacttttcgttaacattgaccaagaactgcatggacctaacattcgat.ctcattcagtattaaaggg
  {MMJUNC}  tggacttttcgttaacattgaccaagaactgcatggacctaacattcgat.ctcattcagtattaaaggg
  {RSJUNAP1} tggacttttcgttaacattgaccaagaactgcatggacctaacattcgat.ctcattcagtattaagggg
  {HUMJUNA} tggac.tttcgttaacattgaccaagaactgcatggacctaacattcgat.ctcattcagtattaaaggg
  Consensus  --GA--TT---TTAACATTGACCAAGA-CTGCATGGACCTAACATTCGAT--TCATTCAGTATTAAAGG-
```

Figure 2. Conserved c-*jun* 3'UTR sequences aligned by *pileup*. A multiple lined-up sequence from *pileup* with a consensus sequence added by *pretty* for the region 291 to 360 of the lined-up sequences. The command used was: % pretty – in=pileup.msf{*} –begin=291 –end5=360 –consensus. As I wanted the sequence to look tidy I included the switches: –line=130 –block=130 which put all the sequence on one line with no spaces to break it up.

2.6 Secondary structure searches of sequence databases

If you think you have identified a possible functional element, look to see if it occurs anywhere else. I have constructed a file containing a block of complementary sequences flanking a known loop-region. The file can be edited, but inserting or removing columns is difficult with most editors. I only know of the (old) IBM Personal Editor (for PCs) that has such a facility. The first few lines of the file are shown in *Table 4*.

Then use *findpatterns* to scan the databases:

```
%findpatterns -dat=ire1.seq -in primate:* -batch.
```

This is an effective and surprisingly specific way of looking for other occurrences of this structure in the databases. Carrying out the search yielded less than 100 matches from Primate sequences in GenEmbl. More than half of these were in non-transcribed regions; most of the rest were previously known iron-response elements. This method takes a long time to run!

Table 4. ire1.seq1[a]

	..			
ire1	0	caaaaacagtgnttttt	0!	the conserved loop
ire2	0	caaaaycagtgngtttt	0!	sequence CAGTGN
ire3	0	caaaagcagtgnytttt	0!	is contained in
ire4	0	caaaatcagtgnatttt	0!	sequences which
ire5	0	caaaayacagtgntgttt	0!	can form a stem
	etc			

[a] IRE1.seq used to search for possible occurrences of a stem–loop structure containing the conserved elements of the iron-response element: a 5′ C residue (unpaired in the IRE, but not tested here) and a loop sequence 5′ CAGUGN.

(A copy is freely available by e-mail request, k.vass@compserv.gla.ac.uk.)

---

3. Summary

The techniques described are not strict prescriptions for how you should proceed. The identification of novel functional elements is axiomatically a chancy business, but it is sensible to examine your data carefully. The possession of the same sequence from more than one species might greatly improve your analytical power. Carefully scan the literature: new examples of post-transcriptional control appear regularly.

References

1. Program Manual for the GCG Package, Version 7, April 1991, Genetics Computer Group, 575 Science Drive, Madison, Wisconsin, USA.
2. Altschul, S. F., Gish, W., Miller, W., Myers, E. W., and Lipman, D. J. (1990). *J. Mol. Biol.*, **215**, 403–10.
3. Jaeger J. A., Turner, D. H., and Zuker, M. (1989). In *Methods in enzymology*, Vol. 183 (ed. R. F. Doolittle), pp. 281–306. Academic Press, London.
4. Gutell, R. R. (1993). *Curr. Opin. Struct. Biol.*, **3**, 313–22.

13

Forecasting protein function

T. C. HODGMAN

1. Introduction

The area of protein function determination has expanded considerably in the past decade, with a huge expansion in the software and databases available, and a change in the kinds of questions being asked. There are now not only proteins but sequence families with unassigned functions; and other protein families whose general activities are known but whose detailed mechanisms and functional parts are the subject of study. Major changes in working styles have occurred, with a move towards distributed systems and use of computer networks, and soon the majority of the software described here will be accessible through World Wide Web connections. It is now impossible, therefore, to cover all the relevant aspects of this topic in one chapter.

This chapter aims to introduce readers to the general methodology, directing them (when appropriate) to other texts and describing the most recent software. These techniques have been validated so often that it is now frequently better to talk in terms of function or structure *forecasting* rather than prediction. Many of the procedures involve sequence comparison and alignment. Therefore, readers are advised to become acquainted with Chapter 7 of this book and ref. 1 which deal with this area. An appreciation of protein structure is also required and various books can be recommended (2–4).

As the language of protein function forecasting can be very technical, various terms are defined now for convenience.

(a) **Query sequence** refers to the sequence under investigation, about which information is required.

(b) **A matching sequence** (or alignment) is a sequence (or alignment) that has been found as a result of a database search.

(c) **Candidate matches** are those matching sequences (or alignments) which look promising and warrant further investigation.

(d) **A domain** is that part of a protein which is structurally distinct, may maintain its tertiary structure when other parts are removed and can still carry out its function in the absence of these other segments.

(e) **A supersecondary structure** is a specific consecutive set of secondary structures, such as the Rossman Fold which consists of β strand - α helix - β strand conformation.

(f) **A subdomain** is a supersecondary structure which carries out part of the overall function of the protein, for example a 'Zinc finger' (5).

(g) **PAM** is the pseudoacronym for the accepted point mutations that can correspond to the divergence between two sequences (6). The higher the PAM value, the more divergent the sequences, with PAM 250 corresponding to about 20% amino acid identity.

2. Structure/function relationships

The sequence of a protein governs its final tertiary structure and function, and sequence analyses are almost all concerned with structural comparison at differing levels. So although protein structure determination is not the main aim of this chapter, an appreciation of protein structural features is required when assessing the significance of sequence matches for functional determination. Briefly, the following features need to be borne in mind.

(a) Globular functional domains are very compact with predominantly hydrophobic residues in the interior and a variety of residues on the surface.

(b) Each amino acid has a different propensity for being on the surface or buried.

(c) Fully functional domains are normally 100–200 residues in length and larger sequences either play a structural role or contain more than one function; for example, virus polyproteins and multifunctional enzymes like fatty acid synthase have over 2000 residues.

(d) The peptide backbone may adopt a variety of ordered secondary structural conformations: helices ($\geqslant 3$ residues long) and β-strand (about two to seven residues long).

(e) The loops connecting them are usually at the surface and are variable in length; some loops fall into recognizable groups (β-turns).

(f) These structures pack together into one of a limited set of supersecondary and hence tertiary structures.

(g) In view of the previous point, natural protein sequences have an amino acid distribution far from random and care is required when using statistical data; the overall distribution of scores in a database search may appear Gaussian but is actually skewed towards the query sequence by weak structural similarities.

(h) Amino acids can be grouped in different ways on the basis of various biophysical criteria (such as size, hydrophobicity, charge, or potential for producing turns).

(i) Analogous proteins maintain active site residues in the same three-dimensional orientation with sequence differences being accommodated by shifts in the packing of secondary structures.

(j) In an alignment, the kinds of amino acids found at a given position might fall into one of the classes described above and, therefore, provide a clue to its position in the tertiary structure.

(k) There is a hierarchy of conservation: active site residues are essential, hydrophobic packing residues are a highly conserved class, and surface residues have low conservation requirements.

(l) Insertions and deletions rarely occur in secondary structures (as this will disturb the packing) but are usually found in their connecting loops.

(m) The linear distribution of packing residues at successive alignment positions can indicate the type of secondary structure found in that region.

(n) Active site residues are often buried and in loops, and their adjacent residues are also often well conserved.

3. General strategy

Sequence comparison and alignment continue to be the principal approaches to determining the possible function of protein sequences. *Figure 1* gives an outline of the general strategy, with the rest of the chapter providing details. *Protocol 1* describes the series of steps that can be taken to determine related candidate matches, each of which has a decreasing level of confidence. These matches must then be tested by checking the literature. When a function has finally been assigned to a query sequence, it is still a formality to carry out experimental work to confirm the forecasted function. The techniques for determining activities in transcription, translation, and bioassays, however, fall well outside the scope of this chapter.

Protocol 1. Strategy for protein function forecasting

1. Break large sequences into overlapping 'domain-sized' fragments, so that a strong match to one region does not obscure the search for weaker matches elsewhere. Structural proteins, such as keratin, fibroin, or collagen, are usually very large and have repetitive sequences. Their different fragments often match the same database entries. Such 'low complexity' sequences can also be embedded in otherwise globular proteins and frequently yield distracting or biologically unhelpful results. However, they can be masked from certain *BLAST* sequence comparisons using *XNU* and *XBLAST* filters (7).

2. Pairwise sequence searches provide the best indicator of domain function (Section 4). However, matches are often weak or dubious.

Protocol 1. *continued*

3. If this is the case, then alignment methods can offer support since structurally and functionally important residues will be conserved (Section 5).

4. When these methods have not yielded any related candidate sequences, then matches to functional subdomains or structure profiles (identified and specified by motifs or patterns) may be tried (Section 6).

5. Comprehensive background information is required on the candidate sequences identified, so that the credibility of the matches may be decided. It is increasingly common to find that a query sequence belongs to a family whose function is unknown.

6. When this is the case, motifs can be defined and have been successfully used to search the sequence databases for analogous regions in other protein families (Section 7).

7. If no matches have been found by these methods, then analysis of the sequence itself may yield useful information, though it will be a long way from determining its function (Section 9).

8. The quality and credibility of the candidate matches are finally tested by comparison with the known information on the candidate's sequences and their relatives. This involves literature surveys, and the match must be realistic in biological terms.

4. Pairwise domain matches

Four programs are commonly used for this type of study: *FASTA, BLAST, PROSRCH*, and its recent variant, *MPSRCH*; the latter two will be considered together. *DFLASH* is another program currently under development. They may all be used by e-mail servers across networks.

All compare the query sequence with each entry of the chosen database. SWISS-PROT and PIR are the most commonly used, though others (such as OWL) might contain old sequences that did not find their way into the major databases. All the programs keep a record of comparison scores, evaluate statistics, and then rank entries in descending order of significance. Despite the potential problems of using statistics, because sequences are far from random character strings, this is still the best way of excluding the vast majority of irrelevant entries while picking up the biologically significant matches. They also generate alignments between the query and each matching sequence for further investigation.

FASTA is the oldest and perhaps the best known. However, *BLAST* is now more widely used because it has the same sensitivity as *FASTA*, is much quicker, and arguably has more helpful output. *PROSRCH, MPSRCH*, and

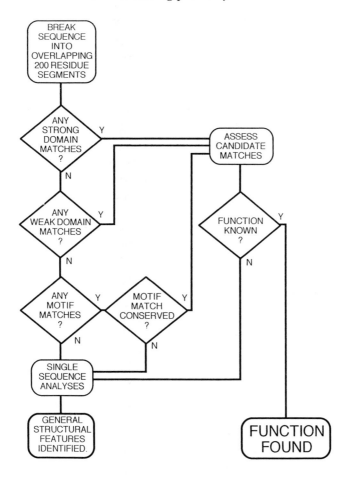

Figure 1. Flowchart of the procedure described in *Protocol 1*.

DFLASH carry out more thorough comparisons using various parallel processors. The first two employ the Smith and Waterman algorithm (8) and usually search SWISS-PROT, whereas the latter uses its own algorithm for searching PIR. They can be used with high PAM values to identify sequences with the same tertiary structure, but not necessarily the same function, and low PAM values to identify common structural/functional motifs. Details of such fine tuning are described in Chapter 7.

4.1 *FASTA* and *BLAST*

For *FASTA*, the input may take a variety of formats. The output includes a histogram, then mean and standard deviation scores of the best (*init*1)

```
                                                              Smallest
                                                              Poisson
                                                     High     Probability
Sequences producing High-scoring Segment Pairs:      Score    P(N)      N

ARGR_ECOLI P15282 ARGININE REPRESSOR.                 770     1.0e-101   1
AHRC_BACSU P17893 ARGININE HYDROXIMATE RESISTANCE PROTEIN.    58   0.053  2
POL_HV2D2  P15833 POL POLYPROTEIN (PROTEASE (EC 3.4.23.-)...  56   0.30   2
POL_HV1RH  P05959 POL POLYPROTEIN (PROTEASE (EC 3.4.23.-) ... 45   0.91   2
METH_ECOLI P13009 5-METHYLTETRAHYDROFOLATE--HOMOCYSTEINE ...  49   0.95   2
```

```
>AHRC_BACSU P17893 ARGININE HYDROXIMATE RESISTANCE PROTEIN.
            Length = 149

  Score = 58 (26.3 bits), Expect = 2.4, P = 0.91
  Identities = 13/34 (38%), Positives = 20/34 (58%)

Query:     82 LKNLVLDIDYNDAVVVIHTSPGAAQLIARLLDSL 115
              L + + ID   ++V+ T PG AQ I  L+D+L
Sbjct:     80 LMDAFVKIDSASHMIVLKTMPGNAQAIGALMDNL 113


  Score = 46 (20.9 bits), Expect = 0.055, Poisson P(2) = 0.053
  Identities = 11/31 (35%), Positives = 16/31 (51%)

Query:    121 ILGTIAGDDTIFTTPANGFTVKDLYEAILEL 151
              ++GTI GDDTI         + +  +LEL
Sbjct:    118 MMGTICGDDTILIICRTPEDTEGVKNRLLEL 148


  Score = 36 (16.3 bits), Expect = 0.21, Poisson P(3) = 0.19
  Identities = 9/32 (28%), Positives = 15/32 (46%)

Query:     41 INQSKVSRMLTKFGAVRTRNAKMEMVYCLPAE 72
              + Q+ VSR + +   V+        Y LPA+
Sbjct:     36 VTQATVSRDIKELHLVKVPTNNGSYKYSLPAD 67


>POL_HV2D2 P15833 POL POLYPROTEIN (PROTEASE (EC 3.4.23.-); REVERSE
            TRANSCRIPTASE (EC 2.7.7.49); RIBONUCLEASE H (EC 3.1.26.4)).
            Length = 1058

  Score = 56 (25.4 bits), Expect = 6.8, P = 1.0
  Identities = 9/45 (20%), Positives = 24/45 (53%)

Query:     77 TTSSPLKNLVLDIDYNDAVVVIHTSPGAAQLIARLLDSLGKAEGI 121
              T S P  N+++D  Y  ++   +   + ++A++++ + K E +
Sbjct:    689 TDSEPQVNIIVDSQYVMGIIAAQPTETESPIVAKIIEEMIKKEAV 733


  Score = 47 (21.3 bits), Expect = 0.36, Poisson P(2) = 0.30
  Identities = 15/44 (34%), Positives = 17/44 (38%)

Query:     76 PTTSSPLKNLVLDIDYNDAVVVIHTSPGAAQLIARLLDSLGKAE 119
              PT    K  +  ID DA   I  P   Q A L S+  AE
Sbjct:    299 PTRQVAEKRRITVIDVGDAYFSIPLDPNFRQYTAFTLPSVNNAE 342
```

Figure 2. Partial output from a *BLAST* search using the *E. coli* arginine repressor. Each item of the ranked list includes the database entry name, accession number, brief description, actual score, and smallest poisson probability P(*N*). The aligned fragments that follow have associated mathematical data and show the positions of residue identity and similarity. Thus compared with itself, the score is very high and P(*N*) is very low. The *Bacillus* arginine repressor (SW:AHRC_BACSU), however, has three approximately collinear blocks that result in about 27% residue identity when the entire sequences are taken into consideration. Two different parts of the HIV Type 2 polyprotein match the same region of the query sequence, and may reflect some structural similarity but probably not more than that.

and total (*init*n) matching segments. Entries with *init*1 scores greater than the mean score plus four standard deviations should be retained for further examination. *BLAST* also accepts various input formats and entries with a smallest poisson probability of 0.1 or less (*Figure 2*) should be retained. Probabilities less than 0.01 correspond to particularly strong matches.

4.2 *MPSRCH* and *PROSRCH*

The parallel-processor programs require suitably written program-control commands (defining the comparison matrix, choice of database, and other search parameters) before being presented with the sequence. However, software exists to simplify the construction and mailing of suitably for-matted messages. *MPSRCH* provides a great wealth of statistical data (*Figure 3*) and the ranked entries show the strength of the similarity in terms of a score, percentage match, and predicted number. The latter is the number of entries expected by chance to have a score greater than or equal to that of the matching sequence. Hence, the lower the number, the more significant the result. Any entry with an expected number less than 0.1 should be considered further as a weak match, and values less than 10^{-4} as strong matches. The criteria for assessing *PROSRCH* output are the same. However, the 'expected number' is shown above each sequence alignment (*Figure 4*).

```
Parameters:     swissprot (33329 seqs, 11484420 residues)
                PAM 200;  Penalty 13;  Perfect Score 3758;  Align 50

Predicted No. is the number of results expected by chance to have a score
greater than or equal to the score of the result being printed, and is
derived by analysis of the total score distribution which gave:

Statistics:     Mean 48.236;  Variance 90.033;  scale 0.536
```

No.	Score	%Match	Length	ID	Description	Pred. No.
1	3758	100.0	643	CR72_BACTI	72 KD CRYSTAL PROTEIN (DEL	0.00e+00
2	343	9.1	633	CR72_BACTK	70 KD CRYSTAL PROTEIN (DEL	0.00e+00
3	334	8.9	633	CR71_BACTK	70 KD CRYSTAL PROTEIN (DEL	0.00e+00
4	175	4.7	823	CRYW_BACTA	130 KD CRYSTAL PROTEIN (DE	9.04e-13
.						
.						
.						
23	104	2.8	1136	CRYS_BACTI	130 KD CRYSTAL PROTEIN (DE	9.47e-02
24	103	2.7	1276	BXD_CLOBO	BOTULINUM NEUROTOXIN TYPE	1.28e-01
25	98	2.6	587	ASO_CUCSA	L-ASCORBATE OXIDASE PRECUR	5.64e-01
.						
30	91	2.4	1135	CRYU_BACTI	130 KD CRYSTAL PROTEIN (DE	4.05e+00

Figure 3. Partial output from *MPSRCH* using the bacterial toxin sequence SW:CR72_BACTI. The parameters shown in the ranked list are self-explanatory. The top 23 entries are all known homologues. The match with *Botulinum* toxin, position 24, could reflect a local structural similarity but nothing more. The remainder are in the zone of spurious matches. However, an exception is at position 30. The values in the %Match column are unfortunately not very helpful.

```
No.   1.    DMD_HUMAN          3685 Amino-acids
DYSTROPHIN.
   Score= 133 Quality   12.963
    35 IDs;    61 CONs;    79 MisMatches;  11 Gaps.  Exp. No. 0.1860E+00 SD   14.3

  Ratio(Found/Expected) Identities  0.776;  Positives  0.834;  Negatives  1.270


          .  .  **.  .     .***  .*  .* *.     **  .    ..    .    * . .
  1882  YKRQADDLLKCLDDIEKKLASLPEPRDERKIKEIDRELQKKEELNAVRRQAEGLSEDGA    1941
   103  FDNEQSDLVHRISS-DKKLEEIPKYKDLLKLF-TTMELMRWSTLVEDYGVELRKGSSETP    160


          *  *  ..  .   ***... *.  .   .   .    *  **. *... . *  .   . ..*
  1942  AMAVEPTQIQLSKRWREIESKFAQFRRLNFAQIHTVREETMMVMTEDMPLEISYVPSTYL    2001
   161  ATDVFSSTEEGEKRWKDLKSRVVE-HNIRIMAKYYTRI-TMKRMAQLLDLSVD-ESEAFL    217


          ...  *.....  *..*   ..  .** ..*... . *..... **  . ... *
  2002  TEITHVSQALLE-VEQLLNAPDL-CAKDFEDLFKQ-EESLKNIKDSLQQSSGRIDIIHSK    2058
   218  SNLV-VNKTIFAKVDRLAGVINFQRPKDPNNLLNDWSQKLNSLM-SLVNKTTHL-IAKEE    274


          .  **
  2059  KTAALQ     2064
   275  MIHNLQ     280
```

Figure 4. The best matching sequence to filamentin (21) using *PROSRCH* at PAM 200. Scoring data appear between the title and the alignment. Although the match was 14.3 standard deviations from the database mean, the level of this match could be expected 0.18 times in that database and has only 20% amino acid identity. On this evidence, the match would remain weak and dubious.

4.3 *DFLASH*

This program is still under development, therefore the reader should obtain the help file to discover what stage of sophistication the program has reached. The length of matching sequence (NRes) and percentage of exact residue identities (Ex%) are taken for consideration by the criteria in Section 4.4. Candidate matches having a comparatively low percentage identity often have Tot% (the percentage of identical plus conservative changes) values approaching twice that of Ex%.

4.4 Assessing retained sequences

The next stage of these studies involves looking at the alignments of the query sequence with the selected entries. Pairwise alignments might only show a short matching segment, in which case dot matrix plots (Section 9.1) are required to show the full extent of the match. *BLAST* shows all the aligned segments it considers significant, and from their positions it becomes readily clear where these can be joined together to produce a longer alignment (see *Figure 2*). Where greater than 25% amino acid identity is found over lengths of 100 or more residues, then a firm candidate match has been identified. For shorter lengths, *Figure 5* can be used to assess whether or not the sequences probably have the same fold. Sequences greater than 50 residues with the same fold can be considered as candidate matches. Shorter

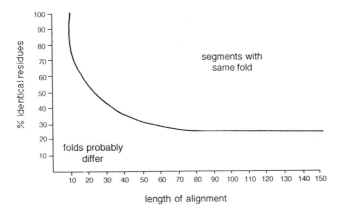

Figure 5. Graph to determine reliable structural similarities based on sequence length and percentage residue identity, redrawn from ref. 29.

segments may indicate structural but not necessarily functional similarity, and methods in the next section are required.

5. Weak domain matches

5.1 General points

The database searches previously described often show up tantalizing weak matches that fall short of the necessary criteria because either the segment length is too short or the percentage identity is too low. The credibility of this weak match can be increased, however, by determining supplementary information about the matching sequence.

(a) Take the DNA sequence encoding this match (details can be obtained from the database entry annotations) and check translations of the alternative reading frames to see if parts of them correspond to the query. This will show if there are any frameshift errors in the query or matching sequences—a growing problem in the databases. Alternatively, this could be indicating that the query sequence is a pseudogene.

(b) When the matching sequence has no obvious homologues in the databases, dot matrix plots and structure forecasts (described below) might show that there is an extended similarity or a common secondary structure profile in the areas of weak sequence similarity, which can justify making the match a candidate sequence.

(c) Alignment of a matching sequence with others that are closely related reveals which residue positions are important for the structure and function of the matching protein family. If these alignment features are also found at the equivalent positions in the query sequence, then a firm candidate match has been found.

239

Considerable progress has been made to assist the latter and several programs are now available. They each have slightly different approaches which will be considered in turn.

5.2 *SBASE*

The program *SBASE* (9) carries out *BLAST* searches on a database of protein sub-sequences that correspond to functional domains only. In this way, the searching is confined to the important sub-sequences. It has the same input requirements and results are interpreted in the same way as *BLAST* output (see above).

5.3 *PRODOM*

PRODOM is an e-mail server that supplies information about related sequences by referring to a predetermined database of domain families. The command Askdom followed by a SWISS-PROT entry name returns the Domain Code numbers for that sequence. Fetchdom followed by the code numbers returns multiple sequence alignments for the regions concerned. In this way, the study of sequence families is made simpler because the alignments have already been carried out. This considerably speeds up the process of comparing a query sequence with the family of a weak match. The server can also carry out *BLAST* searches to initiate the sequence comparison studies of novel proteins. Because *PRODOM* is a derivative of the main sequence databases, its contents will always lag behind them to some extent.

5.4 *PLSEARCH*

This is probably the oldest software in this category. Homologous sequences from the PIR database have been aligned and condensed into character strings that include amino acid ambiguity codes. They are specific for the protein family concerned, and this preprocessing has greatly simplified the process of finding weak matches. The program requires the sequence to be in a particular format, but statistics are evaluated for each motif that provide substantial selectivity.

Figure 6 shows typical output. The histogram gives a rough guide to the significance of a given family match. The crucial value is the SDAM_QUERY, which refers to standard deviations above the mean for the query sequence. Values above four are candidate matches whereas those above six almost certainly have the same fold and general function. Weak matches also occur in the range of values two to four, but cannot easily be separated from spurious matches by this program alone. Alignments of the query sequence to the character strings are also provided and information on these matching patterns can be retrieved by examining a documentation file. This program has the disadvantage, of course, that only the families of the most recently processed database are examined.

```
SDAM
QUERY  NUM
 -3.0    0 |
 -2.5    0 |
 -2.0    0 |
 -1.5    1 |=
 -1.0   31 |================================
 -0.5  220 |===================================================================
  0.0 2812 |===================================================================
  0.5 1460 |===================================================================
  1.0  477 |===================================================================
  1.5  142 |===================================================================
  2.0   34 |================================
  2.5   12 |============
  3.0    3 |===
  3.5    2 |==
  4.0    0 |
  4.5    1 |=
  5.0    2 |==
  5.5    0 |
  6.0    0 |
  6.5    0 |
  7.0    0 |
  7.5    1 |=
  8.0    0 |
  8.5    0 |
  9.0    0 |
  9.5    0 |
 10.0    1 |=
```

NOTE: The above histogram is the distribution of SDAM_QUERY values
      (NOT the distribution of raw scores).

```
        SDAM   SDAM   RAW
   #    QUERY  POS   SCORE
*  1    10.44  37.78  59.43  Pat 1324 [TYPE 1 FIMBRIAE REGULATORY PROTEIN FIME FIMB)]
*  2     7.64  47.47  34.29  Pat 1078 [INTEGRASE / E2 INT) PROTEIN E2) RECOMBINASE]
*  3     5.46  33.77  41.70  Pat 1521 [INTEGRASE TNPB INT) PROTEIN]
*  4     5.29  42.90  32.00  Pat 1200 [HYPOTHETICAL PLASMID REQUIRED PROTEIN IN ]
*  5     4.86  31.98  80.73  Pat TNPI$BACTH [TNP I RESOLVASE (GENE NAME: TNPI).]
*  6     3.93  33.17  28.32  Pat FLMA$ECOLI [F-PLASMID MAINTENANCE PROTEIN A]
*  7     3.62  22.28  13.72  Pat 533 [STAGE REGUL. SPORULN. SYNTHESIS PROTEIN]
*  8     3.46  35.38  70.20  Pat TNPA$STAAU [TRANSPOSASE (TRANSPOSON TN554)]
   9     3.19  21.67  14.64  Pat 598 [50S RIBOSOMAL PROTEIN L29 RPMC)]
  10     3.06  14.74   8.89  Pat 4-F [TRANSCRN. ZN PROTEIN Y-CHROMOSOMAL TESTIS TDF]
  11     2.97  30.21  16.57  Pat 363 [ANTIFREEZE PEPTIDE PROTEIN RD  PRECURSOR]
  12     2.96  28.99  17.70  Pat VG65$BPPZA [EARLY PROTEIN 16.5 (GENE NAME: 16.5).]
 *13     2.84  35.35  66.80  Pat RCI$ECOLI [SHUFFLON-SPECIFIC DNA RECOMBINASE]
  14     2.76  26.45  28.16  Pat 1509 [50S RIBOSOMAL PROTEIN L15 L9]
  15     2.66  22.51  12.44  Pat 173-F [RETINOIC RETINOL-BINDING PROTEIN]
 *16     2.66  32.11  66.55  Pat VINT$LAMBD [INTEGRASE (GENE NAME: INT).]
  17     2.62  28.06  16.73  Pat 572 [NODULATION PROTEIN OF F NODF A) OR HSNA)]
  18     2.61  36.14  15.59  Pat DYRD$ECOLI [DIHYDROFOLATE REDUCTASE TYPE I]
  19     2.61  40.92  26.82  Pat VE8$BPV2 [PROBABLE E8 PROTEIN.]
  20     2.54  33.29  14.32  Pat SECR$CHICK [SECRETIN.]
  21     2.53  40.69  28.90  Pat PP15$HUMAN [PLACENTAL PROTEIN 15 (PP15).]
  22     2.53  26.82  13.29  Pat SODM$STRMU [SUPEROXIDE DISMUTASE (MN-FE)]
  23     2.48  16.71  13.28  Pat 31-F [NEURAMINIDASE (EC 3.2.1.18) (FRAGMENT)]
  24     2.41  51.26  15.07  Pat 563 [FERREDOXIN LIKE PROTEIN FIXX)]
 *25     2.37  37.34  59.79  Pat VINT$BPPH8 [INTEGRASE (GENE NAME: INT).]
```

Figure 6. Partial output from *PLSEARCH* using *E. coli* XerC protein. The histogram shows that some patterns fall a long way from the main distribution and may be considered very seriously. Beside the ranked list, asterisks show which patterns correspond to families with the same function.

5.5 *BLAST3*

This program employs a slightly different approach (10) and is a way of finding domains like those in *PRODOM*. A pairwise match with a score too low be worth considering can become significant when a third matching sequence is taken into account. This program, therefore, initially carries out a *BLAST* search but retains these weak matches as well. Every pair of entries that match the same region of the query sequence are used to generate three-way alignments whose significance can be evaluated. The format of the input is the same as for *BLAST* and *Figure 7* shows part of the program's output. The cut-off scores and expected numbers are sequence sensitive and readers should refer to the original paper for details (10), though a crude guideline is to define candidate matches as those that have an expected number < 0.1.

6. Motif matches

6.1 Sources

Motifs have been determined from various sources. Crystal structure data have provided the most reliable motifs because the significance of each residue position can readily be seen. Functional subdomains have also been identified by *in vitro* mutagenesis studies, assays of proteolytic fragments, chemical crosslinking, and using inhibitory monoclonal antibodies. However, the majority of functionally specific motifs are now defined on the basis of sequence alignments, because of the large number of sequences determined. Owing to the non-randomness of primary sequences, biological knowledge is often a more reliable guide than naive probability data for assessing the significance of any matches.

6.2 Definitions

Motifs have been defined in three ways: using computer plots, helical wheels, and sequence comparison. The first two will be described in Section 9. Motifs based on sequence similarity are defined in two ways:

Figure 7. Partial output from *BLAST3* using clone 1 from the laboratory of S.-J. Richards (30). The sequence region 222–277 has many weak matches some of which are shown in the block marked 1. Below the query sequence is the sub-sequence 90–143 of GBB_YEAST. The three-way alignment of clone 1, GBB_YEAST and TUP1-YEAST (marked below by asterisks) has a score of 179 and is expected 7.1×10^{-5} times in the database. The other alignments in this region match less well, but are all credible matches. A second region of the query sequence has matching alignments in the block marked 2. They are all weak, with the poorest shown at the foot of the figure marked by '!' This is an example of a spurious match. Above each three-way alignment are details of the pairwise statistics.

```
For 2-alignments K = 0.137 and Lambda = 0.322     Expect( S >   71 ) = 0.061
For 3-alignments K = 0.153 and Lambda = 0.255     Expect( S >= 136 ) = 4.3

Number of diags with score between 44 and 71 inclusive:  529.
Constructing 3-alignments with score of at least 136.
3-Way Search.....................................done

1 clone1.mo,          222  FRGHTGAVFSVDYSDELDILVSGSADFAVKWALSAGTCLNTLTGHTEWTKVVLQ   277   S    E
1 [>TUP1_YEAST   628]  90       GHNNKISDFRWSRDSKRILSASQDGFMLIWDSASGLKQNAIPLDSQWVLSCAIS  143  179  7.5e-05
1 [>GBB_YEAST    90]  628       GHKDFVLSVATTQNDEYILSGSKDRGVLFWDKKSGNPLLMLQGHRNSVISVAVA  681  179  7.5e-05
1 [>PWP1_YEAST   300]  201             DLLASAGTDKVIKIWDVKIGKCIGTVS                        227  169  0.00096
1 [>YCW2_YEAST   398]   51  LRGHLAKIYAMHWATDSKLLVSASQDGKLIVWDSYTTNKVHAIPLRSSWVMTCA  104  166  0.0021
1 [>AAC3_DICDI   316]   63  KHTDSVFAIGHHPNLPLVCTGGGDNLAHLW                          92  157  0.020
1 [>YCD9_YEAST   676]  274         ITSVAFSKSGRLLLGGYDDFNCNVWDVLKQERAGVLAGHDNRVS          317  154  0.044
1 [>HIR1_YEAST    23]  308  LAGHDNRVSCLGVTEDGMAVCTGSWDSFLKIWN                       340  148  0.20

2 clone1.mo,          305  PIGREINCKCLKTLS        319   S    E
2 [>IL8_RABIT     27]   26  PIANELRCQCLQTMA         40  142  0.94
2 [>MIP2_MOUSE    30]   28  IGTELRCQCIKTHS          41  142  0.94
2 [>IL8_RABIT     27]   36  PLATELRCQCLQTLQ         50  141  1.2
2 [>IL8_RABIT     27]    2  PVANELRCQCLQTVA         16  138  2.6

* 3-score: 179    E: 7.5e-05      E23: 0.81
* E12: 0.084     E13: 15.        Diag23:  64
* Diag12:  71    Diag13:  55     Scor23:  59
* Scor12:  67    Scor13:  53
*

* clone1.mo,          224  GHTGAVFSVDYSDELDILVSGSADFAVKWALSAGTCLNTLTGHTEWTKVVLQ   277
* >TUP1_YEAST  628  GHKDFVLSVATTQNDEYILSGSKDRGVLFWDKKSGNPLLMLQGHRNSVISVAVA  681
* >GBB_YEAST    90  GHNNKISDFRWSRDSKRILSASQDGFMLIWDSASGLKQNAIPLDSQWVLSCAIS  143
!

! 3-score: 138    E: 2.6
! E12: 1.4e+02   E13: 1.4e+02    E23: 1.4e+02
! Diag12:  48    Diag13:  48     Diag23:  48
! Scor12:  46    Scor13:  48     Scor23:  44
!

1 clone1.mo,          305  PIGREINCKCLKTLS  319
1 >IL8_RABIT     27  RIGTELRCQCIKTHS   41
1 >GRO_RAT        2  PVANELRCQCLQTVA   16
```

(a) **Regular expressions**. These are character strings where the list of permitted residues (or sometimes the residues specifically not permitted) at each position is defined. These motifs are very common owing to the widespread availability of functionally analogous sub-sequences, and are also sometimes called motifs defined by membership of set.

(b) **Weight matrices**. These define arithmetically the importance of each amino acid at each position of the motif, and a successful match involves exceeding some mathematically or statistically defined limit. They are much more subtle at distinguishing functional from merely structurally equivalent motifs, but rely on having many (say >15) examples of the motif. Such matrices are also sometimes called profiles.

Motifs may be combined using Boolean logic and distance constraints to produce **patterns**, and patterns defined on the basis of tertiary structures are usually called **templates**.

6.3 *PROSEARCH*

The largest and most widely used database of motifs is called PROSITE (11). It contains over 300 entries and gives a wealth of detail on the motifs as well as cross-referencing the original sequences in the SWISS-PROT database. The entries may be used to scan a query sequence using a program locally, such as *PROSEARCH* or GCG *MOTIFS*, or submitting the sequence to an e-mail server for remote analysis. Careful interpretation of matches is required because some motifs are intentionally vague, for example phosphorylation sites by protein kinases or *N*-glycosylation signals, and are often false positives having no biological significance. Two approaches to improving the specificity of the database have been developed in recent years.

6.4 *BLOCKS*

Henikoff and colleagues have taken the parent sequences for each PROSITE entry in turn, aligned them, and made profiles of the conserved segments (referred to as **blocks**). This has expanded the length of sequence being studied, bringing more information to bear when searching, and used the subtlety afforded by weight matrix methods. The program to search a sequence with these profiles is called *BLOCKS*. It is most easily used via the e-mail server, but users are very strongly recommended to read the program's help file and the paper describing this work (12). The paper also contains a useful graph relating the score of a block to its likelihood of being significant.

A candidate match has been found either when several blocks from a given family occur in the query sequence in the same (alphabetical) order at similar distances, or a particular block scores highly (say >1400). An example of *BLOCKS* output is shown in *Figure 8*. Detailed information about particular blocks can be obtained from the same mail server, so that biological considerations can be made.

```
Query=clone1.mo  Length: 422  March 28, 1994  10:25  Type: P  Check: 5029  .. ,
    Size=422 Amino Acids
    Database=blocks.dat, Blocks Searched=2679


1.---------------------------------------------------------------------------
Block     Rank Frame Score Strength Location (aa) Description
BL00678     1    0   1166  1530       160-    173 Beta-transducin family Trp-As
BL00678     6    0   1126  1530       240-    253 Beta-transducin family Trp-As
BL00678    31    0    982  1530       198-    211 Beta-transducin family Trp-As

1166=75.60th percentile of anchor block scores for shuffled queries
P not calculated for single block BL00678x
BL00678x     <->x (28,705):160
PWP1_YEAST 302    LASTSADHTVKLWD
                   |  | |  ||||
clone1.mo  161    LcTGSdDlSaKLWD


2.---------------------------------------------------------------------------
Block     Rank Frame Score Strength Location (aa) Description
BL00347A    41    0    959  4269       291-    345 Poly(ADP-ribose) polymerase z
BL00347K   182    0    870  3164       137-    168 Poly(ADP-ribose) polymerase z
BL00347P     2    0   1154  3722        24-     66 Poly(ADP-ribose) polymerase z
BL00347P   360    0    822  3722       127-    169 Poly(ADP-ribose) polymerase z
BL00347Q   327    0    829  3017       348-    379 Poly(ADP-ribose) polymerase z
BL00347V   239    0    850  3447        41-     77 Poly(ADP-ribose) polymerase z
BL00347W    66    0    923  3952        -1-     52 Poly(ADP-ribose) polymerase z

1154=70.20th percentile of anchor block scores for shuffled queries
P not calculated for single block BL00347P
                         |---  423 amino acids---|
   BL00347 AAA::BBBCC:::DDEFFGG:HHIIJJ:::KKLLMMNNOOPPPQQRRRSTTUUVVWWW
clone1.mo                                         :PPP
clone1.mo <                   AAAQQ
clone1.mo <         KK
clone1.mo <         PPP
clone1.mo < VV
clone1.mo <WW


BL00347P     <->P (707,715):24
PPOL_CHICK 710    LNEVQQAVSDGGSESQILDLSNRFYTLIPHDFGMKKPPLLSNL
                   ||           |   |||   ||   ||   |  ||
clone1.mo   25    kNEtldhLislSgavQlrhLSNnleTLlkrDFlkllPleLSfy

-----------------------------------------------------------------------------
```

Figure 8. Partial output from *BLOCKS* using the same query sequence as shown in *Figure 7*. The best ranking block (BL00678) is the only one defining this family and was found three times in clone 1. It shows the similarity to G protein β subunits, though its scores are weak and unconvincing. The program also generates an alignment of the query sub-sequence found with the best matching individual sub-sequence from the set used to define the block (which in this case is PWP1_YEAST). The second best ranking block is one of 23, which are distinguished by suffixes, that define the poly(ADP-ribose) polymerase family. Apart from the low scores, the other matching blocks are not in the correct order, therefore, the query sequence does not belong to this family.

6.5 *BLA*

The other recent approach (13) is executed by a program called *BLA*. Upon submitting a sequence in PIR format, it first carries out a *BLAST* search on a specified database to find other sequences that are significantly related. Then it searches the query and related sequences with the PROSITE database.

The false positive motifs are often filtered out because they are not conserved in the related sequences, and this approach has been successfully used in identifying functions for open reading frames from yeast chromosome 3. It can also process more than one sequence in an input file, and will probably be of most value in assigning functions to new sequence families. The program may also be given the *BLAST* search results so that it can skip directly to the PROSITE searches.

6.6 LUPES

Finally, the Leeds University Protein Engineering Suite (LUPES) contains a database of weight matrices that correlate structure and function, often taking advantage of crystal structure data. The package is resident on the SEQNET system where the contact is `uig@seqnet.daresbury.ac.uk`.

7. URF alignments

With genomic sequencing continuing apace, it is increasingly common to find families of unidentified reading frames (URFs)—protein families for which no function has been assigned. Alignments of these begin to show areas of functional significance by virtue of their patterns of conservation. The greater the number of sequences in the alignment, the easier it is to identify the conserved, and hence important, regions. These may correspond to active site regions or structural features crucial to maintenance of the fold. Motifs of these regions can be defined in the ways described above and used to search for equivalents in the protein sequence databases. The specificity of a motif is revealed by how few clearly unrelated matches are found in the search.

When motifs from the query sequence family are consistently found in the same database entry, then confidence in the match begins to grow. This is particularly enhanced when motifs lie in the same order and are also found in homologues of the matching sequence. A successful example of this approach was the assignment of helicase activity to a conserved plant RNA virus domain (14). It is increasingly common that motifs identify areas of similar secondary structure which bear no relation to functional activity. On the other hand, a broad appreciation of molecular biology is most useful because searches may identify a set of diverse proteins that could have some detailed feature in common, such as metal or other ligand binding, which does not yet have a PROSITE entry.

Although it is worth trying to align a matching sequence with the query URF alignment, it is unlikely to be satisfactory because domain-sized matches would probably have been detected by earlier methods. Furthermore, these approaches can detect active site residues in common that are otherwise surrounded by differing secondary and tertiary structures. Even after confidently assigning candidate matches by these methods, they remain questionable until biochemical experiments have supported the computer-based findings. Thus, these techniques are often more useful for directing the efforts of biochemical research than determining functions on their own.

Four packages, of varying degrees of sophistication, are in widespread use for defining motifs with a view to searching databases: *PTNSRCH*, *PIPL*, *PROFILESEARCH*, and *SCRUTINEER*.

7.1 *PROFILESEARCH*

This program is part of the GCG package. It searches the chosen database with profiles defined by a separate program within the package.

7.2 *PIPL*

This is more flexible and can define patterns on the basis of exact or percentage matches, or by a score matrix (for example PAM tables) to a given string, by weight matrices, membership of set, or for the presence of direct repeats.

7.3 *PTNSRCH*

This package is different in that regular expressions are compared with sequences on the basis of PAM matrices and insertions can also be considered by applying appropriate penalty scores. This computationally demanding procedure is efficiently carried out using a DAP.

7.4 *SCRUTINEER*

This is the most versatile program (15). It works interactively and may define complex patterns involving:

(a) Regular expressions.

(b) Weight matrices defined in various ways.

(c) Mathematical functions using tables or matrices many of which are supplied with the program, though user-defined data may also be used. In a sense these produce and analyse plots internally. These functions can also detect putative amphiphilic helices and β-strands.

(d) PROSITE patterns.

The amino acid classes (especially the hydrophobics) found at successive sequence positions may help in identifying secondary structural elements. Thus:

(a) Surface β-strands are about four to nine amino acids of alternating hydrophobic and hydrophilic amino acids.

(b) Buried β-strands are four to six hydrophobic amino acids.

(c) α-helices have hydrophobic patches identified by helical wheels or nets.

This procedure, sometimes called **hydrophobic cluster analysis** (16), gains greater credibility when the same structure is forecast by other prediction schemes in all the aligned members over the same region. The distribution of elements can identify domains and their boundaries and may suggest functional activities. For example, the general activity of *Bacillus* endotoxins is understood but a study of their sequence alignment (17) indicated the presence of a hydrophobic helical bundle domain (which was presumably involved in forming transmembrane pores) and β-sheet domains (thought to mediate binding to a surface glycoprotein). Subsequent work has substantiated these observations (18).

8. Assessing candidate matches

Having identified related sequences, the next step is to find out as much as possible about them to discover what bearing, if any, they have on the sequence under investigation. Database annotations provide a valuable first source of information including literature citations, and various programs are available to search these (19). The Sequence Retrieval System (*SRS*) interactively combines sequence comparison with textual searches and is available on the Norwegian EMBnet node. A thorough investigation of published work is required because the minute details may be the deciding factor in assessing the credibility of a candidate. Incomplete domains or the loss of critical active site residues could well explain observed phenotypic differences or diseases, or indicate that the sequence belongs to a pseudogene.

Two examples shall be presented. First, the putative protein A87 of plasmid pBR322 shows highly significant similarity to the resolvase of plasmid mini-F (which is required for its segregational stability). However, A87 has one of the catalytic residues mutated and an active site motif is missing, so that this sequence match is revealing more about the prehistory of pBR322 than it is about present functional activities (20). On the other hand, the query sequence in *Figure 4* had only doubtful matches using all the above software. However, papers on the structural proteins dystrophin, actinin and spectrin describe repeats of three-helix bundles that have a common sequence motif. This signature is also present in the query sequence (now called filamentin) and immunocytochemistry has confirmed that it is also a cytoskeletal protein (21).

Literature surveys have become much easier in recent years because the information in literature archives such as *Index Medicus* and the *Science Citation Index* have been transferred to CD-ROMs and are, therefore, accessible online to the general user. The details of these facilities vary from country to country, so users should contact the librarian or information officer of their establishment for the procedures involved. Steps are also underway to produce unified packages which combine searches of the sequence databases with those of the literature archives. One example of this is the package *ENTREZ*, which combines sequence database searching with the relevent subset from *Medline*. An enormous amount of information is now generally accessible through WWW and Internet systems (see Chapters 2, 3, and 4).

Various subsets of the general databases have been compiled that highlight particular aspects of research. For example, NRL-3D is a sequence database that only includes proteins whose tertiary structures have been determined, so that any matches automatically provide structural information. Information on new motifs and sequence families are being reported regularly in *Trends in Biochemical Sciences* and the new resource *Protein Profile* (Academic Press, London, UK) publishes precompiled information on a variety of sequence families liable to be of general interest.

9. Single sequence analyses

If everything above has proved inconclusive, one is left with the sequence itself and any features it may have. These fall into three areas: repeats, biased amino acid composition, and secondary structure forecasting.

9.1 Repeats

Dot matrix plots of the sequence against itself show up repeats as parallel diagonal lines (see *Figure 9*). They are usually found in proteins that have some structural role to play, either in the interactions between multiprotein complexes or with organelles. Individual plots may be drawn using *COMPARE* then *DOTPLOT* in the GCG package, or generated interactively using SIP. Coiled-coil structures can also be found by using this method (19). If there are enough repeats, it may be possible to define a motif and search the protein sequence databases for analogues in other proteins. At the very least, alignment may help to identify secondary structural elements in the repeat structure by the methods described above.

9.2 Biased composition

Since amino acids can be classified by various biophysical criteria, it is possible to scan the sequence for particular features on that basis, often depicting the results as a plot. Thus, a cluster of 20 consecutive hydrophobic amino acids is strongly suggestive of trans-membrane or membrane-interacting sequence. Such a cluster near the amino terminus could very well be a signal peptide sequence, and it is worth searching for a peptidase cleavage site. *SIGPEP* is a well known program for this (22), and *SIGNAL* is a similar program. If these have been identified, then a putative cytoplasmic tail of more than ten residues could be indicative of a membrane protein that is interacting with other molecules either in a receptor or other membrane-bound

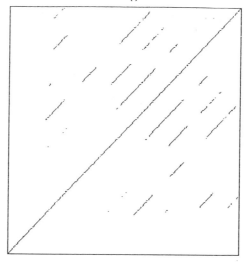

Figure 9. A dotplot of the human G-protein β subunit SW:GBB1_HUMAN compared with itself. The lines parallel to the main diagonal suggest that part of the sequence is repeated at least four times.

signalling pathway. An example of this is the herpes simplex virus glycoprotein US7 which was shown to be part of the immunoglobulin receptor of the virus particle (23, 24).

Clusters of acidic residues could be involved in binding metal ions, whereas clusters of basic residues may show regions which bind nucleic acids. This description also applies to general amino acid composition, and an abundance of other amino acids may have a bearing on its structure and function. Certain amino acids are especially common between compact folding domains, and these criteria can be used in plots for defining domain boundaries (25).

The regular spacing of particular amino acid classes may also provide information about structure/function relationships. The protein/protein interactive domains called leucine zippers were originally identified in this way (26) and highlight a motif that is defined by the general position of amino acid classes on the surface of an α-helix. Hydrophobic moment plots (27) and helical wheels (see *Figure 10*) are used to identify such regions.

9.3 Secondary structure

Owing to the growing number of known protein structures and increasing sophistication of software (for example through the use of neural networks), secondary structure forecasting of intracellular proteins is becoming increasingly reliable. Several programs are available, but one of the most convenient is *PREDICTPROTEIN*, also called *PHD* (28), because it may be used via an e-mail server. It supplies copious output which includes multiple sequence

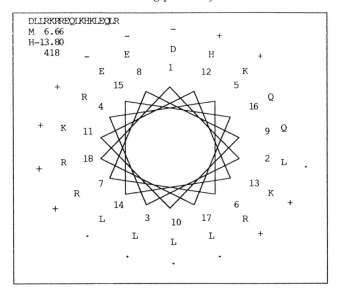

Figure 10. A helical wheel plot (31) of the C-terminal region of human Myc protein. The residues constituting a leucine zipper are marked and clustered on the lower face of the plot while the other residues on the other face are charged. This gives the region a very strong hydrophobic moment of 13.80.

alignments, confidence levels of the structure forecasted at each sequence position, solvent accessibility predictions and scans for transmembrane segments, enabling the user to see immediately which parts of the forecast may be taken seriously. The linear distribution of secondary structural elements provides an overall impression of the protein type and may provide clues for its tertiary fold by analogy with known structures. The *SCOP* package may be helpful here. However, sequences with biased amino acid compositions usually yield poor structure predictions. Thus, such analyses require considerable care and fall outside the scope of this chapter, but could provide pointers to the function of the query sequence.

10. Software sources

The programs mentioned in this chapter and a means to obtain/use them are listed in *Table 1*. Many are accessed via e-mail servers, which are discussed in detail in Chapter 4. However, an increasing number are available through the World Wide Web, and require 'browser' programs, such as Mosaic or Netscape, for easy access. Since the detailed web addresses (URLs) may change, one should connect to a general service in the first instance, such as http://www.ebi.ac.uk/ or http://www.ncbi.nlm.nih.gov/.

Table 1. Software sources

Software	Source	Address
BLA	anon. FTP	`ncbi.nlm.nih.gov`
BLAST	server	`blast@ncbi.nlm.nih.gov`
BLAST3	server	`blast@ncbi.nlm.nih.gov`
BLOCKS	server	`blocks@howard.fhcrc.org`
DFLASH	server	`dflash@watson.ibm.com` (type 'dflash' on the mail subject line and 'help' in the message itself)
ENTREZ	anon. FTP	`ncbi.nlm.nih.gov`
FASTA	server	`fasta@embl-heidelberg.de`
CG	anon. FTP	`ftp.embl-heidelberg.de`
MOSAIC	anon. FTP	`src.doc.ic.ac.uk`
MPSRCH	server	`blitz@embl-heidelberg.de`
PIP	mail	`rs@mrc-lmb.cam.ac.uk` for details
PIPL	mail	`rs@mrc-lmb.cam.ac.uk` for details
PLSEARCH	anon. FTP	`ftp.embl-heidelberg.de`
PREDICTPROTEIN	server	`predictprotein@embl-heidelberg.de`
PRODOM	server	`prodom@toulouse.inra.fr`
PROSEARCH	anon. FTP	`ftp.embl-heidelberg.de`
PROSRCH	server	`dapmail@biocomp.ed.ac.uk`
PTNSRCH	server	`dapmail@biocomp.ed.ac.uk`
SBASE	server	`sbase@icgeb.trieste.it`
SCOP	WWW	`http://www.bio.cam.ac.uk/scop/`
SCRUTINEER	anon. FTP	`ftp.embl-heidelberg.de`
SIGNAL	anon. FTP	`ftp.embl-heidelberg.de`
SIP	mail	`rs@mrc-lmb.cam.ac.uk` for details
SRS	WWW	`http://www.sanger.ac.uk/srs/`

Acknowledgements

I would like to thank David Judge for the use of his computing facilities throughout the preparations for this work, Roman Tatusov and Eugene Koonin for supplying the code for *BLA* prior to its publication and my wife, Jane, for various assistance throughout the preparation of this text.

Note added in proof
Two particular developments in motif analysis are worth bringing to your attention. There is a new motif database to complement *PROSITE*, called *PRINTS*, which is accessible through the World Wide Web at `http://www.biochem.ucl.ac.uk/bsm/dbbrowser/PRINTS/PRINTS .html`. Several tools have become available for finding motifs in sets of unaligned sequences. One that seems to work quite well is called *ASSET* (32).

References

1. Doolittle, R. F. (1990). *Molecular evolution: computer analysis of protein and nucleic acid sequences. Methods in enzymology*, Vol. 183. Academic Press, London.
2. Creichton, T. E. (1984). *Proteins. Structures and molecular properties.* Freeman, New York.
3. Lesk, A. M. (1991). *Protein architecture: a practical approach.* IRL Press, Oxford.
4. Fasman, G. D. (1990). *Prediction of protein structure and the principles of protein conformation.* Plenum Press, New York.
5. Fairall, L. *et al.* (1993). *Nature*, **366**, 483.
6. Dayhoff, M. O. *et al.* (1978). *Atlas of protein sequence and structure*, vol. 5, supplement 3. NBRF, Washington, DC.
7. Claverie, J. M. and States, D. J. (1993). *Comput. Chem.*, **17**, 191.
8. Smith, T. F. and Waterman, M. S. (1981). *Adv. Appl. Math.* **2**, 482.
9. Pongor, S. *et al.* (1993). *Nucleic Acids Res.*, **21**, 3111.
10. Altshul, S. F. and Lipman, D. J. (1990). *Proc. Natl. Acad. Sci, USA*, **87**, 5509.
11. Bairoch, A. (1992). *Nucleic Acids Res.*, **20**, 2013.
12. Henikoff, S. and Henikoff, J. G. (1991). *Nucleic Acids Res.*, **19**, 6565.
13. Tatusov, R. L. and Koonin, E. V. (1994). *Comput. Appl. Biosci.*, **10**, 457.
14. Hodgman, T. C. (1988). *Nature*, **333**, 22, 23, 78.
15. Sibbald, P. R. and Argos, P. (1990). *Comput. Appl. Biosci.*, **6**, 279.
16. Henrissat, B. *et al.* (1990). *Comput. Appl. Biosci.*, **6**, 3.
17. Hodgman, T. C. and Ellar, D. J. (1990). *DNA Sequence*, **1**, 97.
18. Visser, B. *et al.* (1993). In *Bacillus thuringiensis, an environmental biopesticide: theory and practice* (ed. P. F. Entwistle, J. S. Cory, M. J. Bailey, and S. Higgs), pp. 71–88. Wiley, New York.
19. Hodgman, T. C. (1992). In *Microcomputers in biochemistry: a practical approach* (ed. C. F. A. Bryce), pp. 131–58. IRL Press, Oxford.
20. Hodgman, T. C. and Summers, D. K. (1994). *Plasmid*, **32**, 333.
21. Richards, S.-J. *et al.* (1996). Submitted for publication.
22. Von Heijne, G. (1987). *Protein Seq. Data Anal.*, **1**, 41.
23. Hodgman, T. C. and Minson, A. C. (1986). *Virology*, **153,** 1.
24. Bell, S. *et al.* (1990). *J. Virol.*, **64**, 2181.
25. Vonderviszt, F. and Simon, I. (1986). *Biochem. Biophys. Res. Commun.*, **139**, 11.
26. Hurst, H. C. (1994). *Protein Profile*, **1**, 123.
27. Eisenberg, D. (1984). *Annu. Rev. Biochem.*, **53**, 595.
28. Rost, B. *et al.* (1993). *Comput. Appl. Biosci.*, **10**, 53.
29. Sander, C. and Schneider, R. (1991). *Proteins*, **9**, 56.
30. Duff, K. *et al.* (1992). *DNA Sequence*, **3**, 213.
31. Schiffer, M. and Edmundson, A. B. (1967). *Biophys. J.*, **7**, 121.
32. Neuwald, A. F. and Green, P. (1994). *J. Mol. Biol.*, **239**, 698.

14

DNA and RNA structure prediction

ERIC WESTHOF, PASCAL AUFFINGER, and CHRISTINE GASPIN

1. Introduction

An understanding of the functional mechanisms of a biological macro-molecule requires the knowledge not only of its precise molecular organization in space but also of its internal dynamics. Molecular modelling attempts to construct the three-dimensional (3D) structure of a macromolecule on the basis of a mixture of theoretical and experimental data. Hence, prediction methods range from the most mathematically oriented ones, relying solely on computer algorithms, to the most pragmatical and operational one in which insights come alternatively from theory and experiment. Our contention is that modelling and simulation are most interesting in molecular biology when they possess a high predictive power.

Thus, we view modelling as a heuristic tool which should help in the rationalization of experimental observations but also, and most importantly, should suggest new relations between the various components of the modelled molecule. Without a 3D model, mutagenesis of a macromolecule will be, by necessity, somewhat random and, not always informative. In the absence of a 3D model able to organize the data at a higher level, mutagenesis experiments performed under such conditions will mainly confirm an available secondary structure (2D) of a RNA molecule. Such experiments can be useful, however, for bootstrapping a 3D structure which will serve as a framework for organizing existing data and suggesting new mutagenesis. Further, the history of structural discovery shows that there is no correlation between either accuracy or precision and predictive power. For example, molecular biology was born with the 1953 paper by Watson and Crick on the DNA double helix (1), but the structure, although accurate, was not precise by present standards.

The power of visualizing 3D relations is such that models need not always be detailed. On the contrary, extremely precise and detailed X-ray structures can be of no use for uncovering or understanding the function of a crystallized molecule without prior or further biochemical exploration and characterization. In the end, the validity and the accuracy of the model obtained will depend on the nature of the experimental observations collected. However, a mathematical proof guaranteeing the correctness of the derived model is only possible with crystallographic methods (the Fourier theorem). Otherwise, the best that can be achieved is a network of evidence converging on the spatial contacts and relations embodied by a model.

The experimental observations used for deriving a 3D structure can be of quite different nature depending on the techniques employed and on the chemical nature of the macromolecule: from biophysical methods (partial X-ray diffraction data, NMR couplings, or NOEs, and other spectroscopic methods like UV, RAMAN, or circular dichroism), to biochemical approaches (chemical probing or enzymatic attack), and biological data (sequences, phylogenies). High-resolution X-ray crystallographic analysis (diffraction data at 1.5–1.0 Å resolution) yields a wealth of unequalled 3D information. However, this requires not only the crystallization of the macromolecule but also the solution to a phase problem. Generally, with biological macromolecules, the problem is compounded by their size and complexity. Besides, nucleic acids are very difficult to crystallize, since they are highly charged macromolecules which, in the case of RNA molecules, can undergo spontaneous cleavages. In addition, large, nucleic acids and especially RNAs, often exchange between various base pairings and foldings. Recently, NMR methods have proved their usefulness in this area. Chemical and enzymatic probing of nucleic acids in solution yields important information on the stability of the structures and on those bases protected from chemical or enzymatic attack. However, such experimental approaches will not reveal the nature of the interacting partners. Cross-linking experiments have the potential to give that information, but the cross-linking reactions take place in an assembly of molecules generally not all in the same state, and it is difficult to prove that the reactions occurred solely on functional molecules. Sequence data are extremely rich in potential 3D information, since they result from adaptative evolution over millions of years. Thus, if the function is identical and the sequences are sufficiently diverse, the noise level (or covariations resulting from contingencies) will be decreased by sequence comparisons. However, the extraction of 3D content from sequences is difficult and the method will strongly depend on the type of macromolecule under study. For example, self-splicing autocatalytic group I and group II introns, which require only water and ions to function, are more amenable to sequence comparisons than the catalytic RNase P RNA in ribonucleic particles which contains the history of its evolution with the tRNA substrate and with the protein co-factor.

The former experimental approaches (2–4) will not be discussed here. However, it should be kept in mind that the methods described in this chapter range from those in which the incorporation of experimental data is restricted to physical chemistry to those which use and exploit biological information. Molecular mechanics and dynamics belong to the first category.

RNA secondary structure prediction is simplified and on firmer ground with the incorporation of biological and chemical information, and successful RNA 3D modelling is best achieved on the basis of sequence comparisons and chemical probing.

2. Molecular mechanics and molecular dynamics methods

Molecular mechanics (MM) minimizes a particular energy function for a molecular system. The energy function contains steric and geometric terms as well as terms related to atomic interactions. A specific force-field is associated with a given energy function. Molecular dynamics (MD) simulations use similar force-fields and energy functions but, by integration of Newton's equation of motion, allow one to generate time-dependent trajectories of chemical or biochemical systems (5–7). These methods are usually used to add a dynamical perspective to systems for which time-dependent experimental knowledge is scarce and, most importantly, they are also used to process and refine crystallographic (8) or NMR data (*AMBER* (9), *Xplor* (10, 11)), or to calculate free energy differences between related systems by perturbation methods (12, 13). The advantages of MM are its easy implementation and short computing times. The main drawback is that the system might become locked in false minima which depend on possible inaccuracies resulting from the construction of the initial coordinate set or on the choice of the starting conformation. One way to relieve these undesirable effects is to minimize several starting conformations by varying one or more internal coordinates (torsion angles, for example). On the other hand, MD simulations, combined with energy minimizations, are well adapted to the sampling of the conformational space and the localization of local or global energy minima. As it is impossible to recommend, at the actual level of the technique, any definite protocol that one could follow in order to obtain physically meaningful MD simulations, we choose to discuss in this chapter general methodological details with, as guideline protocols, those that we apply in our laboratory on simulations of hydrated DNA and RNA fragments which include the aqueous environment and the counterions (14–16). Other details on simulations of nucleic acids can be found in two reviews (6, 17).

2.1 The potential energy function

The potential energy function, which describes in a simplified way the interactions between the atoms constituting the system, is central to the problem of molecular mechanics and molecular dynamics. A general form of this function, used in the *AMBER* MD package (9), is given by the equation:

$$ E = \sum_{bonds} k_d (d - d_0)^2 + \sum_{angles} k_\theta (\theta - \theta_0)^2 + \sum_{dihedrals} \frac{V_n}{2}[1 + \cos(n\phi - \gamma)] $$

$$ + \sum_{nonbonded} \frac{A_{ij}}{r_{ij}^{12}} - \frac{B_{ij}}{r_{ij}^6} + \sum_{nonbonded} \frac{q_i q_j}{\varepsilon r_{ij}} $$

where the first three terms represent the interactions between atoms separated by less than three bonds, the fourth and fifth term corresponding respectively to the van der Waals and electrostatic interactions occurring between non-bonded atoms. In the electrostatic term, the dielectric parameter ε is either a constant or a function of the distance between the charges.

In addition to the classical terms mentioned above, there is a great variety of additional terms describing for example 10–12 hydrogen bonds (18–20), or mixed terms which couple bond length and bond angle vibrations (21). Other specifics (like choice of parameters, of options, of functions, etc.) can be found in the *AMBER* (18, 19), *CHARMm* (20), *GROMOS* (22), or *OPLS* (23, 24) force-fields. It should be noted that the use of united atom force-fields, where the CH, CH_2, and CH_3 groups are represented by large hydrophobic atoms, compared with the all atom force-fields, is no longer justified, either for simulations *in vacuo* or for simulations taking into account a solvent environment, since the gain in computer time is not worth the approximations introduced in the system.

The choice of a set of partial atomic charges is of particular importance. Next to the classical ways of extracting charges from quantum mechanics calculations, charges derived from experiment were published (25) and recently tested in our laboratory on a simulation of the anticodon arm of tRNA[Asp]. They were shown to give better agreement with known experimental structures than the standard *AMBER* set of charges. Other methods like multipole distributions, in which partial charges are no longer restricted to the atomic positions, have to be considered in the future to increase the accuracy of the electrostatic representation. Ultimately, with adequate computational power, a full electrostatic treatment taking into account the atomic polarizability will be necessary (7). A choice has also to be made concerning the water model to be used. Improving the classical SPC, TIP3P, or TIP4P models (26), the SPC/E model (27) is known to reproduce the diffusion coefficients of water and, therefore, should give more reliable time-dependent quantities.

The treatment of long-range electrostatic interactions (proportional to $1/r_{ij}$) is an issue of great concern. Because of computational limitations, it is very difficult to calculate electrostatic interactions up to a distance greater than a given cut-off value, usually 8–10 Å. This truncation method is not very satisfactory and some authors have shown that they introduce non-negligible artefacts in the calculations (28, 29). To ameliorate the straight truncation of electrostatic forces, a wide range of switching and shifting functions has been employed and discussed (30, 31). Other approaches are possible, like the Ewald summation method used in simulations of a rigid DNA duplex with various counterions and co-ions (32) and of a rigid DNA triple helix in a 1.0 M NaCl aqueous solution (33).

2.2 Molecular dynamics simulation protocols

The following protocols are of course not unique and have to be adapted to each particular system and to the available computational means.

2.2.1 Construction of the system

All available experimental knowledge should be used in order to choose reasonable and interesting starting configurations. They can be extracted from the NDB (Nucleic Acid Data Base) (34) which contains most of the published crystallographic nucleic acid structures as well as structures derived from NMR experiments. Some of those structures are also contained in the PDB (Protein Data Bank) (35). Subsequently, the molecule needs to be solvated. This can be achieved at various levels of approximation. Partial solvation can be performed by putting a shell of water around the entire solute, or only around a site of particular interest (complexation or catalytic sites). There are different ways of constraining the solvent molecules located at the surface of the solvation shell, but some researchers let the water move freely in their simulations. We chose to use periodic boundary conditions which try to mimic an infinite system by replicating images of the simulation shell around the central box. However, the truncation distance used for computing long-range forces limit the range of the 'infinity' of the model.

Next, counterions are placed around the solute. Two methods are generally used for nucleic acids. The first consists in placing the ion along the bisector of the OPO angle at a distance of 4.5–6 Å from the phosphorus atom (9), and the other consists in replacing water molecules with the highest electrostatic potential by counterions until neutrality or the desired total charge is obtained (36). Various counterions have been used such as Na^+, K^+, NH_4^+, Ca^{2+}. To our knowledge, no simulations using the high structuring Mg^{2+} ion have been undertaken so far. The choice and positioning of counterions can be circumvented by reducing, according to the Manning theory of counterion condensation (17), the charges on the phosphate groups, and omitting explicit representation of the ions. However, this leads to values for the charges on the phosphate group below those of some polar atoms in the bases, and therefore alters considerably, and perhaps unrealistically, the water–phosphate interactions.

2.2.2 Equilibration and thermalization

In order to produce stable MD simulations, a good equilibration protocol which avoids early deformations of the solute originating from strong and unfavourable solute–solute, solute–solvent, and solvent–solvent interactions, is essential. Usually, at the beginning, a few hundred steps of energy minimization is used to relieve the main unfavourable constraints from the starting configuration, and, afterwards, the system is brought to equilibrium in several stages (*Figure 1*). First, the solvent alone is allowed to move around the fixed solute and counterions at constant temperature (300 K) and volume. Then, the constraints on the counterions are removed and 1 psec of dynamics at constant temperature and pressure (1 atm.) is performed at respectively 100, 200 K, followed by 5 psec at 300 K. Finally, the whole system is thermalized by a gradual increase of the temperature at each psec from 50 to 300 K by steps of 50 K. Subsequently, the heating step is followed by 5 psec of equilibration at 300 K (16).

Some authors have used constraints to maintain the base pairing of the starting structures during the equilibrium step, or even during the whole simulation. Our recent results proved that this is not always necessary. Breaking of base pairs can result from insufficient equilibration as well as from inaccurate force-field or simulation parameters.

2.2.3 Vacuum simulations

Simulations using no solvent have the advantage of being extremely fast. This allows one to conduct longer simulations and to sample more extensively the

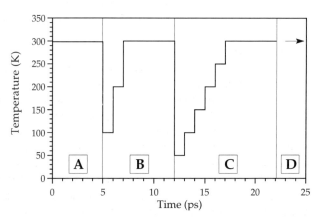

Figure 1. Equilibration protocol for a molecular dynamics simulation of a solvated nucleic acid with counterions. During part A, the solvent alone is allowed to move at 300 K; during part B, constraints are relieved from the counterions in steps of 100 K; during part C, no constraints are applied on the system but after cooling the system down to 50 K, it is warmed up in steps of 50 K; part D corresponds to the subsequent production phase at constant temperature and pressure.

configurational space. To compensate for the absence of solvent and counterions, various dielectric functions have been proposed, but they can give only approximate results since it is well known that specific local interactions with water are necessary to maintain the three-dimensional structure of nucleic acids (37).

2.3 Modelling large nucleic acids

For large nucleic acids, the all atom approach is no longer feasible. In order to be able to simulate such systems, Malhotra *et al.* (38) have developed models of varying resolution ranging from one pseudoatom per helix to one pseudoatom per nucleotide. This allows them to obtain useful but, consequently, much less precise information on the structure of these molecules.

2.4 Analysis of the trajectories

The analysis of the results of the calculations is the last but not the least important part of MD simulations. For nucleic acids, the 'Curves' (39) procedure for helical analysis has been used in a computer graphics utility called 'Dials and Windows' (40) which can monitor and display the time evolution of all the conformational and helical parameters in a DNA oligonucleotide.

3. Fine structure and the search for specific regions in DNA

DNA is not solely a storage medium for genetic information. Any sequence also contains control regions directing the binding of specific proteins as well as regions with static curvature or thermal lability. The prediction of the fine structure of DNA, i.e. the effects of base sequence on 3D structure, is the subject of an enormous literature (41). Here, we will refer more specifically to those methods which possess documented softwares. For small systems (up to 200 base pairs), molecular mechanics methods, as developed in programs such as *AMBER*, *GROMOS*, or *JUMNA* (42) have been used, especially in conjunction with NMR data. *JUMNA* is particularly well adapted to nucleic acids with helical periodicity, either DNA or RNA with between one and four strands in parallel or antiparallel orientations. The study of small systems either by MM (41) or MD (15), allows the extraction of the behaviour of more global parameters (like the average twist angle between two given base pairs or the average roll and tilt angles of a given base pair, i.e. the rotation about the long, respectively short, axis of a base pair, see *Figure 2*). Those parameters can then be inserted in programs using schematic and non-atomic representations of base pairs. Such programs are especially useful for visualizing the path of the helical axis as a function of intra- or interbase pair parameters. Four programs have been extensively used for the prediction

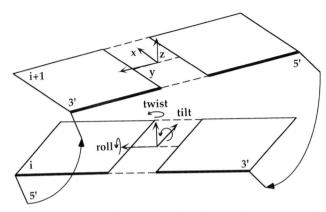

Figure 2. The six parameters relating a base pair to the next one in a double-stranded helix: the three translations along x, y, z (shift, slide, rise) and the three rotations about the z axis (twist angle or rotation angle between base pairs), the y axis (roll angle), and the x axis (tilt angle). For a complete discussion, see Dickerson *et al.* (113).

and display of bent DNA fragments. Bending results from curvature in the plane of the helical axis (controlled mainly by the roll angle) and torsion out of the plane (controlled by variations in the twist angle). The tilt angle is never large because of the resulting compression of the sugar-phosphate backbone. In two programs, *CURVATURE* (43) and that of De Santis *et al.* (44), a given set of those parameters or a mixture of them is used to compute the DNA path. In two other programs, *AUGUR* (45) and *DNA* (46), the user can choose among several sets of parameters or even introduce their own set.

4. RNA secondary structure prediction

Folded 3D RNA molecules are stabilized by a variety of interactions, the most prevalent of which are stacking and hydrogen bonding between bases on strands oriented in antiparallel directions. The 2D structure gives a subset of those interactions represented by Watson–Crick canonical (C–G, G–C, A–U, and U–A) and wobble (G–U and U–G) pairs of bases in double-stranded helices. Such a 2D fold provides an important constraint for determining the 3D structure of RNA molecules (47, 48). Therefore, the determination of the 2D structure is an essential step in the study of the structure–function relationships. Another task associated with RNA 2D structures concerns the automatic identification of specific RNAs in genomic DNA sequences (49, 50).

The determination of a 2D structure results generally from the combination of several approaches, each one using specific knowledge depending on the presence of a set of homologous sequences or of only a single sequence. This section is mainly devoted to the theoretical description of current methods of RNA 2D folding and the associated available programs which are in use today.

4.1 Representation

A 2D fold can be represented on a circle graph where the N nucleotides of the sequence are represented as vertices (dots) and are connected by edges representing the phosphodiester bonds between consecutive nucleotides (along the circle) and hydrogen bonds between bases (across the circle). A valid 2D structure is usually defined as a structure for which the graph contains only edges which do not cross each other. Such a graph is a planar graph. In a more conventional representation, computed with programs such as *Squiggle* (51), *LoopViewer* (52), and *Rnasearch* (53), where bonds are represented as edges of nearly the same size, the folding gives rise to characteristic secondary structural elements which are usually divided into six different types (*Figure 3*): helices, single-stranded regions, bulges, internal loops, hairpin loops, and multibranched loops.

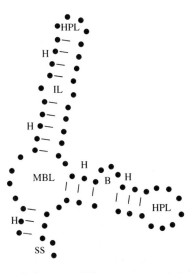

Figure 3. Secondary structural elements. HPL represents a hairpin loop which is formed when an RNA strand folds back on itself. IL represents an internal loop. At least one base is unpaired on each strand of the loop separating two paired regions. A mismatch is a special type of internal loop for which only one nucleotide on each strand is not Watson–Crick paired. B represents a bulge. A bulge has unpaired nucleotides on only one strand. The other strand has uninterrupted base pairing. H represents a helix. A helix is a region of consecutive pairs of bases. MBL represents a multibranched loop or junction. A multibranched loop occurs when double-stranded regions separated by any number of unpaired nucleotides come together. SS represents an unpaired region.

4.2 Data necessary for folding RNA molecules

When a set of homologous sequences (homologous sequences have common ancestry and function) is available, one can search for compensatory base changes which maintain base-paired helices with the help of an available alignment. When only one sequence is available or when RNAs are not conserved among a sufficiently diverse set of organisms, theoretical models of predictions have to be associated with experiments. The related knowledge is based on a set of constraints, the thermodynamic model and the available experimental data on the molecule.

4.2.1 The constraints

Models of prediction generally include the following restrictions on the folding of an RNA into a secondary structure:

(a) Pair restriction forbids all the non-canonical pairings allowing only A–U (two hydrogen bonds), G–C (three hydrogen bonds), and G–U (two hydrogen bonds) pairs.

(b) Uniqueness restriction allows at most one pairing for each base.

(c) Pseudoknot restriction forbids pseudoknots. Two base pairs numbered (i, j) and (k, p) form a pseudoknot if $i < k < j < p$ or $k < i < p < j$. Pseudoknots (*Figure 4*) result from Watson–Crick base pairing involving a stretch of bases in a loop between paired strands and a distal single-stranded region (which could belong itself to a hairpin loop or a bulge). Thus a pseudoknot is akin to a special case of 3D base pairing rather than a structural 2D element. Because efficient programs are essentially based on the ability to decompose a structure into substructures, which is not possible if pseudoknots exist, pseudoknots are usually taken care of in a second step.

(d) Stereochemical restriction requires that at least three ribonucleotides separate two paired strands of ribonucleotides because the chemical linkages cannot stretch beyond a certain distance.

(e) Length restriction affects the length of a helix (the number of base pairs) and allows only helices with a length greater than a given value (usually two).

These restrictions lead to the determination of what is commonly called a valid secondary structure although the restrictions are not always well-founded. In bulges, non-Watson–Crick pairing, such as U–U, A–A, and A–G pairs, are often observed (e.g. 5S rRNA) (2). Also, the existence of unusually stable tetraloops (54), like -GNRA- or -UNCG-, with a non-Watson–Crick pair between G and A (or U and G) shows that hairpin loops can be made with only two unpaired bases (55). Finally, pseudoknots are also extremely frequent in structured RNAs (e.g. group I introns) (56) as in control regions of mRNAs and lead to ambiguities in the 2D definition (57).

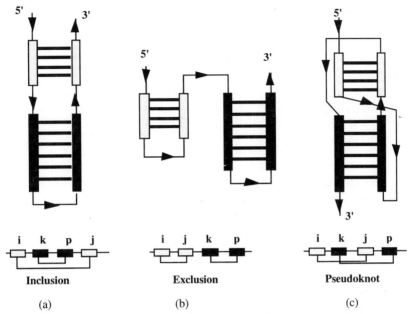

Figure 4. The three possible relationships between two double-stranded helical regions: in (a) and (b) two standard helices and in (c) a pseudoknot. In (a), the paired strands (i,j) are included between the paired strands (k,p) giving a long hairpin interrupted by an internal loop. In (b), the paired strands (i,j) and the paired strands (k,p) form two adjacent hairpins. In (c) the paired elements alternate along the sequence $(i < k < j < p)$, leading to a pseudoknot structure (see ref. 56 for a detailed description of the 3D structures of pseudoknots and of their functions).

4.2.2 The thermodynamic model

The thermodynamic stability of structural elements has been studied to evaluate their probability of formation. These values, computed from experiments on short sequences of nucleotides, give an estimation of the stabilizing free energy of base stacking as well as the destabilizing free energy of single strands. Based on such a set of parameters, several thermodynamic models exist. From the simplified values of Tinoco (58) to Ninio's sophisticated model (59), dedicated to the 5S rRNAs and tRNAs, the most used model nowadays is that of Turner (60). However, it has not been possible to obtain experimental values for each elementary motif, which makes the thermodynamic model rough and uncomplete. For example, until recently, all the loops were considered as destabilizing elements whereas some tetraloops have recently been shown to be very stable (54). Most of the 2D folding programs now take into account the complete model of Turner as well as the parameters associated with the tetraloops (61, 62).

Moreover, the thermodynamic model assumes the Tinoco–Uhlenbeck postulate which states that the free energy of the whole structure is the sum of the free energies of its secondary structural elements. The assumption that the energy of a position in the folded structure is affected only by its nearest neighbours is certainly not correct but the additivity assumption works well and is essential to all prediction algorithms.

4.2.3 Available experimental data

Various enzymes and chemicals are available for probing the solution structure of RNA thus giving detailed data at the nucleotide level. Thus, a real map of the single- or double-stranded regions in the molecule can be established. The mechanisms of action of the probes, the limitations of the technique, and the methods for detection of cuts or modifications are described elsewhere (2, 3). With these data, a number of potential structural elements can be eliminated from consideration in the calculation of folding.

When several mutually exclusive secondary structures exist, site-directed mutagenesis can be used to test for compensatory base changes in the potential helices. Because experiments are time consuming and because precise probing of each nucleotide is difficult, theoretical models of prediction try to incorporate, whenever possible, available data on the studied molecule. The incorporation of this information in 2D folding programs is actually the only way to produce a correct structure.

4.3 Methods of prediction

4.3.1 Sequence comparisons

Comparative analysis of nucleic acid sequences has been widely used for the detection and evaluation of similarities and evolutionary relationships. With RNA molecules, sequence alignments and RNA 2D prediction are intimately related. Comparative analysis is based on the biological paradigm that macromolecules are the product of their historical evolution and that functionally homologous sequences will adopt similar structures. The sequences are first aligned and then searched for compensatory base pair changes. If, during evolution, a base has been modified in a strand of a potential helix (mutation), then this modification must have been compensated on the complementary strand in order to maintain the structure. The presence of several compensatory changes (two or more) in a potential helix allows one to assert the existence of the helix in the structure. Several secondary structure models have been generated by using comparative analysis: tRNA (63), 5S RNA (64), 16S RNA (65), 23S RNA (66), RNase P RNA (4), group I and group II self-splicing introns (67, 68). The method requires that the molecules compared must be sufficiently different to provide enough instances of sequence variations with which to test pairing possibilities but that the molecules do not differ so much that homologous regions cannot be aligned with confidence.

i. Alignment

The objective is to juxtapose related sequences so that homologous residues in each sequence occupy the same column in the alignment (*Figure 5*). Since the 1970 program of Needleman and Wunsch (69), programs to align more than two sequences have been put forth using different strategies including reduction of the problem to three sequences (70), application to closely related sequences (71), the help of a predetermined evolutionary tree (72), the search for common subsequences (73), or the selection of the best

SG	N°	P7.1　　　　P7.1'　P7.2　　　P7.2'	
IA2	73	GUCUUCG------ GACGUAGGGUCAAGCGACUCGA	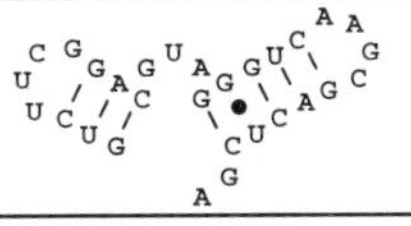 (RNA secondary structure diagram)
	74	UCCCUGAU[7]AGGGAGUAGGGUCAAGCGACCC GA	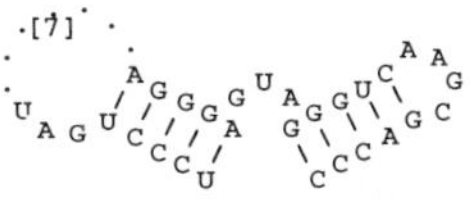 (RNA secondary structure diagram)
	75	UCCCUUUG---- GGGAGUAGGGUCAAGUGACUCGA	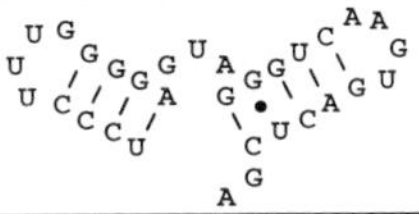 (RNA secondary structure diagram)
	76	UCGAAAC[51]GUAGAGUACCUUA[15]UAGGGG A	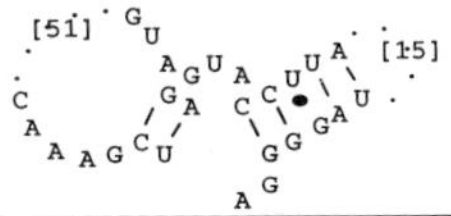 (RNA secondary structure diagram)
IA3	77	UUCUU--GAAAGAGAAAG-AGGUG[9]CGCCUAA	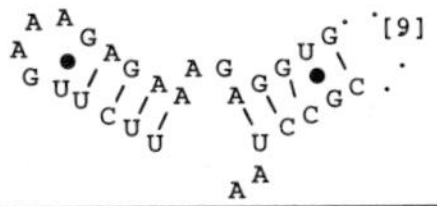 (RNA secondary structure diagram)
	78	AAUC---GAAA-GAUGAG-AGUUU[12]AAGCUAA	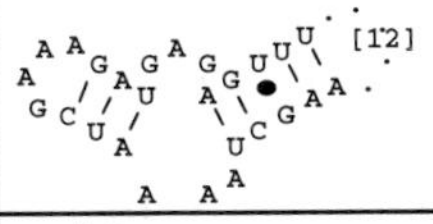 (RNA secondary structure diagram)
	79	UGU[44]GAAACGGCAGG-AUAAC[38]GUUAUAA	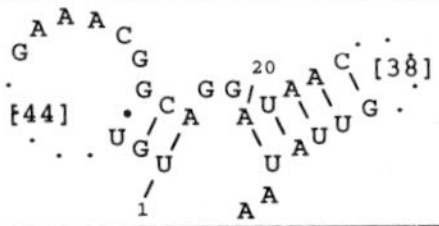 (RNA secondary structure diagram)
	80	UAUAAA[69]UUUAUAGG-AUAUU[16]AGUAUAA	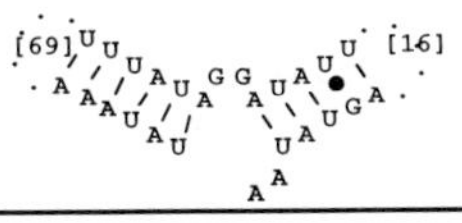 (RNA secondary structure diagram)

Figure 5. Part of the alignment of group I introns corresponding to the structural elements P7.1 and P7.2 of two subgroups IA2 and IA3. (Extracted from the appendix of ref. 48.) The paired sequences are underlined. The numbers correspond to the sequence numbering of ref. 48.

pairwise alignments to gradually align sequences by using an order of incorporation of sequences into the final alignment (*PileUp* (51), *CLUSTAL* (74), and *MultAlin* (75)). Other programs dedicated to the alignment of RNA sequences allow the user to manipulate interactively the proposed alignment (*DCSE* (76), *ALIGNOS* (77)). They offer functions dedicated to secondary structures as well as an interactive environment for manipulating the alignment.

Other recent and interesting programs automatically reconsider the alignment by taking into account new sequences and pre-existing knowledge of the secondary structure (78, 79). Indeed, with the growing number of sequences, specific RNA databases are created and new sequences have to be quickly added to structured databases of homologous RNA molecules. In such databases, it is very desirable that sequences be aligned in accordance with the conserved secondary structural features. Because, in an alignment, optimal structural elements can be misaligned, the program *RNAlign* makes it possible to align a group of aligned sequences with a new sequence, using positions of high sequence conservation and common secondary structures the group as a guide for determining the secondary structure of the new sequence. Thus, *RNAlign* does not suppose that the related sequences are correctly aligned but instead reconsiders the alignment. *RNAlign* was used to build a structured database of RNA from the large ribosomal subunit. The other method of multiple alignment (79), which differs from all those described above, uses stochastic context-free grammars (80) to build a statistical model during, rather than after, the process of alignment and folding. Such an approach was applied to the multiple alignment of tRNA.

ii. Comparative analysis

Given an ordered sequence alignment, comparative analysis can begin. Most computerized approaches to comparative analysis are based on the number of varying positions in base pairs of Watson–Crick helices (81–83). Han and Kim (83) propose a very simple algorithm that builds a covariation matrix where one can visualize, by different characters and for each possible pair of positions, a complementary base change (for each sequence, the base in column i can form a Watson–Crick pair with the base in column j), an exact match (no variation in both columns i and j), a wobble pair (in most sequences the base in column i can form a G–U pair with a base in column j), an inexact pair (a base i does not form a pair with a base in column j for each sequence and the number of pairs is greater than a threshold value) or a mismatch (a base i does not form a pair with a base in column j for each sequence and the number of pairs is lower than a threshold value). In this matrix, possible helices (diagonals of characters) are combined in order to compute valid common secondary structures.

However, all these programs rely on an available alignment which may not be unique, especially when the sequences are highly divergent in primary structure. Moreover, in an alignment optimal for classical scores, the preserved secondary structural elements can be misaligned. Therefore comparative analysis programs such as those presented above have to be used with caution.

4.3.2 Energy minimization

The usual criterion for computing the RNA secondary structure of a single sequence is to minimize the free energy of the folded molecule. Several types of algorithms, among which are *Fold* (84) and *CRUSOE* (85) have been used to find the optimal secondary structure. These methods have been described extensively (60, 84, 86) and will not be described here. Instead, we will describe the main principles of each one and, whenever they exist, the extensions that have been realized in order to compute more appropriate secondary structures.

i. Dynamic programming approaches

The most commonly used algorithm is based on dynamic programming, first used by Nussinov and Jacobson (87). The main advantage of this type of algorithm is speed and thus the ability to fold large molecules. However, they compute only one optimal structure. These algorithms work by first computing optimal structures for fragments of five nucleotides then extending the fragments one nucleotide at a time in both directions until the fragment becomes the whole sequence.

Instead of computing the minimum free energy structure, the partition function of all possible structures and the pairing probability for every possible pair can be calculated, using a dynamic programming algorithm described by McCaskill (88). This program, which is available in the Vienna package (62), allows one to process base pair probabilities through a postscript dotplot where each base pairing probability is represented by a square of corresponding value in the upper part of the matrix. The lower part of the matrix contains the minimum free energy structure according to Zuker's method. In these programs, the temperature at which the base pairings are computed can be varied, as can the choice of the set of energy parameters related to the various elementary structural elements.

ii. Combinatorial approaches

The second type of algorithm usually called a 'combinatorial' approach, works in two steps. It first generates all the possible helices that can be formed from the sequence and then combines them into valid structures (85, 89, 90). This approach, however, is generally limited to molecules with less than 200 bases because of the exponential number of possible combinations.

4.3.3 Extensions of dynamic programming approaches

These algorithms are ultimately limited by our partial understanding of the parameters necessary for the calculation of the free energy. Accordingly, optimally folded structures may not represent the actual base pairing relationships found in the RNA molecule, either because several folded structures with very similar free energies are possible, or because other cellular elements stabilize active RNA structures that otherwise would be thermodynamically less stable. A partial solution is to extend folding programs to allow for the calculation of a range of possible structures that take into account a given set of biochemical data.

Suboptimal folding is the process of determining a set of possible folded structures that have very similar free energy minima but different foldings. Combinatorial approaches can easily compute a set of suboptimal structures. For the case of dynamic programming, several approaches have been developed. The most popular of these is that of Zuker (91), but there are others (92, 93). In the extension developed by Zuker (91), which is an adaptation of the optimal folding method, the result of the suboptimal folding is a series of structures that have similar free energy minima. It is based on the observation that a fold containing a pair (b_i, b_j) divides the structure into two parts: a folding of the included fragment b_i to b_j and a folding of the excluded fragment from b_j to b_i. The two quantities $V(i,j)$ and $V(j,i)$ are computed, $V(i,j)$ representing the minimum folding energy of the included fragment and $V(j,i)$ representing the minimum folding energy of the excluded fragment. In order to compute suboptimal secondary structures, the strategy consists in identifying all bases pairs for which $V(i,j) + V(j,i)$ is close to E_{min}, the energy of an optimal folding of the sequence from 1 to N. In this extension, a P-optimal base pair is defined so as to be contained in at least one folding within P percent of the minimum free energy. Optimal and suboptimal foldings can be generated either automatically or by selecting a base pair. In the first case, optimal and suboptimal foldings are sorted by energy. In the second case, optimal or suboptimal foldings contain the chosen base pair.

In the original package (94), analysis of suboptimal structures is aided by two ways of visualizing the RNA fold: the energy dotplot and a plot of the number of possible different base pairs versus nucleotide position in the sequence (P-num graphs). The program is able to consider various constraints on the folding such as locations of single-stranded sites, double-stranded sites or known helices. It is also possible to force regions to pair together, one region to pair anywhere, one region to be single-stranded or two regions not to pair together. The energy parameters used are those of Turner (60) with additional values for tetraloops.

4.3.4 Interactive computer assisted approaches

Approaches which provide an environment in which the experimentalist can participate in the computational folding of the RNA molecule are called 'interactive' approaches. The strength of interactive approaches lies in their ability to test different structural constraints without modification of the folding program. Structures can thus be continuously modified according to new biochemical information and the user is free to compare biochemical constraints according to intuition. This type of approach (95) is supported by a computer program which allows:

- the examination of as many of the possible substructures as desired
- the use of filtering to incorporate information on pairing length, pairing and stacking energies, experimental data, user assumptions
- the incorporation of related sequences
- user selection and evaluation.

A dotplot matrix, in which the sequence is compared to its reverse complement, allows the visualization of potential helices for selection by the user. A secondary structure is not calculated with this approach. Instead, helices are chosen, then analysed with respect to two criteria such as the energy of the helix and chemical/enzymatic data. Cedergren *et al.* (96) have incorporated the same approach into an RNA folding editor. The program, called *RNASE*, consists of two main units: a helix editor and a structure editor. The user may select desired helices in the secondary or tertiary structure among a list of computed helices. These helices are verified for overlap before being combined into a secondary structure.

The interactive approach we have developed (97) incorporates restrictions from the length of helices and the available data before the step selection and is able to take into account all the usual constraints. In this way, only the possible pairings can be chosen during the selection step. Moreover, the formalism used and the associated algorithms allow one to consider other types of constraints as well as secondary structures with pseudoknots, by adding or removing appropriate constraints. In the selection step, selected elements are not helices but individual pairs of bases. Moreover, energetic criteria encoding the free energy of the molecule is not necessarily taken into account in the search procedure. However, such a criterion can be considered through a selection probability matrix of pairing like that proposed by McCaskill (88) in which the selected pairs become the most probable pairs in accordance with the thermodynamic criteria.

4.3.5 Sequential folding

This type of method relies on the simulation of the folding process (98–102). In these methods, the folding is considered to be a stepwise process where intermediate structures evolve into the native one by subsequent addition of preferred stems. Generally, the programs start to fold the sequence by adding the most stable stems assuming that these are kinetically favoured and act as nucleation centres for local RNA folding. In one method (98), a competition between helices is performed by using random structure generation. The consideration of folding during synthesis is performed by calculating several cycles of folding determination for each incomplete RNA sequence and increasing the sequence after each cycle. In these programs, pseudoknots are allowed to be nucleation centres.

4.4 Limits

4.4.1 Complexity of algorithms

The time complexity of optimal folding methods increases at least approximately with the cube of the length of the sequence, even with a simplification hypothesis, which constitutes a potential limitation. One way to calculate the folded structure of a large RNA is to fold consecutive subregions of the molecule (61), keeping in mind that dynamic programming methods tend to favour pairing of the 5'- and 3'-extremities.

4.4.2 Significance of folded structures

Without experimental data, assessment of the significance of a folded structure is very difficult. Several strategies have been used. For example, alternative foldings can be calculated for a sequence by first using suboptimal folding methods or by varying parameters. Results are then compared and those foldings in which motifs appear systematically may be considered as significant. It is also possible to refold the molecule in successively overlapping pieces, to compare the motifs that arise, and to keep as significant only those that are reproducible (99). A third method is to fold several random sequences that have the same base composition and compare the folding energies (100).

   Moreover, computed minimal energy structures may not be biologically relevant. The problem does not lie merely in the uncompleteness of the thermodynamic parameter sets, the naivety of simple additive models or the fact that input thermodynamic values were derived under conditions that may not truly mimic *in vivo* situations. The ultimate difficulty is rather that many natural RNAs are likely to require helpers (proteins or other RNAs) which control their folding into biologically active forms.

5. RNA tertiary structure construction

Construction of the tertiary structure of an RNA molecule always starts from a given secondary structure. Insights about tertiary contacts can be gained through chemical modifications (which give the relative importance of specific atomic positions) or probing (some protections cannot be explained by the 2D structure), by cross-linking experiments (which directly indicate the partners, assuming a single conformer in solution) and, most efficiently, by careful sequence comparisons (48). The approaches divide themselves into those which rely on mathematical objectivity and automation to those which exploit partial and potentially biased human decisions. In the first category is included the distance geometry method (103) although there are problems choosing the correct chiralities and for avoiding knots in the structures. Another method, *YAMMP* (104), exploits a pseudoatom approach with either one pseudoatom per helix or one pseudoatom per nucleotide. The use of spherical pseudoatoms, however, leads to a loss in the asymmetry of the RNA fragments and, most importantly, all fine interactions which control RNA folding are not modelled. A third approach is based on a constraint satisfaction algorithm. The program, *MC-SYM* (47, 105) searches conformational space such that, for a given set of input constraints (secondary pairings, tertiary pairs, distances), all possible models are produced. With this methodology, Major *et al.* (106) managed, for a tRNA sequence, to generate 26 solutions which displayed the broad features of canonical tRNA structure. Our own approach involves an extensive use of known structures. The framework of those structures is held in a database which is used by the program *FRAGMENT* for inserting the appropriate sequence (106, 107). The fragments produced are then assembled manually on a graphics screen using any modelling software (*FRODO, INSIGHT, PRO-EXPLORE*). The resulting structure is then refined by restrained least-squares minimization programs (*NUCLIN/NUCLSQ*) (108). Molecular mechanics or molecular dynamics could also be employed at this stage. The manipulations on the screen imply some human judgements which depend on the knowledge of 3D structure and the personal bias of the modeller. However, the human mind can quickly exclude sets of solutions and take into account experimental data. The solvent accessibilities of the final model can be easily computed (e.g. *ACCESS*) (109) to validate the structure against experimental reactivities of specific positions to chemical reagents.

6. Conclusions

Table 1 is a compilation of the programs discussed in the present chapter together with the address of the contacting author or distributor. The programs are classified according to the main topics of the chapter. Unfortunately, the programs are often dedicated to some specific machine or system and it is not always convenient to go back and forth between the requested or produced input/output files. At the present time, there is no comprehensive package able to deal with the various aspects of nucleic acid modelling. The development of such packages is in dire need.

Acknowledgements

This research was supported by the GIP-GREG (92H0906), the Ministère de l'éducation nationale, the GDR-1029 'Informatique et Génomes', and the CM2AO program of ORGANIBIO (P.A.).

Table 1. Overview of the programs discussed in the chapter with the address of author or distributor

Program	Key words	Source/reference
Alignments and comparative analysis		
ALIGNOS	Alignment editor	(77)
DCSE	Alignment editor	(76)
RNAlign	Reconsideration of alignment—databases	(78) fcorpet@toulouse.inra.fr
Klinger and Brutlag	Comparative analysis	(82)
Han and Kim	Comparative analysis	(83)
COVARIATION	Comparative analysis	(110) FTP site: iubio.bio.indiana.edu
2D folding programs		
CRUSOE	(Sub)optimal—combinatorial	(85) mgouy@evomol.univ-lyon1.fr
RNASE	Editor, interactive, computer assisted	(96) Montréal University, Canada
McCaskill	Partition function	(62) FTP site: ftp.itc.univie.ac.at
Abrahams	Sequential folding	(99)
MFOLD	(Sub)optimal—dynamic programming	(61) [a]
2D and 3D drawing programs		
Drawna	Automatic 3D ribbon drawings	(40) westhof@ibmc.u-strasbg.fr
Rnasearch	Automatic drawing without overlapping	(53) gaspin@toulouse.inra.fr,
Squiggle	Automatic drawing	[a]
LoopViewer	Automatic drawing	Indiana University, Bloomington, USA Don.Gilbert@IUBio.Bio.Indiana.Edu FTP site: 129.79.224.25
Molecular mechanics and molecular dynamics packages		
AMBER	Free energy calculations, structure refinements,...	(18, 19) amber@cgl.ucsf.edu
CHARMm	Free energy calculations, structure refinements,...	(20)
GROMOS	Free energy calculations, structure refinement	(36)
Xplor	Molecular dynamics and structure refinement	(10) Yale University, New Haven, CT, USA
Quanta/Charmm	Interactive graphics based on CHARMm	Polygen Corporation, Waltham, MA, USA
Insight/Discover	Interactive graphics and molecular mechanics	BIOSYM Technologies, San Diego, CA, USA
Macromodel	Interactive graphics and molecular mechanics	(111) Columbia University, New York
PROSIMULATE PROEXPLORE	Interactive graphics based on GROMOS	Oxford Molecular, The Magdalen Centre, Oxford OX4 4GA, UK
FRODO	Interactive graphics, construction and manipulations of 3D structures	(112)
TURBO FRODO	Interactive graphics, construction and manipulation of 3D structures	Biographics, Marseille, France turbo@lccmb.cnrs-mrs.fr
JUMNA	Junction minimizations of nucleic acids (and nucleic acid–ligand complexes)	(39) IBPC, 13 rue Pierre et Marie Curie, Paris, France
YAMMP	MM on large nucleic acids	(104)
MC-SYM	Conformational search program	(105) major@tremblant.nlm.nih.gov
Nuclin/Nuclsq	Least squares structure refinement	(108) westhof@ibmc.u-strasbg.fr

[a] Genetics Computer Group, Inc., University Research Park, 575 Science Drive, Suite B, Madison, Wisconsin 53711- Help@GCG.Com (51).

References

1. Watson, J. D. and Crick, F. H. C. (1953). *Nature*, **171**, 737.
2. Ehresmann, B., Ehresmann, C., Romby, P., Mougel, M., Baudin, F., Westhof, E., *et al.* (1990). In *The ribosome, structure, function, and evolution* (ed. W. E. Hills, A. Dahlberg, R. A. Garett, P. B. Moore, D. Schlessinger, and J. R. Arner), pp. 148–59. American Society for Microbiology, Washington D.C.
3. Krol, A. and Carbon, P. (1989). In *Methods in enzymology* (ed. J. N. A. Simon and M. I. Simon), Vol. 180, p. 212. Academic Press, London.
4. Woese, C. R. and Pace, N. R. (1993). In *The RNA world* (ed. R. F. Gesteland and J.F. Atkins), p. 91. Cold Spring Harbor Laboratory Press, Cold Spring Harbor, NY.
5. Allen, M. P. and Tildesley, D. J. (ed.) (1987). *Computer simulation of liquids.* Clarendon Press, Oxford.
6. McCammon, J. A. and Harvey, S. C. (ed.) (1987). *Dynamics of proteins and nucleic acids.* Cambridge University Press, Cambridge.
7. Van Gunsteren, W. F. and Berendsen, H. J. C. (1990). *Angew. Chem. Int. Ed. Engl.*, **29**, 992.
8. Gros, P., Fujinaga, M., Mattevi, A., Vellieux, F. M. D., Van Gunsteren, W. G., and Hol, J. (1989). In *Molecular simulation and protein crystallography (Proceedings of the Joint CCP4/CCP5 Study Weekend) SERC* (ed. J. Goodfellow, K. Henrik, and R. Hubbard), p. 1. Daresbury Laboratory, UK.
9. Pearlman, D. A., Case, D. A., Caldwell, J. C., Seibel, G. L., Singh, U. C., Weiner, P., *et al.* (1991). *AMBER 4.0.* University of California, San Francisco.
10. Brünger, A. T. (1990) XPLOR. Yale University, New Haven, CT.
11. Brünger, A. T. (1990). In *Molecular dynamics: applications in molecular biology* (ed. J. M. Goodfellow), pp. 137–78. Macmillan Press, London.
12. Beveridge, D. L. and DiCapua, F. M. (1989). *Annu. Rev. Biophys. Biophys. Chem.*, **18**, 431.
13. McCammon, J. A. (1991). *Curr. Opin. Struct. Biol.*, **1**, 196.
14. Fritsch, V. and Westhof, E. (1991). *J. Am. Chem. Soc.*, **113**, 8271.
15. Brahms, S., Fritsch, V., Brahms, J. G., and Westhof, E. (1992). *J. Mol. Biol.*, **223**, 455.
16. Westhof, E., Rubin-Carrez, C., and Fritsch, V. (1995). In *Computer modelling in molecular biology* (ed. J. M. Goodfellow), pp. 103–31. VCH, NY.
17. Beveridge, D. L., Swaminathan, S., Ravishanker, G., Whithka, J. M., Srinivasan, J., Prevost, C., *et al.* (1993). In *Water and biological macromolecules* (ed. E. Westhof), Vol. 17, pp. 165–225. Macmillan Press Ltd., London.
18. Weiner, S. J., Kollman, P., Case, D. A., Singh, C. U., Ghio, C., Alagona, G., *et al.* (1984). *J. Am. Chem. Soc.*, **106**, 765.
19. Weiner, S. J., Kollman, P. A., Nguyen, D.T., and Case, D. A. (1986). *J. Comput. Chem.*, **7**, 230.
20. Brooks, B. R., Bruccoleri, R. E., Olafson, B. D., States, D. J., Swaminathan, S., and Karplus, M. (1983). *J. Comput. Chem.*, **4**, 187.
21. Dauber-Osguthorpe, P., Roberts, V. A., Osguthorpe, D. J., Wolff, J., Genest, M., and Hagler, A. T. (1988). *Proteins*, **4**, 31.
22. Hermans, J., Berendsen, H. J. C., Van Gunsteren, W. F., and Postma, J. P. M. (1984). *Biopolymers*, **23**, 1513.

23. Jorgensen, W. L. and Tirado-Rives, J. (1988). *J. Am. Chem. Soc.*, **110**, 1657.
24. Pranata, J., Wierschke, S. G., and Jorgensen, W. L. (1991). *J. Am. Chem. Soc.*, **113**, 2810.
25. Pearlman, D. A. and Kim, S. H. (1990). *J. Mol. Biol.*, **211**, 171.
26. Jorgensen, W. L., Chandrasekhar, J., and Madura, J. D. (1983). *J. Chem. Phys.*, **79**, 926.
27. Berendsen, H. J. C., Grigera, J. R., and Straatsma, T. P. (1987). *J. Phys. Chem.*, **97**, 6269.
28. Schreiber, H. and Steinhauser, O. (1992). *Biochemistry*, **31**, 5856.
29. Schreiber, H. and Steinhauser, O. (1992). *J. Chem. Phys.*, **168**, 75.
30. Smith, P. E. and Pettitt, B. M. (1991). *J. Chem. Phys.*, **95**, 8430.
31. Kitson, D. H., Avbelj, F., Moult, J., Nguyen, D. T., Mertz, J. E., Hadzi, D., *et al.* (1993). *Proc. Natl. Acad. Sci. USA*, **90**, 8920.
32. Forester, T. R. and McDonald, I. R. (1991). *Mol. Phys.*, **72**, 643.
33. Mohan, V., Smith, P. E., and Pettitt, B. M. (1993). *J. Phys. Chem.*, **97**, 12984.
34. Berman, H. M., Olson, W. K., Beveridge, D. L., Westbrook, J., Gelbin, A., Demeny, T., *et al.* (1992). *Biophys. J.*, **63**, 751.
35. Bernstein, F. C., Koetzle, T. F., Williams, G. J. B., Meyer, E. F., Brice, M. D., Rodgers, J. R., *et al.* (1977). *J. Mol. Biol.*, **112**, 537.
36. Van Gunsteren, W. F. and Berendsen, H. J. C. (1987). *Groningen Molecular Simulation (GROMOS)*, Library Manual Biomos, Groningen.
37. Westhof, E. and Beveridge, D. L. (1990). *Water Sci. Rev.*, **5**, 24.
38. Malhotra, A., Gabb, H. A. G., and Harvey, S. C. (1993). *Curr. Opin. Struct. Biol.*, **3**, 241.
39. Lavery, R. and Sklenar, H. (1988). *J. Biomol. Struct. Dynam.*, **6**, 63.
40. Ravishanker, G., Swaminathan, S., Beveridge, D. L., Lavery, R., and Sklenar, H. (1989). *J. Biomol. Struct. Dynam.*, **6**, 669.
41. Lavery, R. (1994). In *Advances in computational biology* (ed. O. V. Hugo), Vol. 1, p. 69. JAI Press Inc., Greenwich, Connecticut.
42. Lavery, R. (1988). In *Structure and expression* (ed. W. K. Olson, R. H. Sarma, M. H. Sarma, and M. Sundaralingam), Vol. 3, p. 191. Adenine Press, New York.
43. Shpigelman, E. S., Trifonov, E. N., and Bolshoy, A. (1993). *Comput. Appl. Biosci.*, **9**, 435.
44. De Santis, P., Fuà, M., Palleschi, A., and Savino, M. (1993). *Biophys. Chem.*, **46**, 193.
45. Tan, R. K. Z., Prabhakaran, M., Tung, C. S., and Harvey, S. C. (1988). *Comput. Appl. Biosci.*, **4**, 147.
46. Treger, M. and Westhof, E. (1987). *J. Mol. Graph.*, **5**, 178.
47. Major, F., Turcotte, M., Gautheret, D., Lapalme, G., Fillion, E., and Cedergren, R. (1991). *Science*, **253**, 1255.
48. Michel, F. and Westhof, E. (1990). *J. Mol. Biol.*, **216**, 585.
49. Lisacek, F., Diaz, Y., and Michel, F. (1994). *J. Mol. Biol.*, **235**, 1206.
50. Fichant, G. A. and Burks, C. (1991). *J. Mol. Biol.*, **220**, 659.
51. Genetics Computer Group (1991). *Program Manual for the GCG Package, Version 7.* Madison, WI.
52. Gilbert, D. (1990). *LoopViewer Package*, Bloomington.
53. Muller, G., Gaspin, C., Etienne, A., and Westhof, E. (1993). *Comput. Appl. Biosci.*, **9**, 551.
54. Antao, V. P. and Tinoco, I. (1992). *Nucleic Acids Res.*, **20**, 819.

55. Westhof, E., Romby, P., Romaniuk, P., Ebel, J.-P., Ehresmann, C., and Ehresmann, B. (1989). *J. Mol. Biol.*, **207**, 417.
56. Westhof, E. and Jaeger, L. (1993). *Curr. Opin. Struct. Biol.*, **2**, 327.
57. Westhof, E. and Michel, F. (1994). In *RNA–protein interactions: frontiers in molecular biology*. pp. 25–51. IRL Press, Oxford.
58. Tinoco, I., Uhlenbeck, O., and Levine, M. (1971). *Nature*, **230**, 362.
59. Papanicolaou, C., Gouy, M., and Ninio, J. (1984). *Nucleic Acids Res.*, **12**, 31.
60. Turner, D. H. and Sugimoto, N. (1988). *Annu. Rev. Biophys. Biophys. Chem.*, **17**, 167.
61. Jaeger, J. A., Turner, D. H., and Zuker, M. (1989). *Proc. Natl. Acad. Sci. USA*, **86**, 7706.
62. Hofacker, I., Fontana, W., Stadler, P. F., Bonhoeffer, L. S., Tacker, M., and Schuster, P. (1994). *Monatshefte für Chemie.* **125**, 167.
63. Holley, R. W., Apgar, J., Everett, G. A., Madison, J. T., Marquisee, M., Merrill, S. H., Penswick, S. H., and Zamir, J. R. (1965). *Science*, **147**, 1462.
64. Fox, G. E. and Woese, C. R. (1975). *Nature*, **256**, 505.
65. Woese, C. R., Magrum, L. J., Gupta, R., Siegel, R. B., Stahl, D. A., Kop, J., *et al.* (1980). *Nucleic Acids Res.*, **8**, 2275.
66. Noller, H. F., Kop, J., Wheaton, V., Brosius, J., Gutell, R., Kopylov, A. M., *et al.* (1981). *Nucleic Acids Res.*, **9**, 6167.
67. Davies, R. W., Waring, R. B., Ray, J. A., Brown, T. A., and Scazzocchio, C. (1982). *Nature*, **300**, 719.
68. Michel, F., Jacquier, A., and Dujon, B. (1982). *Biochimie*, **64**, 867.
69. Needleman, S. B. and Wunsch, C. D. (1970). *J. Mol. Biol.*, **48**, 443.
70. Murata, M., Richardson, J. S., and Sussman, J. L. (1985). *Proc. Natl. Acad. Sci. USA*, **82**, 3073.
71. Bains, W. (1989). *Comput. Appl. Biosci.*, **5**, 51.
72. Sankoff, R. J. and Cedergren, G. L. (1976). *J. Mol. Evol.*, **7**, 133.
73. Martinez, H. (1988). *Nucleic Acids Res.*, **16**, 1683.
74. Higgins, D. G., Bleasby, A. J., and Fuchs, R. (1992). *Comput. Appl. Biosci.*, **8**, 189.
75. Corpet, F. (1988). *Nucleic Acids Res.*, **16**, 10881.
76. De Rijk, P. and De Wachter, R. (1993). *Comput. Appl. Biosci.*, **9**, 735.
77. Neurath, H. and Wolters, J. (1992). *Bioinformatics*, **1**, 22.
78. Corpet, F. and Michot, B. (1994). *Comput. Appl. Biosci.*, **10**, 389.
79. Sakakibara, Y., Brown, M., Underwood, R.C., Mian, I.S., and Haussler, D. (1994). In *Proceedings of the 27th Hawaii International Conference on System Sciences*, Hawaii.
80. Searls, D. B. (1992). In *Artificial intelligence and molecular biology* (ed. L. Hunters), p. 47. AAAI Press/The MIT Press, Cambridge, MA.
81. Winker, S., Overbeek, R., Woese, C. R., Olsen, G. J., and Pfluger, N. (1990). *Comput. Appl. Biosci.*, **6**, 365.
82. Klinger, T. M. and Brutlag, (1993). In *Proceedings of ISMB93* (ed. L. Hunter), p. 225. Bethesda, Maryland.
83. Han, K. and Kim, H.-J. (1993). *Nucleic Acids Res.*, **21**, 1251.
84. Zuker, M. and Stiegler, P. (1981). *Nucleic Acids Res.*, **9**, 133.
85. Gouy, M. (1987). In *Nucleic acid and protein sequence analysis: a practical approach* (ed. M. Bishop and C. J. Rawlings), p. 259. IRL Press, Oxford.

86. Zuker, M. (1989). In *Methods in enzymology* (ed. J. E. Dahlberg. and J. N. Abelson), Vol. 180, pp. 262–88. Academic Press, London.
87. Nussinov, R. and Jacobson, A. B. (1980). *Proc. Natl. Acad. Sci. USA*, **77**, 6309.
88. McCaskill, J.S. (1990). *Biopolymers*, **29**, 1105.
89. Pipas, J. M. and McMahon, J. E. (1975). *Proc. Natl. Acad. Sci. USA*, **72**, 2017.
90. Studnicka, G. M., Rahn, G. M., Cummings, I. W., and Salser, W. A. (1978). *Nucleic Acids Res.*, **5**, 3365.
91. Zuker, M. (1989). *Science*, **244**, 48.
92. Yamamoto, K. and Yoshikura, H. (1985). *Comput. Appl. Biosci.*, **1**, 89.
93. Williams, A. L. and Tinoco, I. (1986). *Nucleic Acids Res.*, **14**, 299.
94. Jaeger, J. A., Turner, D. H., and Zuker, M. (1990). In *Methods in enzymology* (ed. R. F. Doolittle), Vol. 183, p. 281. Academic Press, London.
95. Auron, P. E., Rindone, W. P., Vary, C. P. H., Celentano, J. J., and Vournakis, J. N. (1982). *Nucleic Acids Res.*, **10**, 403.
96. Cedergren, R., Gautheret, D., Lapalme, G., and Major, F. (1988). *Comput. Appl. Biosci.*, **4**, 143.
97. Gaspin, C. and Westhof, E. (1995). *J. Mol. Biol.*, **254**, 163.
98. Gultyaev, A. P. (1991). *Nucleic Acids Res.*, **19**, 2489.
99. Abrahams, J. P., Van Den Berg, M., Van Batenburg, E., and Pleij, C. (1990). *Nucleic Acids Res.*, **18**, 3035.
100. Martinez, H. M. (1984). *Nucleic Acids Res.*, **12**, 323.
101. Le, S.-Y., Chen, J.-H., Currey, K. M., and Maizel, J. V. (1988). *Comput. Appl. Biosci.*, **4**, 153.
102. Le, S.-Y. and Maizel, J. V. (1989). *J. Theoret. Biol.*, **138**, 495.
103. Hubbard, J. M. and Hearst, J. E. (1991). *Biochemistry*, **30**, 5458.
104. Malhotra, A., Tan, R. K. Z., and Harvey, C. (1990). *Proc. Natl. Acad. Sci. USA*, **87**, 1950.
105. Major, F., Gautheret, D., and Cedergren, R. (1993). *Proc. Natl. Acad. Sci. USA*, **90**, 9408.
106. Westhof, E., Romby, P., Ehresmann, C., and Ehresmann, B. (1990). In *Theoretical biochemistry, and molecular biophysics* (ed. D. L. Beveridge and R. Lavery), Vol. 1, p. 399. Adenine Press, New York.
107. Westhof, E. (1993). *J. Mol. Struct. (Theochem)*, **286**, 203.
108. Westhof, E., Dumas, P., and Moras, D. (1985). *J. Mol. Biol.*, **184**, 119.
109. Richmond, T. J. (1984). *J. Mol. Biol.*, **178**, 63.
110. Brown, J. W. (1991). *Comput. Appl. Biosci.*, **7**, 391.
111. Mohamadi, F., Richards, N. G. J., Guida, W. C., Liskamp, R., Lipton, M., Caufield, C., *et al.* (1990). *J. Comput. Chem.*, **11**, 440.
112. Jones, T. J. (1978). In *Computational chemistry* (ed. D. Sayre), p. 303. Oxford University Press.
113. Dickerson, R. E., Bansal, M., Calladine, C. R., Diekmann, S., Hunter, W. N., Kennard, O., *et al.* (1989). *J. Mol. Biol.*, **205**, 627.

15

Phylogenetic estimation

NICK GOLDMAN

1. Introduction

Researchers in different fields of biological research often want to know the evolutionary relationship between species or to know about the genetic processes which led to their evolution. To this end, the DNA[*] from representative organisms can be sequenced in the belief that this may hold information regarding evolutionary relationships, or databanks searched to find such data. Or, perhaps more often, a researcher has molecular sequence data from their own study organism(s) that they wish to relate to sequences from other organisms, in which case a databank search can be performed to find a set of sequences suitable for comparison. The selection and alignment of a set of DNA or amino acid sequences is covered elsewhere in this book, and we proceed from the assumption that the researcher has a number of aligned sequences of which they would like to know the evolutionary relationships. This chapter is devoted to advising the researcher on how they may proceed.

Phylogenetic relationships are traditionally depicted on a diagram called a tree, so-named because of the way the branching arrangement of divergent lineages resembles the branches of a tree (*Figure 1*). It is immediately apparent that this representation cannot show the full range of possibilities of evolution; for example, it cannot show 'horizontal' transfers of genetic information such as those caused by hybridization. Nevertheless, the tree diagram has persisted because it is often a good approximation and because it is the best that we can manage computationally. As will become clear in the course of this chapter, the study of phylogeny is a constant trade-off between what approximations are biologically acceptable and what we can work with for practical data analysis.

Corresponding to the true phylogeny is a time scale. Each node, or branching point, represents the time when an ancestral lineage diverged into two (or more) descendants. Current phylogenetic estimation methods are generally unable to estimate the correspondence between an estimated tree and these

---

* The distinction between DNA and RNA is not significant in the rest of this chapter; for convenience the notation of DNA, with bases A, G, C, and T, will be used throughout.

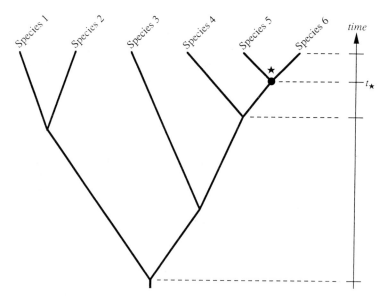

Figure 1. Example phylogenetic tree of six species. Thick lines represent lineages of evolutionary descent (from bottom to top of diagram), diverging at the internal 'nodes' of the tree. The arrow to the right indicates the direction of evolution on the diagram and the true time scale, generally unknown, relating events in all lineages of the tree. For example, the node marked ★ represents a divergence occurring at time t⋆.

times. Many cannot even indicate a tree's orientation in time, i.e. which lineages are ancestors of which others.

Despite this depressing introduction to the inadequacies of current phylogenetic estimation methods, it is indisputably true that molecular sequences do hold information about their evolutionary history. Sequences that are very similar are generally found to be closely related; the more different sequences are, the further apart they are in evolutionary terms. Modern phylogenetic estimation methods exploit this relationship by interpreting sequence similarities or differences, in various ways and according to certain assumptions, to extract some of this information and present it in a form that we recognize. Section 2 of this chapter gives an introduction to principles needed to understand most phylogenetic estimation procedures. Sections 3–6 describe particular approaches, their pros and cons, tips on their use, etc. Three of the main sources of computer software, the *PHYLIP* package (A1) and the *MEGA* (A2) and *PAUP* (A3) programs, each come with excellent documentation and Sections 2–5 are intended to be useful as an introduction to and in conjunction with this. Citations (A1), (A2), etc. refer to computer programs listed in the Appendix to this chapter. Section 5 summarizes differences between methods and makes recommendations—the *PHYLIP* documentation is also particularly valuable in this respect.

Measures of uncertainty or possible error are not well developed in phylogenetic estimation and are widely misunderstood and misinterpreted. As a consequence they are not well represented in the literature, although this does not stop them from being (or perhaps causes them to be) the subject of intense debate at conferences. Some comments on the subject are contained in Section 7.

Phylogenetic estimation methodology is often crude, is subject to the usual statistical errors met in all science, and, worst of all, is often misunderstood. But it is an active field of research, moving faster than most. This chapter is out of date as I write it, and will be more so by the time it is read. In Section 8 I give pointers to allow those interested to follow new research on phylogenetic estimation. As a researcher into these methods, I conclude this introduction with the hope that this chapter will soon need to be superseded by a new practical guide to phylogenetic estimation.

2. Common ground

2.1 Trees

The greatest assumption made in practical phylogenetic analysis is that the evolutionary relationships of molecular sequences may be represented by a branching tree diagram such as that in *Figure 1*. An important point is that there are no closed loops; this implies that there is no transfer of genetic material other than in the direct lineages from ancestor to descendant. There are undoubtedly cases where this will be untrue, but virtually all the (few) methods proposed to deal with this situation consist of *ad hoc* modifications to standard methods.

The true phylogenetic tree relating the sequences will have a root, the position in the tree representing the ancestor of all the sequences, and the time between this point and the sequences (assumed contemporary) must be equal. Many of the methods described below, however, are unable to estimate these. Commonly, the position of the root is unidentifiable. In addition, only an estimated number of base or amino acid substitutions between branching events (nodes of the tree), and not the actual times, can be found (but see Section 2.3.4). This leads to the estimation of a tree with no orientation in time and, if rates of evolution in different parts of the tree cannot be assumed equal, no time scale (see *Figure 2a*). The standard procedure to estimate the direction of evolution within a tree is to include in the analysis one or more sequences whose divergence we know (or are content to assume) to be ancestral to all the others. Once the tree is estimated, a root may be found using this additional knowledge; if one or more sequences' divergence is assumed to be ancestral, then the root of the tree must be in the branch joining those sequences to the rest of the tree (*Figure 2b*). This is known as outgroup rooting, the additional information being introduced by the

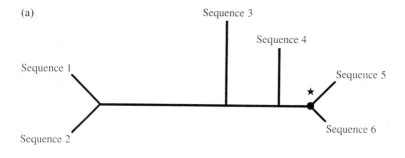

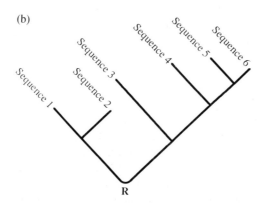

Figure 2. Example 5 of estimated trees of six sequences. (a) This tree might be that estimated from sequences from the six species of the tree of Figure 1. No molecular clock is assumed, so no root of the tree can be estimated and the direction of evolution is unknown. Branch lengths are now typically on a scale of estimated numbers of substitutions per site. The node ★ corresponds to that of Figure 1. (b) The tree of (a) is now rooted using 'outgroup rooting'—here, the assumption that the earliest divergence is between lineages leading to the groups (1, 2) and (3, 4, 5, 6). The estimated root (R) is taken to be somewhere within this branch of the tree (a), which is then redrawn to recreate the direction of evolution from bottom to top of the diagram.

assumption that particular sequences are *out*side the *group* formed by the remaining sequences.

2.2 Data

Molecular sequences form a huge source of data. New information is produced daily, world-wide, by genome analysis projects, commercial agricultural research, medical research, taxonomic research, biodiversity research, and more, and is generally made widely available, via databanks and publication. It is not surprising that molecular sequences are becoming the data of choice for determining evolutionary relationships;

conversely, phylogenetic analysis is an important consideration for all sequences.

All of the major methods for phylogenetic estimation described in this chapter work only for sets of aligned sequences, meaning sets of sequences where each base (or amino acid) in one sequence corresponds to exactly one base (amino acid) of every other sequence. An alignment consists of sequences of nucleotides or amino acids, and gaps. A gap is difficult to interpret in evolutionary terms, and there are no reliable methods which can use the information held in patterns of gaps. Some phylogenetic analyses are able to use the nucleotides or amino acids present at positions where some sequences have gaps; others cannot, and those positions must be discarded from the data to be analysed, even if only one sequence has a gap.

For researchers studying a particular gene, there is no further choice of data to be made apart from the selection of species (often determined by what is available in databanks). For those who are starting with the aim of finding relationships between certain species, and do not mind what sequences they use to achieve this, there may be a choice of sequences to study. I suggest that DNA sequences are always preferable to amino acid sequences. Models of DNA substitution are better developed than amino acid replacement models. More importantly, protein-coding DNA sequences inevitably contain more information than their corresponding amino acid sequences—DNA can always be translated into amino acids, but the amino acid sequence does not contain the information to reconstruct the DNA sequence. A well-constructed phylogenetic estimation method should be able to use the additional information in the DNA sequence. Some further discussion of choice of sequences occurs in Section 2.3.

2.3 Models of evolutionary change

This is a highly controversial subject in the field of phylogenetic analysis. For many years, particularly through the 1970s but sadly still today, many researchers believed it possible to draw inferences about evolutionary histories without having any model describing the patterns that might be seen in the data. This point of view must be hard to understand for any other researcher familiar with statistical methods. It is my belief that phylogenetic estimation is meaningless without some model describing the evolution of the data. Furthermore, recent research shows that the accuracy of phylogenetic estimates depends greatly on the accuracy of models.

The evolution of molecular sequences is highly complex. The simplest acceptable description might be that DNA sequences accumulate errors when copied. These errors may cause alterations of the amino acid sequences coded for by coding DNA; the success of the mutant form, if given the opportunity to spread through a population, will depend on any selective advantage or disadvantage it confers, on the dynamics of the population and on chance. As with almost all such complex systems, from the rolling of dice to the description of chaotic weather patterns, the most successful models are based on probability.

2.3.1 DNA substitution models

For DNA sequence evolution, the most successful and widespread models are based on continuous time Markov chains (1). We assume that each nucleotide site evolves independently of all other sites, and that the evolution of each site in one lineage of a tree is independent of all other lineages. These assumptions are easily criticized on biological grounds; nevertheless, they seem necessary to make the phylogenetic estimation problem mathematically and computationally tractable.

Markov chain models are conveniently defined by a matrix of substitution rates. Every possible substitution (A, G, C, or T changing to any different base) is modelled by a Poisson process (1) with events occurring on average at a certain rate (events per unit time). The rates may be arranged into a matrix, with four rows (*i*) and four columns (*j*), with element (*i,j*) being the rate of substitution from base *i* to base *j*. (The diagonal elements of the matrix are not meaningful in this formulation and can be neglected here.) The most general model would thus have 12 rates, and widely used models comprise different restrictions on these rates. Some of these are described in *Table 1*. These models are very closely related, and this indicates the small range of models that have been found tractable and useful for phylogenetic estimation.

These models do not permit the separation of time and rate of substitution. There is no mathematical difference in average behaviour between a Markov process proceeding at one rate for a certain time, and another process proceeding at twice the rate for half the time. Only the product of rate and time is significant. This is one of the reasons that the roots of trees and directions of evolution cannot be estimated easily, and it means that estimates of time are unattainable. In practice the rate matrices are typically scaled so that the mean rate of substitutions is one per unit time, meaning that evolutionary distances are estimated in substitutions per site.

Rate matrices can be converted into matrices of probabilities so that differences between sequences are apparent, as functions of time (1). These conversions automatically compensate for the fact that if two or more substitutions occur in a lineage at the same site in a sequence, at most one of these will be evident (multiple hits). Substitutions are highly unlikely over very short distances (rate-time products), and increasingly likely over longer distances. Consequently, where there is a choice of sequences to study, that choice is a trade-off between slowly evolving sequences (small rate-time products, leading to few substitutions and little evolutionary information) and faster ones (large rate-time products, leading to the possibility of many multiple hits obscuring evolutionary information). Additionally, longer sequences of course contain more information. In practice, long, fairly slowly evolving sequences should give best results.

Table 1. Some Markov chain models of nucleotide substitution used in phylogenetic estimation

| Model (reference) and description | from | Substitution rates (relative) | | | |
| | | to | | | |
		A	G	C	T
JC69 (2)	A	–	1	1	1
all substitutions have equal	G	1	–	1	1
rate; all bases expected to appear	C	1	1	–	1
with equal frequency	T	1	1	1	–
K80 (3)	A	–	κ	1	1
values of $\kappa > 1$ allow transition/	G	κ	–	1	1
transversion rate bias; bases still	C	1	1	–	κ
appear with equal frequency	T	1	1	κ	–
F81 (4)	A	–	π_G	π_C	π_T
base frequencies $\pi_A, \pi_G, \pi_C, \pi_T,$	G	π_A	–	π_C	π_T
taken from pooled data, allow	C	π_A	π_G	–	π_T
unequal values to be attained	T	π_A	π_G	π_C	–
F84[a] (5)	A	–	$(1+\kappa/\pi_R)\pi_G$	π_C	π_T
combination of properties of K80	G	$(1+\kappa/\pi_R)\pi_A$	–	π_C	π_T
and F81 models (parameter κ has	C	π_A	π_G	–	$(1+\kappa/\pi_Y)\pi_T$
different meaning)	T	π_A	π_G	$(1+\kappa/\pi_Y)\pi_C$	–
HKY85 (6)	A	–	$\kappa\pi_G$	π_C	π_T
combination of properties of K80	G	$\kappa\pi_A$	–	π_C	π_T
and F81 models (more natural	C	π_A	π_G	–	$\kappa\pi_T$
formulation)	T	π_A	π_G	$\kappa\pi_C$	–
TN93 (7)	A	–	$\kappa_1\pi_G$	π_C	π_T
similar to HKY85 but with two	G	$\kappa_1\pi_A$	–	π_C	π_T
different transition/transversion	C	π_A	π_G	–	$\kappa_2\pi_T$
ratios	T	π_A	π_G	$\kappa_2\pi_C$	–

[a] Originally used by Dr Joseph Felsenstein in programs in the *PHYLIP* package, from 1984. $\pi_R = \pi_A + \pi_G$; $\pi_Y = \pi_C + \pi_T$.

2.3.2 Amino acid replacement models

Models operating at the level of amino acid replacement are much less developed, probably because of the relative ease with which DNA sequences can now be obtained. The general form of such models would be as for DNA, with continuous time Markov chain models having a 20×20 rate matrix (since there are 20 amino acids). Two such models are described in refs 8–10; see also Section 4.1.

2.3.3 Other models

Markov models incorporate a meaningful concept of time-dependence of substitution. Other models have in the past been proposed that directly fix probabilities of change. These lose the interpretation of relating to an evolutionary process that is in some sense uniform over evolutionary time, and some also discard dependence on time. This seems undesirable, given a belief in some underlying randomness in the accumulation of substitutions (note that this does not mean randomness in their meaning or in the evolutionary forces creating them, but only in their visible pattern).

2.3.4 Rate variation across branches

It is possible to model overall substitution rates as constant over all branches of a tree, or potentially different in each branch. The restriction that they are constant corresponds to the assumption of a molecular clock (11). If a phylogenetic estimation method assumes the existence of a molecular clock, and assuming all the sequences studied are contemporary, then it is possible to find the point in the estimated tree corresponding to the root. This is another way to estimate rooted trees, suitable for sequences for which the molecular clock assumption is justified (see Section 7.2.2). For such rooted trees, where all divergences can be placed on a single time scale, the entire tree may be 'calibrated' in terms of real time if any one point in the tree can be dated, e.g. using palaeontological data (12).

2.3.5 Rate variation across sites

Generally, phylogenetic methods assume the same overall substitution rate at all sites of DNA sequences. It is becoming increasingly popular to allow for variation across sites by modelling the distribution of rates across sites using a gamma distribution (13, 10): the substitution rate at each site is assumed to be a random variable drawn from a gamma distribution. Models incorporating this assumption are indicated by '$+\Gamma$', e.g. JC69$+\Gamma$ is used to denote the JC69 model (*Table 1*) with additionally the gamma distribution description of rate variation across nucleotide sites.

2.4 Estimation

As with other statistical applications, different methods of estimation may be applied. The method of maximum likelihood is one which seeks to maximize the ability of the model to explain observed data. The method of maximum likelihood has the advantage of a huge statistical literature of supporting results, methods and theorems, and few disadvantages in general. It is applicable to the problems of phylogenetic estimation, and methods based upon its use are described in Section 3.1.

Maximum likelihood methods try to fit models to data using the maximization of probability. Other methods generally try to optimize some other measure of fit between model and data, for instance by minimizing a deviation between some expected outcomes and their observed values. These methods are also used in phylogenetic estimation, and are described in Section 4.

An important question is that of what we can expect to be able to estimate. Clearly we want to estimate a phylogenetic tree, and this is possible. Methods using substitution models which include time-dependence permit, in addition, the estimation of branch lengths, in terms of numbers of base substitutions or amino acid replacements per site. In addition, certain parameters of the models used can be estimated by some methods, either during tree estimation or else by separate procedures, and this can give interesting insights into processes of evolution. Some methods incorporate estimation of more information, and this can cause undesirable effects. In Section 3.2 below, I describe how parsimony analyses estimate almost as much information as is originally present in the data, and not surprisingly a method which tries to estimate as much information as it is given as data has serious drawbacks (Section 7.2.1).

2.5 Heuristics

Finally, we must consider a problem involved in writing algorithms or computer programs to perform phylogenetic analyses. A fundamental difficulty in phylogenetic analysis is the number of possible phylogenetic trees that exist for even a small number of sequences. For example, there are 39 208 unrooted trees for eight sequences; for 12 sequences, this rises to more than 6.9×10^9 trees[*] (14). In practice, it is not possible to evaluate every possible tree and so the search for the best tree (according to whatever criteria) often cannot be complete. Even with clever searching methods, known as **heuristics**, the general problem is that the best tree found may not be the actual best one. In one sense, this is simply a problem for mathematicians or programmers. However, the effects cannot always be ignored. In the following sections regarding specific estimation methods, I point out the problems this introduces and make suggestions as to what can be done about it.

3. Phylogenetic estimation methods based on sequences

3.1 Maximum likelihood methods

Given aligned molecular sequences and a proposed phylogenetic tree, branch lengths and Markov chain model as in Section 2.3, it is possible (4) to calculate the likelihood of the data, i.e. the probability of observing those data, given the model, parameter values etc. (*never* to be confused with the probability of the parameter values being correct given the data). The larger the likelihood is, the better an explanation of the data the model gives. The method of maximum likelihood is to vary all the unknown parameters (tree, branch lengths, etc.) until the highest possible likelihood is attained, and to take these values as parameter estimates (15).

---

[*] 'The principal uses of these numbers will be. . . to frighten taxonomists' (14, p. 27).

The models currently used in maximum likelihood phylogenetic methods are not designed to use information held in patterns of gaps in sequence alignments, and I recommend that positions of an alignment where any sequence has a gap are removed prior to phylogenetic analysis.

Unlike traditional statistical problems, where it is often possible to write down formulae for maximum likelihood parameter estimates that avoid the need ever to use the actual likelihood value, this is not the case in phylogenetic estimation (due in part to the large number of different tree topologies that must be considered). In practice, computer programs make a guess at a likely solution and subsequently vary all the parameters, recording likelihood values and noting the best achieved so far. These are taken as starting points for subsequent searching, until the program can find no improvements. This will generally be a *local* maximum of the likelihood (no small change to any parameter can increase the likelihood), but sadly there is no guarantee that it is the *global* maximum (no changes to any parameters can increase the likelihood).

Maximum likelihood methods have the advantage of a large body of statistical theory behind them (15), making it easier to devise tests and interpret results. They operate on the complete sequence alignment, considering equally every position of every sequence in each likelihood calculation. This makes the best possible use of the information in the sequences, and the method is only restricted by the accuracy of the assumptions it makes and the base substitution or amino acid replacement model it uses (and by the difficulty of finding the globally optimal parameter values). Of course, there is a trade-off, and that is in speed. Implementations differ, particularly in the complexity of models used and the speed of computers, but it is rarely possible to make a detailed study of more than 20 sequences at once. Individual trees can be assessed for 50 or more sequences, but due to the complexity of the phylogenetic estimation problem and the difficulty of finding the global optimum such individual trees are not generally adequate for a reliable understanding of the phylogeny of the sequences.

Because of the need for maximum likelihood methods explicitly to compare likelihood values for many different parameter value combinations, these methods necessarily make very clear the model that is being used. The likelihood value gives a readily available score, with the advantage that it is easy to compare results from different runs, different parameter values etc., in an entirely objective manner. In practice, likelihood values are very small (they are the probability of obtaining the observed data, out of the set of all possible outcomes) and their natural logarithms are used. This can cause some confusion as the logarithms are negative numbers, and we must recall that (e.g.) -1000 is greater than (thus 'better' than) -2000. Because of their many advantages, maximum likelihood methods are the methods of choice whenever they are feasible.

3.1.1 *DNAML* and *DNAMLK*

Two widely available and reliable computer programs that implement the maximum likelihood method for phylogenetic estimation from DNA sequences are the *DNAML* (DNA maximum likelihood) and *DNAMLK* programs (A1) written, maintained and distributed by Dr Joseph Felsenstein. They are general purpose programs which perform maximum likelihood analyses under the substitution model called F84 in Section 2.3 (*Table 1*). Option choices (see below) permit the model to be made equivalent to JC69, K80, and F81, and the programs will perform searches among all possible trees to attempt to find the best estimate of tree and branch lengths. In addition, they can also assess trees of the user's own choice.

The difference between the two programs is that *DNAMLK* assumes a molecular clock. As described in Section 2.3.4, this means that rooted trees are estimated and calibration against a real time scale may be possible. The choice between *DNAML* and *DNAMLK* rests upon whether or not the molecular clock assumption is suitable for the data at hand (see Sections 5 and 7.2.2).

Both *DNAML* and *DNAMLK* permit the user to force the transition/transversion parameter κ of the F84 model to take any value: choosing a value of 0 has the effect of nullifying the parameter, and the model becomes identical to F81. The programs use a fixed value of this parameter (supplied by the user), and do not estimate the value from the data. It is, however, possible to perform this estimation 'by hand', by performing multiple runs with different values and selecting the result with the highest likelihood. Base frequencies may be estimated using averages from the sequences, or can be fixed by the user; there is rarely any reason to do the latter, except to force them all to equal 1/4, in which case F84 becomes K80 (F81 becomes JC69)—this is occasionally useful for comparison with other methods.

Both *DNAML* and *DNAMLK* treat sequence sites where there are gaps in a non-standard manner. The method used is akin to treating each gap as though it were an undetermined nucleotide, but there is no evidence that this is biologically meaningful or gives good results and I recommend that any site with gaps be removed before analysis.

To improve the ability of these programs to find the global maximum likelihood, two options are available. Selecting 'global rearrangements' causes an extended search algorithm to be used, and this is always desirable. There is still no guarantee of finding the best solution and the second option, of selecting a randomized order of input of sequences to the algorithm, allows different local solutions to be found in repeated program runs. This option should also generally be used, especially with larger datasets (more than eight or ten sequences), until the user has confidence that no improved solutions can be found (or at least until they have a feeling for which parts of the phylogenetic tree are reasonably constant, and which are variable, among the best found). Different solutions from different runs are correctly compared simply by comparing their likelihood values: higher likelihoods are always better, and the presentation of a solution of lower likelihood simply represents the inability of the search algorithm always to find the best solution.

DNAML and *DNAMLK* also have an option whereby different sites in the sequences can be taken to have different overall rates of substitution. In principle, this can be used to differentiate between codon positions, between introns and exons, etc. In practice, since the programs require the user to preselect the positions and relative rates of substitution, it is difficult to decide what values to use. Consequently, this option should be considered experimental and should be used with great caution.

3.1.2 *BASEML* and *BASEMLG*

Dr Ziheng Yang has written the programs *BASEML* and *BASEMLG* (A4) for maximum likelihood analysis of aligned DNA sequences. The programs use a variety of substitution models, will search among trees to find estimates of tree and branch lengths or assess trees of the user's choice, and can incorporate a molecular clock (Section 2.3.4) and a gamma distribution of substitution rates across sites (Section 2.3.5). *BASEML* and *BASEMLG* are able to calculate maximum likelihood estimates of parameters of substitution models (e.g. κ of the HKY85 model). As with *DNAML* and *DNAMLK* (Section 3.1.1), choice of substitution model should be determined by suitability for the data at hand (see Section 7.2.2). *BASEML* and *BASEMLG* have not yet been widely tested, but appear reliable and are highly recommended as containing some of the most advanced methods for substitution model fitting and phylogenetic tree estimation.

3.1.3 *PAUP**

Dr David Swofford's *PAUP** program (A3; see also Section 3.2.1), scheduled for release in summer 1996, will incorporate analyses of DNA sequences using maximum likelihood techniques and a variety of DNA substitution models.

3.1.4 *PROTML*

Dr Masami Hasegawa and colleagues have devised and distribute the *PROTML* (protein maximum likelihood) program for analysis of aligned amino acid sequences (A5). The program uses the amino acid Markov chain model described in refs 8, 9 and is otherwise similar to *DNAML*. No molecular clock is assumed and, as with *DNAML* and *DNAMLK*, positions in an alignment where any sequence has a gap should be excluded from the data before analysis. The same cautions also apply regarding the use of multiple program runs to make it more likely that the global maximum likelihood and corresponding estimates are found.

3.2 Parsimony methods

Phylogenetic estimation methods based on the 'principle of parsimony' have a controversial past. The use of the word parsimony comes from their basic criterion for estimation, that the preferred estimate of tree is the one which requires the smallest number of base substitutions or amino acid replacements to have occurred (16). Defences of this criterion have in the past crossed from the belief that estimation can be performed with no models into abstract philosophy, both of which are untenable. 'Parsimony', in the sense of Occam's razor, may be suitable for choosing mathematical models of evolution, but it is unlikely that evolution has followed this principle.

The soundest justifications for the use of parsimony-based methods now rely on their being understood as unusual maximum likelihood methods, or as approximations to the standard maximum likelihood methods (17). The former relies on the use of an implausible model of DNA substitution which assumes that the probability of observing a substitution is independent of time. This derivation also explains how parsimony estimates the sequences at each branching point of a tree and that these estimates can seriously affect the quality of estimates of other parameters as well as being inaccurate themselves. The latter justification is shown by long experience to be true, but the accuracy of the approximation is not measurable in any general way. This means that the interpretation of parsimony analysis results is difficult, in terms of what the likely sources of error are and how serious the errors might be. Many examples have now accumulated in which it is guaranteed that parsimony methods will produce wrong answers (see Section 7.2.1).

Parsimony methods can be applied to alignments of DNA sequences or of amino acid sequences. The parsimony criterion can only generate unrooted trees; outgroup rooting methods are widely used to solve this problem.

It is common practice in parsimony analyses to treat each 'site' in a gap in an alignment as though it represented a single unknown nucleotide or amino acid. An adaptation of the parsimony criterion then permits the assignment of a base/amino acid in place of each gap site in such a way as to minimize the number of hypothesized base substitutions or amino acid replacements. There seems to be neither biological nor theoretical justification for this, and the practice cannot be recommended. As with maximum likelihood analyses, such positions should be excluded from parsimony analyses.

Despite all these drawbacks, parsimony methods remain the most popular in terms of frequency of publication (18). This can be attributed to two main factors: the widespread availability of 'friendly' computer programs which implement the methods and the speed of parsimony algorithms. Although these face the usual problem of having an enormous number of trees to consider, the assessment of each one (the calculation of the tree's score, or minimum required number of substitutions) is very rapid in comparison to a likelihood calculation. This means that much larger problems can be considered—up to 50 sequences cause little trouble, and analyses of over 100 sequences are not uncommon. These advantages have been seen to outweigh the theoretical disadvantages outlined above. This is perhaps understandable for large alignments that cannot be analysed by maximum likelihood methods, but the parsimony methodology does not permit any identification of which results will be most reliable and which least reliable. Parsimony analyses might best be considered a form of exploratory data analysis, whereas maximum likelihood methods give a full statistical phylogenetic analysis of molecular sequence data. There is no excuse for the general use of parsimony-based methods. On the other hand, many journals and their referees still expect to see parsimony analysis of sequence alignments and so in a *Practical Approach* book it is necessary to describe these methods briefly.

3.2.1 *PAUP*

Dr David Swofford's *PAUP* (phylogenetic analysis using parsimony) program (A3) performs virtually any parsimony-based phylogenetic analysis, of aligned sequences of either DNA or amino acids. The simplest analyses permit every possible tree to be evaluated according to the parsimony criterion, with each required base substitution (DNA sequences) or amino acid replacement (amino acid sequences) counting for one unit of 'length' and the 'shortest' tree being considered the best. In this simplest case every evolutionary change (substitution or replacement) is considered equal, but it is also possible to give various weights to different changes. An example of the use of this option is to increase the weight given to DNA transversion substitutions, a crude way of accounting for unequal transition/transversion rates. (Incidentally, it is hard to understand how this practice can be reconciled with the belief that parsimony embodies no model of evolutionary processes for if this were the case, why should any substitution be weighted differently from any other?) Similarly, sequence positions can be given different weights if for any reason they are believed to hold more information about the phylogeny. In both cases, the user has to supply the weights and to decide where they are applied, making the options' use highly subjective. These options should only be used when there is genuine extrinsic evidence that they are justified and that suggests reasonable values for the weights—rarely the case.

PAUP contains some of the most sophisticated algorithms for finding the most parsimonious tree(s) for a dataset. For small numbers of sequences (up to about 15), it is possible to consider every possible tree, or to use an algorithm known as 'branch and bound', to guarantee finding the most parsimonious tree. One or other of these options should always be used if feasible. Failing this, it is best to perform multiple program runs, using a number of different initial trees to initiate the search algorithm (*PAUP* will generate random starting trees if required) and using the most powerful search strategy ('tree bisection-reconnection'). There is no excuse for neglecting these options.

After finding the best tree or trees, *PAUP* has extensive facilities for the manipulation and output of results. Many of these are designed to cope with the regular discovery of multiple equally good (though often dissimilar) trees, and are designed to reduce a large number of incompatible trees to some form of average result. In addition, output options are designed to deal with the fact that the method produces, as well as estimates of the tree itself, estimates of the sequences at each node of each tree. This is a large amount of information and there are many ways to display the estimated sequences, lists of where on the tree substitutions are estimated to have occurred, etc. Because of the uncertain statistical status of these estimates, the accuracy of conclusions drawn from them is entirely unassessable.

3.2.2 Other programs implementing parsimony

A number of other computer programs implement parsimony analyses of DNA and amino acid sequences. These include *DNAPARS*, *DNAPENNY* and *PROTPARS* from the *PHYLIP* package (A1), *MEGA* (A2), and *HENNIG86* (whose author, Dr J. S. Farris, could not be contacted at the time of writing). The implementations are basically the same as in *PAUP*, which is so powerful as to encompass virtually all the options they contain. This means that the only advantage in using a variety of parsimony programs is in their different abilities in heuristic searches. It is not known which programs are best in this respect—probably each has strengths and weaknesses. *PAUP* and *HENNIG86* seem widely accepted; *MEGA* is a newer program and is less well tested.

4. Phylogenetic estimation methods based on distances

A different approach to phylogenetic estimation fits trees to sequences not directly, according to models of the evolution of sequences related by phylogenetic trees, but through an intermediate stage. This approach was devised in pre-computer times, when simultaneous comparison of more than two sequences was not feasible. The alternative procedure is to calculate evolutionary distances between pairs of the aligned sequences, the simplest meaningful comparisons possible, and then fit those distances to a phylogenetic tree. Each stage of these analyses, computation of distances and subsequent fitting of distances to trees, is much quicker than maximum likelihood or parsimony analyses, and this makes analysis of large datasets possible (and accounts for the popularity of distance-based methods). The inevitable drawback, of course, is in the quality of the results. It is certain that pairwise distances contain less information than the original sequences, and the statistical properties of distance matrix-based methods are not as good as those of maximum likelihood methods. However, maximum likelihood methods can be impractical for large datasets and for these cases it is useful to describe some distance matrix methods that may be encountered.

4.1 Sequence distances

Initially, the simplest measure of distance between pairs of aligned DNA or amino acid sequences was employed—the proportion of positions of the sequences where the nucleotides or amino acids differed. However, the proportion of positions showing differences is a poor indicator of the number of changes that have actually occurred: any site at which two or more changes have occurred will show at most one difference, and may show none at all if a change is subsequently reversed or if the same change occurs independently in two evolutionary lineages. The proportion of positions different is necessarily an underestimate of the true frequency of changes, particularly for more distantly related sequences (where multiple hits/reversals are more likely). Markov chain models were introduced, automatically allowing for multiple hits and reversals. The first such model was used by Jukes and Cantor (2), who published the formula for converting the observed proportion of differences between pairs of DNA sequences into the estimated number of substitutions actually occurring per site for the JC69 model. As more complex Markov chain models were developed for DNA sequences (*Table 1*), it has become possible to estimate evolutionary distances according to biologically more accurate models. Models, and corresponding distance estimators, for amino acid sequences are less well developed. The only widely available distance estimator uses the amino acid equivalent of JC69, i.e. the model assuming all replacements occur at equal rates and that all amino acids occur with equal frequency.

Markov chain models of sequence evolution have the advantage that the distances derived from them, and thus the branch lengths of trees estimated from these distances, are measured in units (e.g. of base substitutions per nucleotide site) that may be meaningfully related to the amounts of evolutionary change they represent. The same is not true of other measures that have been proposed (e.g. proportions). It is also worth noting that positions of alignments where sequences have gaps may still be of some use. For each comparison of a pair of sequences, every position where those sequences have no gaps can be included in the distance calculation, even if *other* sequences have gaps in those positions. This utilizes more of the information present in the sequences, when gaps are present. There seems to be no argument against this practice, which is therefore recommended.

The decision to be made by the researcher is the choice of distance measure used. For DNA sequences, distance measures derived from the models described in *Table 1* are the most generally available ones. The original literature contains the appropriate formulae, which are generally not difficult to incorporate into computer programs, and there are a number of readily available programs which will read sequence alignments and calculate distance matrices under a choice of models. Two of the best are *MEGA* and *DNADIST*.

4.1.1 *MEGA*

MEGA (10, A2) is a relatively new program, from the laboratory of Dr Masatoshi Nei. It performs various useful utility functions relating to phylogenetic estimation, and includes a number of distance measure calculations. For DNA sequences, distances derived from the JC69, K80, TN93, JC69+Γ, K80+Γ, and TN93+Γ models are available. The '+Γ' models require the user to supply an appropriate value of the shape parameter of the gamma distribution. For amino acid sequences, the only useful measure available is that derived from the 'all rates equal' model.

4.1.2 *DNADIST*

DNADIST is one of the programs in Felsenstein's *PHYLIP* package (A1). The JC69, K80, F84, and K80+Γ models are available for calculating matrices of pairwise distances from aligned DNA sequences. An important point is that the implementations of the K80 and K80+Γ models do not use the method of Kimura's original publication (3), which implicitly re-estimates the transition/transversion ratio parameter separately for every pair of sequences, but uses an alternative procedure which assumes a single fixed value of the parameter to apply to all sequences. In practice, the difference is generally small (if any reasonable value is specified), but the approach in *DNADIST* has the disadvantage that the user must somehow choose an appropriate value for the parameter. The implementation of the F84 model suffers the same drawback regarding the choice of a value for its transition/transversion ratio parameter, as does the K80+Γ model regarding the choice of its gamma distribution shape parameter.

4.2 Phylogenetic trees from distance matrices

Once a meaningful matrix of pairwise distances between molecular sequences is prepared, it is natural to try to fit a phylogenetic tree to these distances, and a number of methods to do so have been proposed. Some were borrowed from other scientific disciplines, with varying regard for their suitability to the phylogenetic estimation problem, whereas others were specifically designed to be appropriate for phylogenetics, with varying success. All are basically attempting to match the observed distances to a hypothesized tree and associated branch lengths; all those described below calculate some 'weighted least squares' score, either using the score as an explicit criterion of fit or as part of an algorithm that creates a tree. The 'explicit fit' approach has the definite advantage that results are assessable by an objective criterion, and different results using different program options can be compared with each other. The 'algorithmic' methods are very difficult to assess objectively, although some studies have shown them to perform well in terms of estimating the correct tree from simulated data with a known phylogeny.

Some methods can generate branch length estimates less than zero, but this is biologically meaningless and any method that cannot suppress such estimates should be viewed with suspicion.

I now describe some of the more popular methods, including mention of computer programs that implement them.

4.2.1 Fitch–Margoliash method

The method proposed by Fitch and Margoliash (19) uses a least squares fit of data to tree, weighted by the inverse of the squared observed distances. (This weighting is suitable for the situation when the variances of distance estimates are proportional to the square of the estimated distances, as is generally the case for molecular sequence distances.) This criterion allows the assessment of any candidate tree simply by comparing the expected distances, calculated by measuring along the tree between pairs of sequences, with the observed distances. Methods for searching the set of all possible trees are, as usual, appropriate for finding the best estimate of phylogeny; the usual problems apply when the number of sequences is large, but in practice are less severe than with maximum likelihood methods since the Fitch–Margoliash (FM) method is so much faster at evaluating each tree. The FM method permits the use of (unrooted) trees that do not assume a molecular clock, or the trees considered as candidates can be restricted to those which conform to a molecular clock (giving rooted trees, with the possibility of calibration; Section 2.3.4). The best fit is certain to be no worse in the former case; it is not clear how the difference in fit, which might give a measure of the suitability of a molecular clock assumption for a given distance matrix, can be assessed.

The FM method is implemented in the *FITCH* and *KITSCH* programs of Felsenstein's *PHYLIP* package (A1). *FITCH* assumes no molecular clock, whereas *KITSCH* restricts the trees considered to those which conform to a molecular clock. As with the *DNAML* and *DNAMLK* programs (Section 3.1.1), both *FITCH* and *KITSCH* include the 'global rearrangement' option which improves the tree searching strategy and this option, as well as multiple program runs with different sequence input orders, should be used whenever possible. In addition, it is possible to forbid branch lengths from being assigned values less than zero, and this option is also highly recommended in order to restrict the searches to biologically meaningful trees.

FITCH and *KITSCH* also permit the use of weighting schemes other than 'inverse square distance', but this is not recommended for DNA or amino acid sequence distances.

4.2.2 Neighbor-joining and UPGMA methods

These methods use a different weighting of the least squares criterion, and are defined simply as algorithms to produce trees and not in terms of a criterion of goodness of fit. UPGMA (20) assumes a molecular clock and neighbor-joining (NJ) assumes there is no molecular clock (21). In each case, a procedure is defined in which a tree is built from an initial starting point with the decision of how to modify the tree at every stage defined by the current state of the tree and a computed criterion that uses an unweighted least squares criterion. When the final tree is generated, no measure of the 'goodness of fit' between tree and data is available. Because of their algorithmic (rather than search based) nature, UPGMA and NJ are exceedingly fast. Simulation studies have shown NJ to perform well compared to parsimony methods.

However, there are definite drawbacks with the algorithmic approach. The most serious is that there is no way to compare the resulting trees from different runs, as no goodness of fit criterion is defined. Repeat runs, with the order of the sequences in the distance matrix changed, can give different estimated trees and whereas with maximum likelihood, parsimony and FM methods there is an absolute criterion which can be used to decide objectively which of a number of trees is a better estimate, this is impossible with NJ or UPGMA. In addition, the algorithms do not eliminate the possibility of branches being assigned negative lengths and this often leads to estimated trees with incomprehensible biological properties.

The NJ and UPGMA methods of tree estimation are available in the *MEGA* program (A2), and in the *NEIGHBOR* program of the *PHYLIP* package (A1).

5. Comparison of methods

Now that the main phylogenetic estimation methods have been described, together with details of their more widely available computer implementations, it is necessary to make comparisons and give recommendations on their use. Particularly in recent years, there have been a number of studies comparing various methods by measuring their performance when analysing simulated data, for which the 'true' answer is known. I am in agreement with Hillis *et al.* (22), that it is very difficult to draw general conclusions from the results of these highly specific studies and that many are performed by researchers with a particular axe to grind. It should be noted that the conclusions of such papers are rarely at variance with their authors' previous research interests. I encourage the reader to think very carefully about what such papers have shown, and not just to accept sweeping generalizations from the papers' abstracts. Perhaps this should be done before accepting the following recommendations, which are based on my own experience and on my reading of the literature.

As noted above, DNA sequences must contain more information than amino acid sequences and phylogenetic estimation methods based on DNA are generally better developed than for amino acids. Consequently, I recommend the use of DNA sequences whenever the choice exists.

Maximum likelihood is without doubt the method of first choice for phylogenetic estimation. It is the only method which can combine the best models of DNA substitution or amino acid replacement, simultaneous analysis of every position of every sequence, an exceptionally well justified and well tested estimation method, and statistical testing of the estimates made (see Section 7). It permits well known, objective tests comparing different sets of assumptions to see which are most accurate and most likely to give good results, and is additionally of interest to those researching molecular evolution, and not simply phylogeny, as explicit models describing evolutionary processes in molecular sequences can be optimized to provide the best possible fit of model to data, and these models can be tested to see if they provide a statistically acceptable description.

There is no guarantee that the maximum likelihood method will necessarily produce better estimates of phylogenetic trees. Nevertheless, virtually all the theoretical and simulation studies that have compared maximum likelihood methods to others have concluded that there are almost no circumstances in which maximum likelihood methods perform significantly worse, and there are many highly plausible circumstances in which they perform significantly better (23–25). Consequently, if maximum likelihood analysis is possible, it should always be used.

Studies of the accuracy of different models of DNA substitution generally show that the most complex models available are the most accurate (Section 7.2.2), and lead to better phylogenetic estimates. In the past, choice of models was something of a stab in the dark. Most researchers, intuitively feeling that DNA substitution is a complex process, chose to use the most complex models available to them, and recent results (26) have shown that this was the right decision—simpler models are almost invariably seen to be significantly inferior to the more complex models, which should be used whenever possible. Similarly, the more general model permitting different overall rates of evolution in different lineages (i.e. no molecular clock) can be used unless there is particular evidence indicating a reliable molecular clock for a specific DNA alignment (see also Section 7.2.2). It seems likely that the same rules should apply to amino acid data.

As would be expected for an analysis method that is notably superior in many respects, there is a penalty to be paid. This is the time taken to perform analyses; maximum likelihood methods are very slow. On many workstation or mainframe computer systems, the practical limit to the number of sequences that can be analysed is probably in the region of 30–40. The number that can be rigorously analysed, with satisfactory use of multiple program runs to test the success of search algorithms and to compare different models' assumptions and parameter values, is likely to be half this. These numbers are not adequate for the studies that many researchers wish to undertake. There are few solutions to this problem, other than to reduce the number of sequences studied. Is it necessary to analyse all the sequences together, or is there some natural way to break the total number into smaller groups? Although there is no statistically validated way to combine two or more sub-trees, perhaps it will be acceptable to use a number of representative sequences to estimate the overall structure of a phylogenetic tree, and then analyse particular groups of sequences to see the fine detail in different regions of that tree.

If this is not possible or acceptable, another method must be used. My recommendation then is to use the Fitch and Margoliash distance matrix based method. This has fewer drawbacks than other distance methods—there is a strictly defined and objective criterion of fit between model and tree, and unlike some other distance methods it is possible to use versions of the method which prohibit biologically meaningless negative branch lengths. When thought suitable, it is also possible to restrict the candidate phylogenetic trees to those conforming to a molecular clock. Results from simulation studies show the FM method to be successful in a range of realistic circumstances (25). The distances used should again be derived from the most complex models. In particular, the F84, HKY85, TN93, or TN93+Γ models are recommended for DNA sequences.

If speed of program execution is still a problem, the neighbor-joining distance matrix based method may be used. Despite not having a measure of fit of data to tree that is comparable from one program run to another, the method has been found to produce reasonable results under a range of conditions (25, 27). As with the Fitch–Margoliash method, emphasis should be placed on using the most accurate distance measures possible. The UPGMA method is now generally dismissed as inadequate (28).

I do not recommend parsimony methods for any phylogenetic estimation problem. Others (including journal editors and referees) may not see this the same way. In particular, it is often recommended that different methods be used in the same study, for 'comparison of results'. It is not clear how the comparison of results from a statistically better method (maximum likelihood) with those from a less well justified and well understood method (distance or parsimony based) can clarify matters, but nevertheless we are all forced to do this at one time or another.

6. Other phylogenetic estimation methods

Various other methods of phylogenetic estimation are seen occasionally in the literature. Some of these are mentioned briefly here, to allow the researcher a certain familiarity with a few of the remaining most common terms likely to be encountered. These methods are not so widely accepted as those described in Sections 3 and 4. None of the methods below is recommended, being at best of unproven or untested value, other than as experimental techniques potentially interesting to theoreticians.

6.1 Lake's method of invariants

This method is perhaps the simplest to calculate, and barely requires a computer at all (immediately making one suspicious of its efficacy). Dr James Lake proposed that under a fairly general model for DNA sequence evolution, the expected numbers of certain patterns of nucleotides at sites of a four sequence alignment would be related in particular ways if certain trees were true (29). The method applies to unrooted trees, of which there are just three for four sequences. Lake proposed three simple formulae ('invariants') combining the numbers of certain site patterns; one formula corresponded to each tree, and depending on which formulae gave values equal, or not equal, to zero for a given sequence alignment, the tree corresponding to that formula was selected.

A more readable account of Lake's method of invariants is given by Cavender (30). Initially, there was great interest in Lake's method because of its simplicity. Subsequent theoretical study has shown the method to have very poor statistical powers (28), quite apart from the fact that it was only suitable for analysis of four sequences. After a flurry of interest the method is effectively forgotten as a serious method for estimating phylogeny. For historical interest, I note that programs implementing Lake's method include *PAUP* (A3) and the *DNAINVAR* program of *PHYLIP* (A1).

6.2 Hein's method of simultaneous alignment and phylogenetic tree estimation

It has been noted for some time that information is contained in the patterns of shared insertions and deletions visible in a multiple sequence alignment. All the methods discussed so far in this chapter are unable to utilize this information. Dr Jotun Hein devised an analysis in which DNA of amino acid sequence alignments and trees are estimated simultaneously, each contributing to the other. Initial pairwise alignments are used to generate a distance matrix and thence a tree; then that tree is used to create a multiple sequence alignment (31). The whole process may be iterated, and the procedure of alternating between finding trees and modifying the sequence alignment continued until no improvements are evident, at which point an estimated tree and alignment are available.

This method, embodied in Hein's *TREEALIGN* program (A6), has not become widespread. The reason probably is that it incorporates none of the most acceptable alignment methods, distance measures nor tree estimation methods. It does have certain appealing features, in particular the realization that insertion/deletion events are evolutionary processes too, giving information that ideally would be used to improve our ability to estimate phylogeny.

6.3 Minimum message length coding

This is a statistical estimation method in which the model and parameters permitting the most compact recoding of data, with no loss of information, are taken as estimates. The recoding never actually occurs; it is notional, and simply provides the criterion for inference. The method is very closely related to maximum likelihood methods. Minimum message length coding methods have been applied to phylogenetic estimation by Dr Lloyd Allison and colleagues (32), who make available some of their programs (A7). The method is not in the mainstream of phylogenetic estimation, but its highly general formulation allows a consistent approach to be made to a number of molecular sequence analysis problems.

7. Measuring uncertainty

Molecular evolution is highly complex. After many years of debate on the processes controlling it, there is still no agreement even on what are the most important factors—as an example, take the fundamental debate about the relative contributions of random drift and natural selection (11). But there is at least agreement that molecular evolution is not deterministic; that is, it is not wholly determined by past history and equations which, if we but knew and could solve them, would tell us the future. As a consequence, we expect to analyse molecular genetic data using probabilistic models and, even if our models are highly accurate, we expect to observe stochastic fluctuations in our data inexplicable other than as the chance outcomes of unrepeatable circumstances.

In common with all other scientific disciplines, then, phylogenetic estimation requires us to deal with data which include 'errors', i.e. pieces of data which, simply by chance, do not fit our expectations perfectly. This is why the choice of method for phylogenetic estimation is so important. If all data were 'perfect', there is little doubt that all the methods described in this chapter (with the possible exclusion of parsimony analyses: see Section 7.2.1) would produce the correct answer time after time. The 'quality' of a method is really just its ability to cope with stochastic fluctuation and with discrepancies between real data and assumed models.

An important consequence is that it would be useful to be able to give an indication of the accuracy of results, as well as presenting the results themselves. Of course, we cannot know the true accuracy without knowing the true answer, so in practice we are looking to give a measure of estimated accuracy. The arrested development of phylogenetic estimation methods meant that this question was not even posed for a long time. Theoretical advances, increasingly good empirical studies and recognition of good practice have now brought the question to the fore and various techniques for estimating likely errors have been proposed.

Sadly, the complexity of the phylogenetic estimation problem means that few of these techniques are reliable or even do what they are intended to, but some are useful and many should at least be recognized and understood by the phylogenetic researcher, if only to know that their use is not appropriate to a particular problem. Section 7.1 describes a number of techniques designed to estimate measures of uncertainty in results due to statistical fluctuations, and Section 7.2 describes the detection and effects of systematic differences between real data and the assumptions made by phylogenetic estimation methods.

7.1 Statistical fluctuation

7.1.1 Bootstrap analyses

The 'bootstrap' is a statistical technique which uses computing power to make statistical estimates when mathematical analysis cannot provide suitable formulae. Its major use is in providing error estimates in complex problems. This would seem to be precisely what is required in phylogenetic estimation, to which it was introduced by Dr Joseph Felsenstein (33).

The bootstrap works by creating new datasets from the original data. A bootstrap data set is created by repeatedly drawing items of data (for our purposes, sites of a DNA or amino acid sequence alignment) at random (with replacement) from the original data until a data set the same size as the original one is produced. Each site of the original alignment may appear once, twice, or more in a bootstrap data set, or it may not appear at all. Bootstrap data sets are analysed exactly as though each were the original; the procedure is repeated many times (typically in the low hundreds, but see Section 8), and the distribution of bootstrap results gives an estimate of the variability of the original result.

The bootstrap is often used in phylogenetics to attempt to indicate levels of confidence in the groups of sequences in an estimated tree (a group meaning the set of sequences descended from a particular internal node of the tree). Every group of the estimated tree is searched for in each bootstrap data set tree, and the numbers found of each are recorded. It is then possible to indicate at each node of the estimated tree the proportion of bootstrap samples in which these groups are reproduced. These numbers are commonly taken as indications of the confidence we might have in the tree, proportions near to 1 suggesting high confidence that the group is a true one, and proportions near to 0 indicating the opposite.

There are a number of problems with this interpretation. Even taking these proportions at face value they are difficult to interpret statistically. The calculation of a large number of confidence estimates (here, one for each node of the tree) is unlikely to give reliable results as each proportion is calculated independently of all others, while we wish to interpret the tree as a whole. If two groups are each estimated by a bootstrap to have a 95% chance of being correct, and assuming them to be independent, then the chance that both are correct is $(95/100)^2$, or 90.25%; as more and more groups are considered, these probabilities decrease until they are virtually meaningless. A much more meaningful procedure, rarely followed, would be to preselect a single group of interest (33).

An additional problem with the bootstrap is that it confounds the errors introduced by stochastic fluctuations in the data and those introduced by the phylogenetic estimation method used. Different ways of estimating trees can (and often do) give different results for the same data. If a method is used in a situation where it is not appropriate, and has a tendency to produce misleading results (see also Section 7.2), then a bootstrap analysis will simply confirm these errors: the bootstrap replicates will repeatedly re-create the errors of the original analysis and suggest high confidence in the inaccurate estimates.

This illustrates that the bootstrap does not estimate the accuracy of the analysis, but (at best) only the variability inherent in the data. The bootstrap is not capable of indicating whether or not the method of analysis is inappropriate or is producing inaccurate results. This makes any interpretation of its results even more difficult. Nevertheless, some journals seem to require bootstrap analysis in the mistaken belief that it is a general purpose measure of the quality of phylogenetic estimates. Bootstrap analyses can be performed within the *PAUP* (parsimony) and *MEGA* (distance matrix) programs, and using the *SEQBOOT* program of the *PHYLIP* package in conjunction with other phylogenetic estimation programs in that package.

7.1.2 Standard errors of branch lengths

Phylogenetic trees estimated by maximum likelihood or distance matrix based methods have lengths assigned to each branch, representing the estimated numbers of substitutions per site over the period of time represented by that branch. When those branch lengths have been estimated as part of a maximum likelihood analysis, it is also possible to obtain estimates of the standard errors of the branch lengths. This comes from a general result of maximum likelihood theory which relates the standard errors to the curvature of the likelihood function at its maximum. The resulting estimates are generally underestimates of the true standard error, and are valid only in the vaguely defined case of there being 'enough' data, but in practice in other statistical applications where estimates are of a single parameter value they have been found to be accurate and useful.

Problems in the application of this method to phylogenetic estimation are, first, that the existence of a branch depends on having already estimated the tree correctly, and rarely will we know if this is the case (34) and, secondly, that to estimate simultaneously a standard error for each branch leads to problems of interpretation of many independent error measures, as described for the bootstrap in Section 7.1.1. Although the *DNAML* program of the

PHYLIP package will calculate standard errors of branch lengths in this way, they should not be taken too seriously.

An analogous method for estimating the standard errors of branch lengths of trees estimated by the neighbor-joining method is described and implemented in the *MEGA* program (10). This implementation suffers from the drawback of switching from the algorithmic neighbor-joining methodology to an ordinary least squares methodology part way through the analysis, making its statistical properties even more obscure.

If an estimated phylogenetic tree has some very small branches there is a tendency to suspect that these may be more likely to be 'wrong', perhaps because it feels like a smaller change to rearrange a tree at a small branch than at a large one. It has been suggested that if analysis of standard errors indicates a branch to be significantly different from zero then it is real, whereas if it is not significantly different from zero it may be considered somehow of uncertain reality. This has led to further confusion and bogus testing of hypotheses of evolutionary relationship (34). Concentration on standard errors in the cases where branch lengths are near zero is additionally unfortunate; quite apart from the problems described above, this is precisely the case in which the theoretical assumptions leading to the error estimates are most inaccurate. There is in fact no evidence that a branch of small estimated length indicates anything other than an estimate of a small amount of evolution (26).

7.1.3 Standard errors of other parameters

The same standard error estimation derived from maximum likelihood theory is applicable to the more conventional parameters of models of DNA substitution, for example the parameter κ controlling the relative rates of transition and transversion substitutions in the K80 and HKY85 models. Error estimates appear to be fairly reliable in these cases (26). The only available programs incorporating maximum likelihood estimation and which estimate parameters of these more complex models are the *BASEML* and *BASEMLG* programs (A4). The *DNAML* and *DNAMLK* programs of the *PHYLIP* package do not perform estimation of such parameters but instead require the user to provide (fixed) values. It is possible to use multiple runs to estimate standard errors, based on numerical analysis of the likelihood values for parameter values near their maximum likelihood estimator, but the details are beyond the scope of this chapter.

7.1.4 Statistical testing of different trees

Dr Hirohisa Kishino and colleagues have proposed methods (8), again based on maximum likelihood analyses, for testing whether one tree (generally the maximum likelihood estimate) is significantly better than another (generally the tree with second highest likelihood, or some other particular tree of interest). Although the underlying theory for these tests is sound, problems arise with the estimation of the distribution against which the test statistic is compared to perform the test. Kishino and colleagues use approximations of uncertain validity; strictly they apply only to the case that the trees being compared are chosen according to extrinsic criteria, and not to the case of comparing the maximum likelihood tree with the tree of second highest likelihood. The criterion for selecting the trees that are to be compared is too closely related to the methods for testing the trees for the latter not to be dependent on the former. However, the methods appear to perform well in some cases (35). The *DNAML* program of the *PHYLIP* package has an option that supports these tests.

7.2 Systematic errors

The discussion above considers stochastic fluctuations which ensure that the data we want to analyse will not be in perfect accord with expectations. In phylogenetic estimation there is another source of error to consider; that introduced by the use of inadequate estimation techniques. There are various reasons for such inadequacy occurring, some unavoidable and others which can be avoided by suitable choice of methods.

7.2.1 Statistical consistency

A statistical estimation method is said to be 'consistent' if, given sufficient data conforming to the model that the method assumes, it is certain to give a correct estimate. This highly desirable property is not shared by all phylogenetic estimation methods. If a method is not consistent there is always the possibility that, no matter how many data are collected, estimates will be incorrect—often with increasing certainty as the amount of data rises. It is known that maximum likelihood methods are consistent. After some confusion regarding parsimony analyses, caused by the (now discredited) defence that they incorporate no model, it is accepted that under a wide range of realistic conditions parsimony methods are not consistent (36). The status of other methods is not so clear. The neighbor-joining method has been intensely studied in recent years, and there seems to be no evidence that it is not consistent.

A criticism of the consideration of consistency when choosing a method for phylogenetic estimation has been that it is dependent on the assumption of some model of sequence evolution, whereas it is unlikely that any model currently available is accurate. In answer to this, I point our that some models are becoming fairly accurate (see below) and that it is always advisable to perform the best available data analysis, whatever one's reservations about that particular method, in preference to analyses with even more flaws.

7.2.2 Accuracy of DNA substitution models

A statistical test has been devised (37) which makes it possible to test statistically whether or not a model of DNA substitution is giving a good description of the patterns evident in aligned sequence data. The test is very slow to perform, because multiple simulations are necessary to assess test statistics instead of simply looking up values in a table as in traditional statistics. In the small number of studies where this test has been used to date, the general conclusion is that the most widely utilized models are generally inadequate for protein-coding sequences (37, 26). They may be rather more acceptable for non-coding sequences; in addition, more complex models incorporating rate variation across sites of the sequences seem quite adequate for some coding sequences. This has confirmed the long-held suspicion that selective effects and rate variation at different nucleotide sites are the most important of the biological features generally omitted from current models.

An alternative version of the test of (37) permits the comparison of two specific models. Such a test can be used to judge whether a more complex model is significantly better than a simpler one (e.g. testing HKY85 against K80 or HKY85+Γ against HKY85) or whether a molecular clock is indicated by the data (37, 26). Subsequent research has shown that simulations are unnecessary for these tests, which can be performed 'by hand' following maximum likelihood analysis (26, 34). Such tests have confirmed that complex models are almost invariably best; in addition, they have indicated that the molecular clock assumption is reasonable for some sequences. Further work has proposed simpler 'diagnostic' tests (38) that can indicate areas of failure of models for particular datasets; unfortunately, none of the tests mentioned in this section is currently available in distributed computer programs.

When models are shown to be suitable for the analysis of particular sequence alignments, the researcher can have increased confidence that results from analysis techniques that use those models are likely to be good. In cases where models are not adequate, the estimates of trees may well be affected and certainly the accuracy of estimates of branch lengths is greatly reduced (26). It is not possible to place too much emphasis on the importance of the consideration of the applicability of different models.

8. The future of phylogenetic estimation

This chapter has, in a spirit of realism as well as because of the need to highlight differences between methods of phylogenetic estimation, contained criticisms of every method. Even the most highly recommended methods have drawbacks. Improvements seen in this chapter to be necessary, and other developments, are being made and it is my intention in this section to indicate how the reader may remain up-to-date in the field of phylogenetic estimation and what advances they may hope to see soon.

Almost all improvements to phylogenetic estimation methods are first published in the *Journal of Molecular Evolution, Molecular Biology and Evolution, Molecular Phylogenetics and Evolution*, and *Systematic Biology*. Regular scanning of these journals should ensure that no major developments in the field are missed. The journal *Computer Applications in the Biosciences* often contains announcements of newly available computer software, but the other journals should generally be consulted for critical assessment of any new methods.

New methods of phylogenetic analysis do appear from time to time, but unpredictably. Gradual advances in established methods are, however, always available. I include now some pointers towards interesting new research that may have an important impact in the next few years. Methods mentioned in this section are generally too recent for their usefulness to have been established.

The methods most highly recommended in this chapter are those based on maximum likelihood methodology. The main faults of these are that they are too slow to be of practical use to many researchers and that the models they employ are unrealistic. Regarding the basic speed of methods, it is worth noting Dr Gary Olsen's *FASTDNAML* program (39, A8), which is an accelerated version of *DNAML*. Remember also that computer hardware is still developing quickly and problems that were considered virtually impossible in 1994 are routinely tackled in 1996. There is every reason to expect the same improvement from 1996 to 1998.

Regarding the perceived inadequacies of the models used with maximum likelihood methods, research into improvements is already under way. An important factor seems to be the relaxation of the assumption of equal rates of substitution at each site of a sequence. Dr Ziheng Yang has devised a maximum likelihood treatment (13) of models incorporating the gamma distribution of substitution rates across sites (Section 2.3.5)—this is even slower to compute than before, but he is developing an approximate analysis that is much faster (40). Another related method being developed by Drs Gary Churchill and Joseph Felsenstein permits the rate of nucleotide substitution at each site, instead of being constant or drawn from a gamma distribution, to be related to that at neighbouring sites.

Another major criticism of the models is that, for protein-coding DNA, the assumption that each site evolves independently of all others is implausible. The models of Yang and of Churchill and Felsenstein make some improvements in this respect. In addition, Dr Ziheng Yang and I (41) and Drs Spencer Muse and Brandon Gaut (42) have developed new models which operate at the level of the codon (nucleotide triplet), and these have permitted incorporation of genetic code and codon usage information and also of measures of the differences between the amino acids for which the triplets code. Initial results appear to be promising. A similar approach has been adopted by Drs Michael Schöniger and Arndt von Haeseler, who have modelled correlated pairs of nucleotides (43). While in the main designed for maximum likelihood phylogenetic estimation, all of these developments of models are also of use to those forced to use distance matrix-based methods, as these are dependent on the same models.

Distance matrix-based and parsimony-based methods are not advancing at the same rate, and this is again testimony to the powerful generality of maximum likelihood methodology. There has been considerable interest recently in simulation studies that seem to indicate that distances calculated using overly simple DNA substitution models can give more reliable estimates of phylogenetic trees (though not of the actual distances) than more complex and more accurate models, when each is used with distance matrix based tree estimation methods (44–46). It appears, however, that this effect may only be present when very simple substitution models are 'true', which will not be the case for real data. The use of more complex and more accurate models is still advised.

Another highly productive area of research is the study of the bootstrap method in phylogenetics. Recent papers have greatly improved understanding of what the bootstrap may and may not be expected to achieve (47), and the length of sequences and number of bootstrap replicates needed to achieve good results in a reasonable time (47–50). It is likely that there will soon be more results in this area.

It is also likely that 'production line research' using large-scale simulations to compare the relative performances of different methods will continue. Although it is difficult to draw general conclusions from simulations which must assume particular models to apply and particular trees for study (22), these studies are gradually increasing our understanding of how different methods perform under a variety of realistic circumstances.

Two fundamental yet unrealistic assumptions in current phylogenetic estimation methods may soon be relaxed. The first, that sequences are related according to a tree, has been generalized to allow 'networks' containing loops (51). Second, probabilistic models of insertion/deletion processes are being devised (5) which may be suitable for use in maximum likelihood estimation, allowing more information than ever before to be extracted from molecular sequences.

Appendix. Computer programs

A.1 *PHYLIP*

Dr Joseph Felsenstein has developed and distributes *PHYLIP* (phylogeny inference package), a package of computer programs for many aspects of phylogenetic estimation. *PHYLIP* is probably the most useful collection of programs available for phylogenetic estimation. The programs *DNAML*, *DNAMLK*, *DNAPARS*, *DNAPENNY*, *PROTPARS*, *DNADIST*, *FITCH*, *KITSCH*, *NEIGHBOR*, *DNAINVAR*, and *SEQBOOT* have all been mentioned above, and the package also includes utility programs and programs for data other than DNA and amino acid sequences useful to researchers in phylogenetics. The documentation available with *PHYLIP* is excellent, both that relating specifically to the use of each program in the package and also the wider-ranging documents which make a superb introduction to phylogenetic estimation. Even if you do not want, or do not have facilities, to use *PHYLIP* I strongly recommend you get the documentation.

The programs in *PHYLIP* version 3.5 are distributed as C source code, and are easily run on virtually any computer with a C compiler. Executable files are also available for PC and Macintosh computers.

The package is distributed free of charge. It is available by anonymous *FTP* from `evolution.genetics.washington.edu` (128.95.12.41)—start with the file `/pub/phylip/Read.Me`—or by contacting Dr Felsenstein directly at the Department of Genetics SK-50, University of Washington, Seattle, WA 98195, USA. E-mail address: `joe@genetics.washington.edu` (on the WWW, see
`http://evolution.genetics.washington.edu/phylip.html`).

A.2 *MEGA*

The *MEGA* program was developed at the Institute of Molecular Evolutionary Genetics, Pennsylvania State University. *MEGA* contains many utility functions for the analysis of DNA sequence data, including distance matrix calculation using a number of advanced models of DNA substitution and phylogeny estimation using parsimony and distance matrix methods. The *MEGA* documentation gives a good grounding in the methods the program incorporates.

MEGA version 1.0 for DOS is distributed as an executable file plus example datasets. The nominal charge made for *MEGA* may be defrayed for those unable to pay for some reason. An order form for *MEGA* is available from Drs Sudhir Kumar, Koichiro Tamura and Masatoshi Nei, Institute of Molecular Evolutionary Genetics, Pennsylvania State University, 328 Mueller Laboratory, University Park, PA 16802, USA. E-mail address: `imeg@psuvm.psu.edu`.

A.3 *PAUP*

PAUP (phylogenetic analysis using parsimony) is the most powerful widespread program for all forms of parsimony based analyses. It is developed and written by Dr David Swofford. The documentation distributed with *PAUP* forms an excellent introduction to parsimony-based phylogenetic estimation methods.

Distribution of *PAUP* version 3.1 is currently suspended, pending release of version 4.0 (provisionally called *PAUP**) in summer 1996. *PAUP** will come in the form of C source code for Unix and VAX/VMS systems, and executable files for Macintosh and DOS. It will be available for a nominal charge from Sinauer Associates, Inc., 23 Plumtree Road, Sunderland, MA 01375-0407, USA. E-mail address: `orders@sinauer.com`.

A.4 *BASEML* and *BASEMLG*

The *BASEML* (nucleotide BASE maximum likelihood) and *BASEMLG* (*BASEML* with gamma distribution) programs were written by Dr Ziheng Yang of the Department of Integrative Biology, University of California at Berkeley (ziheng@mws4.biol.berkeley.edu) to perform tree estimation and substitution model fitting using maximum likelihood methods.

They and other programs, collectively known as the package PAML, are available as C source code free of charge by anonymous *FTP* from ftp.bio.indiana.edu (129.79.225.25)—see the directories Incoming or molbio/evolve.

A.5 *PROTML*

Drs Jun Adachi and Masami Hasegawa make available their *PROTML* (protein maximum likelihood) program, which estimates phylogenetic trees from aligned protein sequences using maximum likelihood methodology.

PROTML is written in C for Unix-like operating systems. It (and other programs for phylogenetic estimation) is distributed free of charge, and is available by anonymous *FTP* from sunmh.ism.ac.jp (133.58.12.20), in the directory /pub/molphy/. Documentation is also available from the program's authors Drs Jun Adachi and Masami Hasegawa, Department of Statistical Science, The Graduate University for Advanced Study, 4-6-7 Minami-Azabu, Minato-ku, Tokyo 106, Japan. E-mail address: adachi@ism.ac.jp.

A.6 *TREEALIGN*

Dr Jotun Hein's *TREEALIGN* program simultaneously aligns molecular sequences and estimates the phylogenetic tree relating them. It is available free of charge by electronic mail from the EMBL Netserver (send e-mail containing only the word 'help' to netserv@embl-heidelberg.de) or by contacting Dr Jotun J. Hein, Institute of Biological Sciences, University of Aarhus, Building 540, Ny Munkegade, DK 8000 Aarhus C, Denmark. E-mail address: jotun@hardy.pop.bio.aau.dk.

A.7 Minimum message length encoding

Programs using minimum message length encoding, developed by Dr Lloyd Allison and colleagues, are available by contacting Dr Lloyd Allison, Department of Computer Science, Monash University, Victoria 3168, Australia. E-mail address: lloyd@bruce.cs.monash.edu.au.

A.7 *FASTDNAML*

Dr Gary Olsen's *FASTDNAML*, an accelerated version of Felsenstein's *DNAML*, is distributed as C source code. It is available free of charge by 'anonymous *FTP*' from `rdp.life.uiuc.edu` (128.174.228.13)—see the directory `/pub/RDP/programs/fastDNAml`—or by contacting Dr Olsen directly at the Department of Microbiology, University of Illinois, 131 Burrill Hall, 407 South Goodwin Avenue, Urbana, IL 61801, USA. E-mail address: `gary@phylo.life.uiuc.edu`

References

1. Cox, D. R. and Miller, H. D. (1977). *The theory of stochastic processes*. Chapman and Hall, London.
2. Jukes, T. H. and Cantor, C. R. (1969). In *Mammalian protein metabolism* (ed. H. N. Munro), Vol. 3, pp. 21–132. Academic Press, New York.
3. Kimura, M. (1980). *J. Mol. Evol.*, **16**, 111.
4. Felsenstein, J. (1981). *J. Mol. Evol.*, **17**, 368.
5. Thorne, J. L., Kishino, H., and Felsenstein, J. (1992). *J. Mol. Evol.*, **34**, 3.
6. Hasegawa, M., Kishino, H., and Yano, T. (1985). *J. Mol. Evol.*, **22**, 160.
7. Tamura, K. and Nei, M. (1993). *Mol. Biol. Evol.*, **10**, 512.
8. Kishino, H., Miyata, T., and Hasegawa, M. (1990). *J. Mol. Evol.*, **31**, 151.
9. Adachi, J. and Hasegawa, M. (1992). *MOLPHY: programs for molecular phylogenetics, I. PROTML: maximum likelihood inference of protein phylogeny*, Computer Science Monograph No. 27. Institute of Statistical Mathematics, Tokyo.
10. Kumar, S., Tamura, K., and Nei, M. (1993). *MEGA: molecular evolutionary genetics analysis*, version 1.01. The Pennsylvania State University, University Park, PA.
11. Kimura, M. (1983). *The neutral theory of molecular evolution*. Cambridge University Press, Cambridge.
12. Sarich, V. M. and Wilson, A. C. (1967). *Science*, **158**, 1200.
13. Yang, Z. (1993). *Mol. Biol. Evol.*, **10**, 1396.
14. Felsenstein, J. (1978). *Syst. Zool.*, **27**, 27.
15. Kendall, M. and Stuart, A. (1979). *The advanced theory of statistics*. Charles Griffin, London.
16. Felsenstein, J. (1983). *Annu. Rev. Ecol. Syst.*, **14**, 313.
17. Goldman, N. (1990). *Syst. Zool.*, **39**, 345.
18. Sanderson, M. J., Baldwin, B. G., Bharathan, G., Campbell, C. S., von Dohlen, C., Ferguson, D., *et al.* (1993). *Syst. Biol.*, **42**, 562.
19. Fitch, W. M. and Margoliash, E. (1967). *Science*, **155**, 279.
20. Sneath, P. H. A. and Sokal, R. R. (1973). *Numerical taxonomy*. W. H. Freeman, San Francisco.
21. Saitou, N. and Nei, M. (1987). *Mol. Biol. Evol.*, **4**, 406.
22. Hillis, D. M., Huelsenbeck, J. P., and Cunningham, C. W. (1994). *Science*, **264**, 671.
23. Hasegawa, M., Kishino, H., and Saitou, N. (1991). *J. Mol. Evol.*, **32**, 443.
24. Hasegawa, M. and Fujiwara, M. (1993). *Mol. Phyl. Evol.*, **2**, 1.
25. Kuhner, M. K. and Felsenstein, J. (1994). *Mol. Biol. Evol.*, **11**, 459.
26. Yang, Z., Goldman, N., and Friday, A. (1994). *Mol. Biol. Evol.*, **11**, 316.
27. Saitou, N. and Imanishi, T. (1989). *Mol. Biol. Evol.*, **6**, 514.
28. Huelsenbeck, J. P. and Hillis, D. M. (1993). *Syst. Biol.*, **42**, 247.
29. Lake, J. A. (1987). *Mol. Biol. Evol.*, **4**, 167.

30. Cavender, J. A. (1989). *Mol. Biol. Evol.*, **6**, 301.
31. Hein, J. J. (1994). In *Methods in molecular biology* (ed. A. M. Griffin and H. G. Griffin), Vol. 25: Computer Analysis of Sequence Data, Part II, pp. 349–364. Humana Press Inc., Totowa, NJ.
32. Allison, L., Wallace, C. S., and Yee, C. N. (1992). *25th Hawaii Int. Conf. Sys. Sci.*, **1**, 663.
33. Felsenstein, J. (1985). *Evolution*, **39**, 783.
34. Yang, Z., Goldman, N., and Friday, A. (1995). *Syst. Biol.,* **44**, 384.
35. Hasegawa, M. and Kishino, H. (1994). *Mol. Biol. Evol.*, **11**, 142.
36. Zharkikh, A. and Li, W.-H. (1993). *Syst. Biol.*, **42**, 113.
37. Goldman, N. (1993). *J. Mol. Evol.*, **36**, 182.
38. Goldman, N. (1993). *J. Mol. Evol.*, **37**, 650.
39. Olsen, G. J., Matsuda, H., Hagstrom, R., and Overbeek, R. (1994). *Comput. Appl. Biosci.*, **10**, 41.
40. Yang, Z. (1994). *J. Mol. Evol.*, **39**, 306
41. Goldman, N. and Yang, Z. (1994). *Mol. Biol. Evol.*, **11**, 725.
42. Muse, S. V. and Gaut, B. S. (1994). *Mol. Biol. Evol.*, **11**, 715.
43. Schöniger, M. and von Haeseler, A. (1994). *Mol. Phyl. Evol.*, **3**, 240.
44. Schöniger, M. and von Haeseler, A. (1993). *Mol. Biol. Evol.*, **10**, 471.
45. Tajima, F. and Takezaki, N. (1994). *Mol. Biol. Evol.*, **11**, 278.
46. Goldstein, D. B. and Pollock, D. D. (1994). *Theor. Popul. Biol.*, **45**, 219.
47. Hillis, D. M. and Bull, J. J. (1993). *Syst. Biol.*, **42**, 182.
48. Hedges, S. B. (1992). *Mol. Biol. Evol.*, **9**, 366.
49. Lecointre, G., Philippe, H., Lê, H. L. V., and Le Guyader, H. (1994). *Mol. Phyl. Evol.*, **3**, 292.
50. Brown, J. K. M. (1994). *Proc. Natl. Acad. Sci. USA*, **91**, 12293.
51. von Haeseler, A. and Churchill, G. A. (1993). *J. Mol. Evol.*, **37**, 77.

16

Evolution and relationships of protein families

WILLIAM R. TAYLOR

1. Introduction

To identify remote protein sequence and structural similarities it is necessary to understand the processes that have given rise to the current forms. This can be gained partly through a comparative study of the current sequence and structural databanks, but to interpret these interrelationships fully it is necessary to have a model, even if only conceptual, of how the current state arose. The rich variety of protein sequence and structure observed today has resulted from a long process of evolution. Taxonomists, who consider the equivalent problem in the biological world, have a great advantage over their molecular counterparts as they have examples of intermediate forms in the fossil record, in addition to the 'living relics' (like the coelacanth). In the molecular world, however, an evolutionary model must be developed through direct observation of genetic events (such as recombination, splicing, etc.) as there is, effectively, no fossil record*. 'Living relics' can also be found in the protein world (1) but it must be remembered (as in the biological world) that such 'relics' have a lineage equally as long as any other living thing (2). This is emphasized by the observation that the divergence of equivalent proteins among the bacteria is as great as between any bacterium and a 'higher' organism.

The observed shapes and sequences of proteins are a result not only of their history, but also of the physico-chemical constraints imposed by their constituent components (e.g. the strength of covalent and hydrogen bonds), their environment (aqueous or lipid, intra- or extracellular) and the tasks that they are required to perform (catalysis or recognition). It is difficult to separate the forms imposed by these constraints from those that have been inherited, but this is, nevertheless, a problem worth tackling, because, if the physico-chemical constraints could be quantified, then the evolutionary component (being the remainder) would similarly be known. For example, the

* Contrary to the impression created by Hollywood, the amount known about ancient DNA is trivially small.

wings of vertebrates must maintain a certain relationship to the size of the animal to allow flight. This is a purely physical constraint. By contrast, there is no physical law which maintains that feathers are the only material from which wings can be constructed; so finding that all birds use feathers leads us to believe that they had a common ancestor while finding that bats and birds have wings of an equivalent relative size does not inspire the same conclusion. A parallel situation in molecular evolution might involve a general enzymatic reaction that requires a certain juxtaposition of chemical groups (supported by a sufficiently stable framework). If it could be shown that only one protein chain fold is able to achieve this, then no evolutionary inference can be made about equivalent enzymes using this catalytic mechanism (having the same fold). However, if the necessary groups can be supported by, say, fifty different folds, then a group of enzymes with the same fold appears much more likely to be related.

This chapter aims to provide some guidance on how to interpret the alignment of remotely related protein sequences and structures. To do this it is not only necessary to understand the comparison methods themselves but also the underlying principles of structure and evolution on which the comparisons methods are (or should be) based. The content of this chapter will therefore tend to be theoretical, and while pointers will be given to practical comparison methods, the main purpose of the work will be to convey the basic principles. This will consider how conventional methods of sequence comparison can be extended into regions of very remote similarity, using both multiple sequence data and structural constraints. Sequence threading methods (which compare a sequence directly to a structure) will be reviewed continuing on to the comparison of protein structures. The evolutionary mechanisms that underpin (and partly have been deduced using) these comparison methods will then be reviewed, followed by a critique of some of the more theoretical treatments that are relevant to the assessment of the significance of very remote similarities.

2. Sequence similarity

2.1 Pairwise sequence alignment

Alignment is one of the most important and widely used methods of biological sequence analysis. It has revealed many unexpected similarities which have given fresh insight into previously uncharacterized systems. Of the many methods to align sequences, the most general and widely used are based on the dynamic programming algorithm (3–5). This algorithm finds the optimal alignment under a given scoring scheme, providing pairwise matches are independent. (See ref. 6 for a review.)

The basic dynamic programming algorithm is very general and has been used not only for the comparison of one sequence with another but also using structures encoded as sequences. This allows these 'structures' to be compared against each other and directly with an amino acid sequence. (These alternative applications will be described below.) From an evolutionary viewpoint, however, the dynamic programming algorithm is constrained by the inherent assumption that the only mechanisms at work are insertion, deletion, and substitution. Transposition and inversion of segments are forbidden and in nucleic acid analysis (where these events might reasonably be expected) different algorithms are often applied (7).

Although the algorithmic basis of sequence alignment is rigorous, the controlling parameters are, by contrast, poorly characterized. These include the model of amino acid similarity and the penalty imposed for the insertion of gaps (see also Chapter 7).

2.1.1 Gap penalty

There has been much 'experimentation' and theoretical analysis on the best size and form for the gap penalty. The theoreticians have been most concerned that its form should preserve the guaranteed optimality of the dynamic programming algorithm. This is of prime importance for analytic (mathematical) analysis, the results of which enable probability distributions to be deduced so allowing the statistical significance of a match to be calculated (8). An alternative approach is not to rely too much on theory but to derive an empirical gap function. The required analysis of gap lengths and composition has been carried out in some detail using alignments based on known structures (9) and on extensive sequence comparison (10, 11).

Gap penalties can be either simple (one penalty for any gap size) or complex (typically, one penalty to open the gap and another making it dependent on gap size). A form of gap penalty often used is one with an opening cost followed by a linear increase with the size of the insert (generally referred to as an *affine* model) (12). The only general constraint on the gap penalty is that it should increase with length. Some alignment algorithms, however, impose the further condition that once a gap is started its size must become increasingly less important (13) (referred to as a *concave* function as the incremental cost of extending a gap becomes less).

A further general result to emerge from the analysis of complex gap penalties is that in the phase-space of the two gap parameters, the alignments fall into two types. When the penalty for a gap is high the best ungapped (or local) alignment is optimal but as penalties are reduced, a boundary is crossed (a phase transition) into the region of gapped alignments. Interestingly, correct protein sequence alignments seem to lie close to this boundary (14). This analysis quantifies the intuitive feel for the point at which an alignment has the 'right' number of gaps.

The form of the gap penalty reflects an underlying model of protein evolution which, in part, is a reflection of the stability of protein structure. A simple model for the latter is that the core is very sensitive to change whereas the surface, and especially exposed loops, are susceptible to change (15). Indeed, it probably makes little difference to a protein structure whether 10 or 100 residues have been inserted into an exposed loop—providing the insert is in the form of a compact, independently folded, domain. On this basis, the simple gap penalty (no extension penalty) provides an adequate model for remotely related proteins.

Whatever the form of the gap penalty, it is important that its size is appropriate relative to the scoring scheme. For example, a gap penalty (of 20) that gives good results with identities (scored as 10) will have a different effect when used with a full amino acid substitution matrix. Previously, two matches balanced a gap but if the substitution matrix has, on average, the same score (10) for an identity, then the additional scores from non-identical residues will render the gap penalty effectively smaller.

2.1.2 Substitution matrices

When dealing with remote protein similarities, the importance of the form of the gap penalty is dwarfed in comparison to the ill-determined nature of the amino acid substitution model. Good amino acid substitution probabilities are unknown, not through any lack of effort on the part of researchers in the field, but because the individual probability for a particular amino acid at a particular location in one protein mutating to another amino acid in another protein cannot be calculated. To do so it would be necessary to know the structural (and all other) constraints not only on the mutating position but also at that position in all intermediate ('fossil') structures. As remotely related protein sequences may have diverged more than a billion years ago, such knowledge is unobtainable. The best that can be done is to compile 'average' probabilities based on amino acid exchanges over many different structural environments in many different proteins. A substitution table of these average probabilities has, therefore, an unavoidably high inherent error when applied to evaluate any specific change.

One way to avoid the problem outlined above is to consider only very closely related sequences. If there is only a single amino acid difference then one sequence probably evolved directly from the other in a single step or a common ancestor can be deduced through phylogenetic reconstruction (see Chapter 15 and ref. 16). This method, however, only provides a substitution table for the comparison of sequences separated by a single point (accepted) mutation (PAM). To apply the results at 1 PAM to remotely related sequences, it is necessary to extrapolate the PAM table by repeated application, typically a few hundred times, which in practice, is achieved by matrix multiplication (16).

Despite the rigorous derivation of the 1 PAM matrix, the validity of its extrapolation can be questioned as the results differ from a matrix derived directly from distantly related sequences (11). One compromise is to derive a matrix from fairly closely related sequences but without the calculation of ancestors (17), allowing the large current volumes of sequence data to be used automatically. A different compromise is to use more distantly related sequences but only the parts of the alignments that are unambiguous. Since this matrix is derived by summation over conserved blocks it is referred to as the BLOSUM matrix (18). The resulting matrix, however, has the inherent problem that the residues in the blocks will not be typical of the whole sequence (on average) but has the advantage that they will be biased towards the regions most likely to align (typically secondary structures and active sites). Many comparative evaluations of these various matrices indicate that although most of the recent ones are better than the older matrix of Dayhoff (19), there is little practical difference between them (20).

2.1.3 Alignment significance

The alignment of a pair of protein sequences cannot be used to detect remote similarities. ('Remote' is defined here as those similarities that cannot be detected by the alignment of two sequences.) A fundamental limit is approached which is the degree of similarity that would be attained through the alignment of two random sequences.

In practice, this limit is not even attained, since protein sequences are neither random in composition nor in internal organization, giving a broad region of uncertainty which has been graphically referred to as the 'twilight zone' (21). In practice, sequences sharing less than 20–25% identity (when optimally aligned) cannot be treated as related with any confidence. This rough guide depends on the distribution of matches. For example, if the 20% occurred in a single run then the chance of this occurrence (for sequences of non-trivial length) would be small, whereas if the matches were scattered randomly along the alignment then they would be more likely to have arisen by chance. Statistics have been calculated to assess the significance of matched segments of different lengths that do not contain gaps (22). These form the underlying basis of the *BLAST* program (23) (see also Chapter 7). The significance of segment matching has also been quantified through structural correspondence, usefully dividing the space into 'true' and 'random' alignments (24) (*Figure 1*). More commonly, however, several matching segments are encountered with gaps in between and exact statistics are being developed (25, 26).

If one or more of the matched segments corresponds to a known motif, as might be found, for example, in the PROSITE database (27), then confidence in the alignment, and hence the relationship can be boosted greatly (28). However, clumped (motif-like) matches can also arise through the coincidental matching of secondary structures. A particularly misleading occurrence is the matching of periodic leucines (an amino acid pattern common in α-helices and coiled-coils) which is also characteristic of a DNA-binding motif (29). Loop regions can also mislead, and the over-interpretation of glycine-rich loops as nucleotide-binding sites (30) is common.

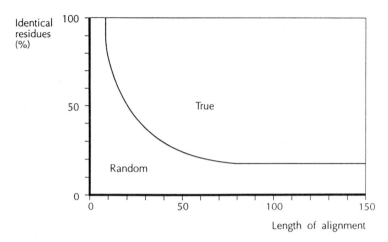

Figure 1. Probable structural correspondence of matched segments. In the 'true' region the secondary structures of the sequence fragments correspond but not in the 'random' region while about the dividing line is a region of uncertainty. For long segments this approaches the 20% barrier obtained with random sequences, while at the other extreme, even identical pentapeptide matches cannot be considered significant. No gaps were allowed in the matched sequences. (After ref. 24.)

Remote pairwise alignments should be treated with great caution, especially if they have been 'discovered' through databank scanning. With the current large size of the sequence databanks (currently around 100 000 protein sequences) chance alignments in the 'twilight zone' are common. A rough guide to assessing alignments of marginal credibility is to look not only at the identities (matched amino acids) and their distribution, but also at the regions between matches. If these retain some remnant of similarity based on the physico-chemical properties of the amino acids (31), then the result might be worth pursuing.

2.2 Multiple sequence alignments

Fortunately, many of the problems described in the preceding section can, increasingly, be avoided as the great volume of (derived) protein sequence data currently available often leads to the situation of having several clearly related sequences that can be aligned with confidence. With such data a whole new aspect of the protein is revealed as it becomes apparent which parts of the sequence are vital (and have been conserved) and which parts, to varying degrees, are dispensable.

There are many methods of multiple sequence alignment (see reviews in ref. 32) but most practical methods are based on the progressive combination of pairwise alignments (33–36). These methods have been incorporated into various packages such as *CLUSTAL V* (37), *CAMELEON* (Oxford Molecular Ltd, using Taylor's program (38)) or *PILEUP* (UWGCG, (39), using Feng *et al.*s program (33)), but with various enhancements and improvements, the current formulations are sometimes difficult to distinguish. They will not be considered in any detail since it is not clear that their performance on remotely related structures is sufficiently different to justify analysis and the limiting factor in all of them is the quality of the data they are given to align.

Typically, six sequences would be sufficient to reveal regions of conservation and variation, providing that they have a wide spread in similarity (six closely related sequences are little better than one). However, with some families (such as the protein kinases, immunoglobulins, or globins), hundreds of members are known in addition to several tertiary structures and a wealth of biochemical data (including site-directed mutants). The analysis and cross-referencing of the information in these data-rich families is currently carried out by 'experts' with specialized knowledge of the field, but as further sequences and structures are determined it will become necessary for the analysis and cross-referencing to be carried out more automatically. To meet this need the various databanks are incorporating cross-references and becoming more accessible (40).

From an evolutionary perspective, one of the most interesting aspects of these extensive analyses is to determine the common cores of ancient proteins that are found across a wide range of phyla, and often in all living organisms (41). These ancient conserved regions (ACRs) provide an indication of the minimal set of proteins required to support basic life functions. The number of such proteins may be as low as 900 and a representative of 600 of these may already be found in the current databases. This is comparable to a similar estimate of 1000 for the number of families based on the analysis of recurrence in a recent influx of sequence data (42) (see below). However, this correspondence may be coincidental (43).

2.3 Structure biased alignment

The information associated with a data-rich protein family can be used, first, to attain the best multiple sequence alignment of the family and secondly to aid in the search for further members of the family. Many of the approaches exploiting these additional data can be applied to either pairwise or multiple alignment but are generally more effective with more than two sequences.

2.3.1 Gap bias

Relative insertions and deletions of sequence (indels or gaps) are much less likely to be found in segments of secondary structure (9). If the structure of the protein is known, a good alignment program should use this information and modify its local gap penalty accordingly to avoid breaking secondary structures and inserting residues in the hydrophobic core (44–48). Although this method is commonly applied when a structure is known, it is less common for the predicted secondary structure or core to be used in the same way. This is perhaps because of the added uncertainty in the prediction, but with multiple sequence data, predictions of useful accuracy can be attained (49, 50). Such an approach overlaps with, but to some extent, complements preferences for certain residues to appear in insertions or at the broken ends of gaps described above.

2.3.2 Substitution bias

Structural information can be used not only to bias the location of gaps, but also to bias the matching of residues. At a simple level, certain amino acid substitutions are more favoured in different secondary structures. For example, I→L is a change more favoured in the α-helical conformation whereas I→V would be more favoured in β-conformation (since L and V strongly favour α and β conformations, respectively). The results of this type of analysis can be encoded into structure specific substitution matrices which are applied according to the type of structure of one (or more than one) of the sequences in the alignment (51). The number of tables depends on the amount of available structural data and with high-resolution structures, quite specific tables can be constructed (52). Structural subdivision cannot, however, be taken too far since the number of well resolved structures is limited. Each additional structural subclass requires that the available data are spread more thinly and care must be taken that there are sufficient data to give reliable values for each table entry.

At worst, the structural data might only be predicted and although this is probably too unreliable for globular proteins, a substitution matrix for transmembrane regions (53) can usefully be applied according to predicted segments (54).

Figure 2. Alignment-based comparison methods using the sequences and structures of two proteins *A* (residues a–p) and *B* (residues 1–14). Each comparison uses the alignment matrix (AM) but with a different table. Comparing two sequences, the AM is filled with similarity values for every pair of residues from *A* and *B*, e.g. for position *j* in *A* (S) and position three in sequence *B* (N), the similarity of S and N is found in Table (a) and transferred into the AM at *j*,3 (full lines). This is repeated for all pairs, so filling the matrix. For sequence-to-structure, Table (b) is used, giving the propensity for each acid (upper-case) to be in a structural environment (symbols), e.g. for position 12 in sequence *B* (P) and position m in structure *A* (say, 'buried and in α-helix', □), the propensity for P in □ is entered into the AM at *m*,12. (dot-dashed lines). For structure-to-structure, Table (c) gives the similarity of one environment (symbols) to another, e.g. for *d* in A (□) and 2 in structure *B* (□) a value is entered into the AM at *d*,2. (dashed lines). The AM is processed in the same way each time, finding the path that 'picks-up' the highest sum of scores (white line) giving the alignment: DGFLQNPVK**SGEVARN (A)
VLNSTVRDKPTPVA (B).

2.4 Sequence threading

The incorporation of structural data can be approached in a different way: rather than align sequences through a structure-specific table (relating amino acid to amino acid), the substructures can be encoded as a sequence and aligned directly with an amino acid sequence using a substructure-to-amino acid table (*Figure 2*). This method was first used to align a sequence encoded as hydrophobicity against a sequence of observed residue exposure (55), and

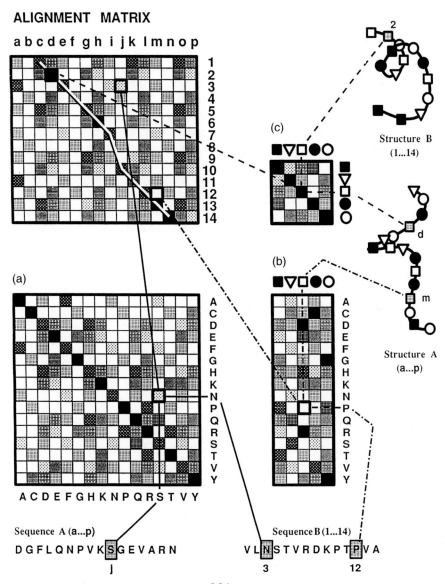

ALIGNMENT MATRIX

(a)

(b)

(c)

Structure B (1...14)

Structure A (a...p)

Sequence A (a...p)

D G F L Q N P V K S G E V A R N

j

Sequence B (1...14)

V L N S T V R D K P T P V A

3   12

has since been elaborated to incorporate a greater variety of substructures (or more exactly structural environments); such as 'buried in α-helix' (56), or finer specifications (57).

The preceding approach captures residue burial and secondary structural conformation, which, being the two most basic features of globular proteins, allow very remote similarities to be recognized; such as the similarity of flavodoxin to a bacterial chemotaxis control protein (cheY) (55), or the similarity of actin to hexokinase (56). Despite their successes at the limits of plausible similarity, it is not obvious that the approach provides a great advance over the older pattern based methods some of which embody structural constraints (58). This approach has attained qualitatively equivalent results both in the globin family (59) and sugar kinases (60).

All methods based purely on sequence comparison (both profiles and patterns) cannot, in principle, distinguish different folds that might give rise to the same patterns of exposure and secondary structure. This situation is more common than might be expected, since proteins use recurring super-secondary structural motifs in different ways (such as the β-α-β unit) (61). Since supersecondary structures are semi-autonomous folding units, they retain approximately the same local structural environments, making recognition by sequence comparison methods difficult. For example, it is difficult to distinguish the alternating β/α (TIM) barrel fold from the Rossmann fold (*Figure 3*).

To avoid this type of ambiguity requires the inclusion of pairwise residue interactions in the comparison method (62–64). When insertions and deletions are considered, however, this leads to a computationally difficult problem which has been solved in different ways (65–67). It is not clear that the additional complexity of incorporating pairwise residue interactions leads to a significant increase in the ability to recognize remote relationships, however, they appear to help resolve ambiguity in the repeating β/α class of protein (68) and occasional examples, such as the recognition of the phycocyanin fold as a globin (65) suggests that they can extend the range of recognition beyond what can be achieved using sequence alignment methods without pairwise interactions (69).

3. Structural comparison

As with sequences, the comparison of a pair of closely related protein structures presents no fundamental problems and can be achieved through the simple superposition of the two structures (70). In general, protein structures are better conserved through evolution than the sequences that determine them. This maxim restates the observation that the 20 amino acids are functionally redundant—probably by factor of four (71). (In other words, five amino acids would be sufficient.) This gives plenty of scope to vary the sequence while still maintaining the structure. As a consequence, the

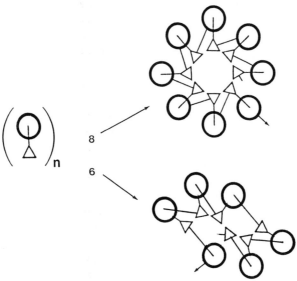

Figure 3. Folding options for tandem β–α units: β-strands are represented by triangles (which, unlike some representations has no directional implication) and all strands are parallel and progress towards the viewer. In reality, the strands are both curved and twisted and this is suggested by their non-linear alignment. α-Helices are represented as bold circles and are in reality about twice as broad as a β-strand. The direction of the chain is indicated by a terminal arrow-head. With eight units the sheet can form a barrel and the different radii at the β and α layers accommodate their different size. The barrel structure is found in many enzymes, typified by triosephosphate isomerase (TIM) and is referred to as a *TIM-barrel*. Six units cannot form a barrel but an inversion in one-half of the sheet allows the helices to be placed both above and below. The resulting arrangement has twofold symmetry and occurs widely among dinucleotide binding proteins. It is typified by the dehydrogenases where it is referred to as a Rossmann fold. In both structures the strands and helices have similar local environments.

comparison of protein structure provides our best probe into the deep evolutionary past (72).

3.1 Recent comparison methods

Methods of structure comparison have moved from the comparison of intact proteins by rigid-body superposition towards methods based more on local structural similarity or (like the sequence threading methods) based on comparing structural environments. This approach can be formulated in terms of the standard sequence comparison algorithm (dynamic programming) (73, 74) so bringing a unity to the comparison of sequence and structure. This formulation of structure comparison allows all the methods previously applied to sequence comparison to be used on structures, such as variants to look for local matches (motifs) or multiple comparison based on consensus (averaged) structures (75).

Some current methods use local structure comparisons but have avoided their subsequent analysis by dynamic programming (this is equivalent to the calculation of a sequence dotplot) (76, 77). These methods have the advantage that they can recognize structural transposition (or even inversion) (78) which can be useful when searching for a conserved arrangement of catalytic groups (independently of the chain fold) (79), or for packing similarities when the chain has been artificially reconnected (80). It is not clear how often the latter events occur in nature and although it may be unwise to exclude them, the generality of the method required for their detection may weaken the recognition of 'true' evolutionary relationships which preserve the sequential order.

3.2 Fold classification

The previous section on structural comparison raises the problem of what defines a fold and what changes can be accepted while still allowing two proteins to be considered related. This problem parallels the ongoing debate in taxonomy of how to define a species and so, not surprisingly, is central to molecular taxonomy.

The initial approach to molecular taxonomy was largely based on expert assessment (81–84). Folds have been classified into general topological categories such as 'singly-wound' or 'doubly-wound', corresponding to the TIM barrel and Rossmann fold, respectively (*Figure 3*). Other more descriptive terms have been employed also; such as 'Greek key' or 'jelly-roll' (85)—often leading to arcane descriptions like '. . . is a β-sandwich protein containing a Greek key partially extended towards a "jelly-roll".'. (NB, this is not a quote from ref. 85) (*Figure 4*).

Such descriptions are relatively unambiguous for certain features but like physiological taxonomy, do not deal easily with 'misfit' creatures that have, say, feathered wings as well as arms and legs. In addition, feature classification does not provide a distance measure between objects without recourse to the complexities of cladistic analysis.

In an attempt to overcome these difficulties, the 'new wave' of structure comparison methods (based on local environments) have been turned to the problem of fold classification (for a short review see ref. 86). These methods still have difficulty bettering the 'expert eye' in remotely related fold recognition (87) but with the increasing size of the protein structure databank, even the dedicated experts, with maybe one exception (88), have trouble keeping abreast of all the latest folds and their similarities. Some automatic comparison methods have now been applied to the pairwise comparison of a representative selection of the protein databank (80, 89–91). These studies have not revealed anything unexpected in the overall classification of structure although they have identified some novel similarities (92, 93), and motifs (94).

The results of these comparisons tend to reveal more about the assumptions in the comparison methods than about proteins. If, for example the

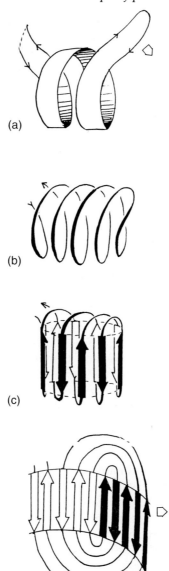

(a)

(b)

(c)

(d)

Figure 4. Various representations of an all-β protein. (a) Emphasizing the double helical (doubly wound) nature of the chain which may have played a role in folding. (b) The final double-wound structure. (c) Hydrogen bonds between vertical strands creates a cylindrical β-sheet (arrows). (d) Opening the sheet (at the left edge) produces a two-dimensional representation emphasizing the spiral that would be seen looking down the helix axis in (a) or (b) from the left. The two white and two black strands in the centre describe a Greek key motif while the extended spiral is referred to as a jelly roll. Most β-sheets are less regular

method is too simple (90) then large groups of structures can be linked through a trivial comparison (for example a small protein with only one α-helix). On the other hand, methods that are too specific fail to connect sub-families that are clearly related (to the expert eye). Between these extremes a reasonable balance can be found but it is disappointing that this critical cut-off must be imposed and does not arise intrinsically.

3.3 How many protein folds?

The number of observed protein folds can be obtained on the assumption of a reasonable cut-off value for structural similarity. Typically this would be chosen to link all the Rossmann fold containing enzymes (dinucleotide binding) with other similar folds such as flavodoxin, cheY (see above), and perhaps even the *ras* oncogene product or ribosomal elongation factor Tu (*Figure 5*). A cut-off in this range gives an estimate of less than 100 distinct folds (91).

Whatever the exact value for the number of known folds, it is surprising how many novel structure (and sequence) determinations reveal a familiar fold (95). One of the most frequent to recur is the $(\beta-\alpha)_8$ barrel (represented by the glycolytic enzyme triosephosphate isomerase, or TIM). Its persistence was remarkable several years ago (96) and there are now over 25 examples, all with little functional or sequence similarity (97, 98). Perhaps the phenomenon of recurring folds was best typified by the systematic sequencing and structural determination of the enzymes on the mandalate pathway (99). The structures of some of these enzymes (including TIM barrels) might have been recognized by remote sequence similarity to known structures.

The recurrence of folds leads to the possibility that the total number of folds (both seen and yet to be seen) might be estimated. Following the logic that the frequency of recurrence depends on the size of the underlying population of folds from which examples are drawn. A calculation was previously attempted along these lines based on the recurrence of exons (see below) (100) but this has now been largely discredited (101, 102). A more recent attempt was based on sequence analysis of the yeast chromosome III data and concluded that typically 1000 folds might be expected (42). However, considering the number of assumptions in such calculations, they must be treated cautiously. Given that the comparison of structure provides a deeper evolutionary probe than sequence comparison, this number would be expected to be less if the calculation were repeated on structural data and a rough estimate of 500–700 folds has been suggested (86).

A problem in all the above calculations is that they assume that each fold is equally represented in the underlying population. From the observed frequencies, this is clearly not valid. Indeed, the systematic analysis of the structural databank has revealed that folds are heavily biased towards a few folds, such as the Rossmann fold, TIM barrel, four-helical-bundle, and Greek key motif (103, 104). Adjusting for this phenomenon in the calculation of the

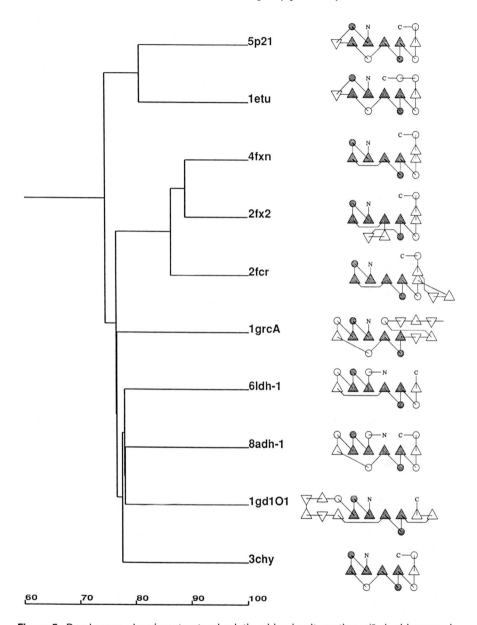

Figure 5. Dendogram showing structural relationships in alternating α/β doubly-wound folds, generated from pairwise structure comparison using the method of Taylor *et al.* (75). A schematic representation is shown adjacent to the corresponding Protein Data-bank code for the structure. These include a bacterial chemotaxis protein (chy), dinu-cleotide binding domains (grc, ldh, adh, gd1), flavodoxins (fxn, fx2, fcr), a ribosomal elongation factor (etu), and the *ras* oncogene product (p21). All the domains bind nucleotide.

total number of folds has the effect of depressing the estimated number. Indeed it is depressed so much that it becomes less than the number of known folds (Orengo and colleagues, personal communication).

This problem might be resolved, however, by adopting a simple model for the underlying frequency of each fold. It would not be unreasonable to assume, either for evolutionary or kinetic reasons, that the frequency of a fold is proportional to its simplicity (as measured by internal symmetry). This assumption would reflect also the observation that the observed folds are dominated by a few super-families, which have high internal symmetry. These common symmetric folds would be followed in the frequency distribution by a long 'tail' of rare complex folds the number of which would be difficult, if not impossible, to estimate. If this hypothesis reflects the true distribution, then unexpected folds may continue to be revealed by the crystallographers for many years to come—much to the consternation of structure predictors!

4. Molecular evolution

As shown by the discussion of structure comparison in the previous section, the distribution of expected protein folds (and hence the significance of similarity) is tightly bound with assumptions of the underlying mechanisms of molecular evolution. For example, how frequent are duplications? can transpositions be tolerated within a fold? Much of the evidence on which our model for these phenomena is based has come from the discovery of events in the relatively recent evolutionary past which have clear sequence similarity. By extrapolation it can be assumed that the mechanisms have remained the same and can be used to account for more distant evolutionary events.

4.1 Genetic algorithm model

Evolution is a process of optimization, generally under continually changing conditions. For simplicity we can assume that the conditions are constant, allowing a simple model of evolution to be constructed in which sequences are selected for propagation according to their evaluation by a fixed fitness function. In situations where the evaluation of this function is too complex for conventional minimization, this approach can be used as a practical (computer) method called the **genetic algorithm** which is especially suited to sequence-based applications (105, 106). The method incorporates a random mutation model to generate diversity in its population of sequences, however, this alone is insufficient to produce good convergence on an optimal solution. The key feature that makes the genetic algorithm effective is its ability to exchange pieces of sequence between members of the population. These exchanges, called **cross-overs**, take place between one or more exchange points and mimic the strategy exploited at the level of the genome in sexual reproduction.

Swopping genetic material in this way allows locally optimal solutions, developed independently in different sequences, to be recombined to produce a super-sequence. The process is wasteful, since for every good/good combination there will be many good/poor and poor/poor combinations but the good combinations survive and multiply, enriching the population for the next cycle. If proteins had to make only stepwise (residue-by-residue) changes, then many wasteful or lethal 'dead-ends' would be investigated that would become increasingly likely to undo the good solutions as they approach the optimum.

The effectiveness of the cross-over strategy, however, relies on the existence of locally optimal solutions and if the stability of the minimum solution were equally dependent on every element of the sequence the cross-over strategy could not work. Proteins lie somewhere between these extremes and whereas the stability of the fold is, to some extent, dependent on the interaction of every amino acid, there is also a clear hierarchy of structure from secondary, through super-secondary to domains, which is ideally suited to genetic algorithm optimization, even in computer simulations (107).

4.2 Gene duplication and fusion

4.2.1 Genetic mechanisms

A related strategy (not exploited in computer algorithms) is to combine independent structures, either two the same or two different, into a single protein. There are many ways to generate duplications and translocations: translocations simply require the incorrect religation of broken double-stranded DNA, while an easy route to generate duplication involves staggered (double) strand damage combined with 'fill-in' repair of the broken (single strand) ends before religation (ref. 108, and following papers). There are many clear examples indicating that duplication (with fusion) has occurred extensively, both in the recent and remote past (for review, see ref. 15). These include single duplication (109, 110), through triple- and double-duplication (111), to multiplication (112) and explosion (113). Indeed, the proteins that do not contain some indication of duplication (or pseudo-symmetry) in their structure or sequence are the exceptions (Flores and Taylor, unpublished results).

In addition to fused genes, stop codons can also be copied, leading to multiple gene copies. These provide an important route to the evolution of new function from existing (vital) proteins (114), and can lead to a variety of unexpected genetic effects (115), which, for simplicity, will be ignored here.

4.2.2 Dimeric precursors

Protein structures that function as dimers would be the most likely candidates for gene duplication into a fused protein as they have already evolved complementary interacting surfaces. A probable example of this process is seen in the aspartyl proteases. The form found in higher organisms has two (remotely related) domains with considerable differences in loop lengths and

subdomain packing but the same chain fold. Each domain, however, con-
tributes an aspartic acid to the active site which in the high-resolution structures can be seen to have an almost exact twofold (180°) relationship (116).
In addition the same twofold axis closely corresponds to the symmetric relationship of the two domains suggesting a precursor molecule which func-
tioned as a dimer (117). An example of such a protein was identified in the
retroviral proteases (118), and modelled as a dimeric enzyme (28)—a result
later confirmed by many crystallographic studies (119).

The dimers most susceptible to duplication and fusion would be those in
which the two ends to be joined (the N-terminus of one subunit with the C-
terminus of its symmetric half) lie close together. Without this, some unwind-
ing of the chain at each terminus would be necessary, or an additional linking
segment would be needed. Both would give rise to new interactions with the
probability of these being unfavourable. A direct implication of this is that
the remaining free ends (now the termini of the fused-gene product) must,
because of the twofold symmetry, lie close together (Taylor and Rippmann,
unpublished deduction). Interestingly, this would explain the common prox-
imity of the termini in protein domains (120), a phenomenon that is largely
unexplained by other effects.

4.3 Introns and exons

The general strategy of strand exchange as a mechanism for generating low-
error diversity could be made more efficient if the recombination (or cross-
over) points, which are a source of added error, avoided the locally optimized
sequence regions. In terms of protein structure this would entail introducing
a bias for cross-over to occur in the regions of sequence coding for surface
loop regions. However, as the main mechanism of recombination (including
sexual) involves random strand breakage (121) it is difficult to envisage any
method of control.

The discovery of genes split by regions of non-coding sequence revealed a
potential mechanism (122, 123). The intervening regions of 'junk' DNA
(introns) are typically much longer than the coding regions (exons) making it
more likely that a random recombination site will occur in an intron. By
selection, the introns have come to lie in regions of the gene corresponding to
the less critical features in protein structure, such as surface loops (124). The
only problem with this simple strategy is that the introns must be removed
(spliced out) before the RNA message is translated and this vital task is carried out by a complex protein/RNA mechanism (125, 126).

Since their discovery, there has been much speculation on the origins of
introns and their significance to evolution both at the level of protein structure (127, 128) and at higher levels of organization (129). A major puzzle was
their absence in the prokaryotes. This was originally explained as removal
due to the pressures of rapid replication, implying that the ancestral organism
had introns (129) (and most of the papers above). The 'old intron' hypothesis

never explained why the clearest examples of exons corresponded to domains found in proteins of relatively recent origin (130–132) and doubts about the fundamental nature of exons were expressed (133), despite the growing dogma. Recently, however, the tables have turned (or been slightly rotated) again with the discovery of introns in two prokaryotes leading to renewed discussion (134, 135).

From evidence that introns can be inserted into coding regions (134, 136), supported by a variety of arguments (102, 137), it is now thought that the well-ordered splicing of introns seen today may be an evolutionary recent mechanism (138). Indeed, Patthy argues that the exon/intron mechanism (if not 'invented') was exploited at the time of the metazoan radiation (or 'big-bang') in the Cambrian (500 MyBP). This recent importance is supported by both the phylogenetic distribution of exon containing proteins and their common occurrence in extracellular proteins associated with cell–cell communication. Although the latter is indirect evidence, the argument that a novel source of protein diversity was needed at this time is persuasive as the complex body forms that arose in the early Cambrian would have required many new cell-surface receptors, both in embryogenesis and function, specific for each new tissue and their interactions. This would have been especially true in the vertebrates where the rapid evolution of complex nervous systems was required to co-ordinate the new body forms and process sensory data.

Although the recent importance of introns can now no longer be doubted, it still does not account for their origin. The modern splicing apparatus (the spliceosome) is assembled from snRNPs which contain a catalytic RNA component (125, 126), implying an origin in the RNA world (3.5 billion years ago—long before the Cambrian). This raises the problem of what the snRNPs were doing before the metazoan radiation. Although uncertain, the origin of introns is probably linked with retroviruses, transposable elements and principally, self-splicing introns,* suggesting an earlier 'selfish' past (125, 139). It is probable that the 'modern' intron mechanism arose in eukaryotes through the co-operative evolution of self-splicing introns which were introduced by the symbiont precursors of mitochondria and chloroplasts (139). This evolution was probably complete by 700 million years ago—closer to, but still distinctly before the Cambrian radiation. It is therefore, unlikely that the evolution of RNA splicing actually causing the explosion in phyla, more likely, the new pressures of tissue differentiation and communication led to the exploitation of this ancient 'background' mechanism.

* The introns discussed above are those found in eukaryotic nuclei which are removed from mRNA by the spliceosome and referred to as spliceosomal introns. The other, non-spliceosomal, introns are self-splicing (autocatalytic RNA) found both in eukaryotic and prokaryotic organisms, not only in mRNA but also (indeed, primarily) in tRNA and rRNA. They are divided into three groups (I, II, and III) depending on splicing mechanism and co-factors and their reliance on a catalytic RNA function strongly suggests an ancient origin in the RNA world. The three groups should not be confused with the three codon reading frames in which spliceosomal introns can be inserted.

4.4 Evolution of function

4.4.1 A simple model

The preceding sections have concentrated on the evolution of protein structure and while it is clear that structure sets the boundaries within which evolutionary exploration is confined, the ultimate selection pressure (at the molecular level) is on protein function. Some of the 'options' available to proteins in different situations in an arbitrary sequence space are illustrated in *Figure 6*. Any move in this space represents a change in sequence, with small moves corresponding to point mutations and large jumps to splicing (or frame shifts). Each area labelled 'Fold' encloses sequences with a distinct protein fold (a super-family)—such as the $(\beta/\alpha)_8$ (TIM) barrel. Although each fold is depicted as a single, smoothly bounded area, it is more likely to

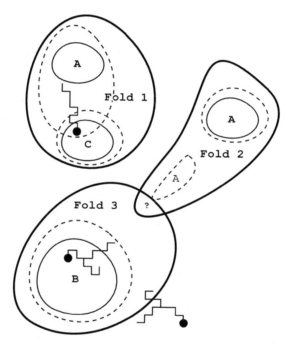

Figure 6. Evolutionary options for proteins. An imaginary sequence space is depicted in which sequences evolve by short steps. Some sequences (●) correspond to folds (Fold 1...), which can have various functions (A, B, ...). Partial functions are enclosed by dashed lines and where these overlap a path is created for the evolution of one function into the other. (See main text for further discussion.) The Fold-1 C-ase enzyme is also a weak A-ase, allowing divergence of a copy towards an optimal A-ase. The Fold-3 B-ase evolves and duplicates with one copy retaining and the other losing function. An unfolded 'random' sequence similarly wanders with one copy attaining a fold (3). The interesting possibility of sequences which can form two different folds is also indicated as '?' and would correspond to the position of the prion protein.

be a discontinuous (probably fractal) multidimensional hypervolume. In addition, the boundaries are not fixed but will move under different physiological conditions (e.g. pH, temperature, etc.).

Within each fold region, sequence subsets are enclosed representing proteins with different functions. There can be different functions with the same fold (e.g. A and C in Fold 1), or different folds with the same function (A in Folds 1 and 2). If, in the latter situation, the two folds perform their function using the same mechanism, they would probably be held up as an example of convergent evolution. There has been much sterile discussion in the literature on whether particular similarities are divergent or convergent evolution. This can be avoided by adopting the simple maxim that, if the folds are the same, the sequences have diverged from a common ancestor but, if the folds are different and the mechanism is the same then they have converged. As there is generally no independent means of verification, all further debate is unscientific (15).

It is also possible to have a function that does not require a folded chain, or a sequence that can adopt more than one fold, but for simplicity, it will be assumed that functional proteins are uniquely folded. A more relevant possibility is that functions might overlap, with their intersection subset representing proteins carrying out both functions. This situation might arise either through an overlap in catalytic specificity at a single active site or through dual (or multiple) sites on the same protein. Each function in the latter situation would probably be associated with distinct domains and as these effectively correspond to smaller proteins, the possibility will be ignored. A more realistic situation for two functions at one site is that one function is dominant over the other. In general, both fold and function would have different degrees of stability and efficiency (respectively), giving a continuously varying (contoured) field over the sequence space.

Onto this model, selection pressure must be applied, which most simply, can be assumed to 'quickly' kill off any sequences without a function. A vital protein will, therefore, be tightly 'confined' within the subset of sequence space where it retains its function. To explore outside this space it is necessary to make a new copy on which to experiment—just as any good programmer would not experiment with their only copy of the computer-code of a working program. This can be achieved by gene duplication, allowing one copy to avoid selection pressure—either completely as an unexpressed pseudogene or partially, as a redundant (expressed) member in a multiple copy gene. Still assuming a relatively quick death for functionless proteins means that mutants can only survive for a limited time away from an 'island' of function. Even if mutations in the protein are now nonlethal, its expression carries some additional load, putting the organism at a selective disadvantage. This implies that the most probable evolutionary paths between functions are across the closest gaps where the functions overlap.

4.2.2 Examples

There are many examples of different enzymes with the same fold but there are not many where the sequences can be shown to be related (to the degree of implying a common ancestor). In these few examples, the proteins retain their key catalytic residues and so support the idea that enzymes can evolve different specificities while retaining the same catalytic mechanism (140, 141). A good example is the similarity of mandelate racemase to muconate lactonizing enzyme. These enzymes have similar structures (TIM barrels) and similar active sites but use different divalent metals. Interestingly, the lactonizing enzyme can function with either metal, which together with other active site differences, suggests how an ancestral enzyme might have had low level racemase activity which could have provided an evolutionary pathway across to the new function (99).

The evolution of a new enzymatic function is useless in isolation as any product would simply accumulate, probably to a lethal extent. This implies the 'co-ordinated' evolution of whole pathways—an unlikely event if the changes had to be made simultaneously. However, the existence of low-level alternative substrates across a number of enzymes allows for many possibilities, as all combinations of steps involving a common product/substrate are potential metabolic pathways. If one combination produces something useful, this will provide sufficient evolutionary pressure for the evolution of a new set of enzymes. Evidence that this can occur comes from further study of the mandelate pathway where it has been shown, either by structure or sequence comparison, that each step in the pathway has evolved from an enzyme in another pathway (99). Such studies open the possibility of reconstructing the evolution of metabolism by 'peeling-off' successive layers of pathways. Alternatively, the comparison of the metabolism of different organisms can give some indication of what was operational in the ancient 'proto-organisms', suggesting a core from which later pathways evolved (142, 143).

The previous examples have illustrated the evolution of one enzyme into another (or one specific function to another specific function). Examples can also be found where an enzyme has moved into an unspecific functional role. This has been observed in the structural eye-lens proteins which include a diverse collection of enzymes (114, 144), some of which retain some functional activity. Although the possible importance of residual function cannot be ruled-out, the prime function of these proteins appears to be to maintain the optical properties of the lens. This structural function does not greatly conflict with enzymatic function and an example has been seen where the same gene is 'shared' without duplication (114). As in the evolution of enzymes, this may represent an early stage in divergence before the conflicting pressures of the combined functions become sufficiently severe to favour duplication and functional uncoupling.

5. Theory

If the prediction of tertiary structure from just sequence data were 100% accurate, then many of the problems raised above concerning the number and symmetry of protein folds could be solved simply by generating all possible sequences, predicting their structure and comparing the results. This would provide not only the number of folds but also their underlying frequencies. Comparison of the distributions to the observed would then distinguish physico-chemical constraints from evolved biases. Unfortunately, the required prediction methods are not remotely accurate enough or, if they were, it is unlikely that computers would be fast enough to apply them to a useful number of sequences.

Despite this pessimistic critique, some limited speculations can be made in this direction by generating random sequences, folding them (in a computer simulation) and then assessing the degree to which the sequences are able to adopt a stable fold. At one extreme, if all sequences can fold, then new proteins would be likely to arise 'spontaneously' (say, from frameshifts or translated introns). Away from this extreme, the dependence on evolution from existing structures would be more important, increasing the persistence of evolutionary relics.

For computational tractability, such calculations are best carried out using a simplified protein model. The sequence is typically a binary string of hydrophobic and hydrophilic residues whereas the chain is confined to lie on a lattice. The lattice has variously been two- (145) or (more commonly) three-dimensional (146, 147), with each residue reduced to a point. A further level of simplification can be attained by adopting fixed secondary structures which can then be reduced to line segments (148). Distance geometry also provides an alternative to the artificial constraints of a lattice, while still avoiding the long calculation times of conventional molecular dynamics, allowing realistic 'random' proteins to be produced (149). In addition to these simulation approaches, the problem has been treated more theoretically, applying the principles of statistical mechanics (150–152).

Despite their widely different basis, most of these studies conclude that many 'random' sequences appear capable of forming stable folds, a conclusion now supported by experimental results (153). This implies that totally new proteins might well appear spontaneously, or that large random insertions might be tolerated, as they would have greater probability of forming an independent folding unit which would not disrupt the existing structure. This latter possibility might easily occur through the mutation of an intron splice site, leading to the translation of the 'random' intron sequence. Such domains of random origin would not require any immediate function to survive, but could 'hitch-hike' on the strength of the function of their attached protein. If they later acquired a function they might even escape their parent (by random processes of intron insertion) and become autonomous proteins (or functional domains). By whatever mechanism they arose, these novel proteins would carry no evolutionary imprint of internal pseudosymmetry and may account for some of the less regular or complex folds.

6. Conclusions

Before the end of the century, the genomes of at least three disparate organisms (bacterium, yeast, and nematode) should be sequenced. These data will provide, for the first time a view of all the essential proteins needed for life. It is likely that after this, there will be few remaining surprises as the proteins utilized by higher organisms (with the exception of some plant proteins) will probably all have evolved from this core. Untangling these lines of descent, both within and between species, will pose a difficult challenge to the molecular evolutionist but will give an unrivalled view of the relationships of proteins.

Many of these comparisons will be difficult, if not impossible, without the aid of structural data. Although 250 protein structures are being determined each year (and doubling every few years), the number of novel folds revealed does not increase as quickly as most structures are close homologues (or trivial mutants) of those already known (154). This places great importance on the methods of structure biased sequence comparison (e.g. threading) to extrapolate this, relatively, limited structural knowledge as far as possible into the new wealth of sequence data.

References

1. R. Cammack, D. Hall, and K. Rao (1971). *New Sci. Sci. J.*, **23**, 696–698.
2. W. Bains (1987). *Trends Biochem. Sci.*, **12**, 90–91.
3. S. B. Needleman and C. D. Wunsch (1970). *J. Mol. Biol.*, **48**, 443–453.
4. P. H. Sellers (1974). *J. Combinator. Theor.*, **16**, 253–258.
5. T. F. Smith and M. S. Waterman (1981). *J. Mol. Biol.*, **147**, 195–197.
6. W. R. Pearson and W. Miller (1992). In *Methods in enzymology* (ed. L. Brand and M. L. Johnson), Vol. 210, pp. 575–601. Academic Press, New York.
7. M. Schoniger and M. S. Waterman (1992). *Bull. Math. Biol.*, **54**, 521–536.
8. M. S. Waterman (1984). *Bull. Math. Biol.*, **46**, 473–500.
9. S. Pascarella and P. Argos (1992). *J. Mol. Biol.*, **224**, 461–471.
10. G. H. Gonnet, M. A. Cohen, and S. A. Benner (1992). *Science*, **256**, 1443–1445.
11. S. A. Benner, M. A. Cohen, and G. H. Gonnet (1993). *J. Mol. Biol.*, **229**, 1065–1082.
12. S. F. Altschul and B. W. Erickson (1986). *Bull. Math. Biol.*, **48**, 603–616.
13. W. Miller and E. W. Myers (1989). *Bull. Math. Biol.*, **50**, 97–120.
14. M. Vingron and M. S. Waterman (1994). *J. Mol. Biol.*, **235**, 1–12.
15. M. Bajaj and T. Blundell (1984). *Ann. Rev. Biophys. Bioeng.*, **13**, 453–492.
16. M. O. Dayhoff, R. V. Eck, and C. M. Park (1972). In *Atlas of protein sequence and structure*, Vol. 5, pp. 89–99. Biomed. Res. Found., Washington DC.

17. D. T. Jones, W. R. Taylor, and J. M. Thornton (1992). *Comput. Appl. Biosci.*, **8**, 275–282.
18. S. Henikoff and J. G. Henikoff (1992). *Proc. Natl. Acad. Sci. USA*, **89**, 10915–10919.
19. M. O. Dayhoff, R. M. Schwartz, and B. C. Orcutt (1978). In *Atlas of protein sequence and structure* (ed. M. O. Dayhoff), Vol. 5, pp. 345–352. Nat. Biomed. Res. Foundation, Washington DC..
20. S. Henikoff and J. G. Henikoff (1993). *Prot. Struct. Funct. Genet.*, **17**, 49–61.
21. R. F. Doolittle (1978). *Of URFs and ORFs: a primer on how to analyse derived amino acid sequences.* University Science Books, California, USA.
22. S. Karlin, P. Bucher, V. Brendel, and S. F. Altschul (1991). *Annu. Rev. Biophys. Biophys. Chem.*, **20**, 175–203.
23. S. F. Altschul, W. Gish, W. Miller, E. W. Myers, and D. J. Lipman (1990). *J. Mol. Biol.*, **214**, 403–410.
24. C. Sander and R. Schneider (1994). *Prot. Struct. Funct. Genet.*, **9**, 56–68.
25. S. Karlin and S. F. Altschul (1993). *Proc. Natl. Acad. Sci. USA*, **90**, 5873–5877.
26. M. S. Waterman and M. Vingron (1994). *Proc. Natl. Acad. Sci. USA*, **91**, 4625–4628.
27. A. Bairoch (1990). *PC/Gene: a protein and nucleic acid sequence analysis microcomputer package, PROSITE: a dictionary of sites and patterns in proteins, and SWISS-PROT: a protein sequence data bank.* PhD thesis, University of Geneva.
28. L. H. Pearl and W. R. Taylor (1987). *Nature*, **329**, 351–354.
29. J. C. Hu, E. K. O'Shea, P. S. Kim, and R. T. Sauer (1990). *Science*, **250**, 1400–1403.
30. M. J. E. Sternberg and W. R. Taylor (1984). *FEBS Lett.*, **175**, 387–392.
31. W. R. Taylor (1986). *J. Theor. Biol.*, **119**, 205–218.
32. R. F. Doolittle (ed.) (1990). *Methods in enzymology.* Vol. 185. Academic Press, San Diego, CA.
33. D. F. Feng and R. F. Doolittle (1987). *J. Mol. Evol.*, **25**, 351–360.
34. W. R. Taylor (1987). *Comput. Appl. Biosci.*, **3**, 81–87.
35. G. J. Barton and M. J. E. Sternberg (1987). *J. Mol. Biol.*, **198**, 327–337.
36. D. G. Higgins and P. M. Sharp (1988). *Gene*, **73**, 237–244.
37. D. G. Higgins, A. J. Bleasby, and R. Fuchs (1992). *Comput. Appl. Biosci.*, **8**, 189–191.
38. W. R. Taylor (1988). *J. Molec. Evol.*, **28**, 161–169.
39. J. Devereux, P. Haeberli, and O. Smithies (1984). *Nucleic Acids Res.*, **12**, 387–395.
40. R. D. Appel, A. Bairoch, and D. F. Hochstrasser (1994). *Trends Biochem. Sci.*, **19**, 258–260.
41. P. Green, D. Lipman, L. Hillier, R. Waterston, D. States, and J.-M. Claverie (1993). *Science*, **259**, 1711–1716.
42. C. Chothia (1992). *Nature*, **357**, 543–544.
43. P. Green (1994). *Curr. Opin. Struct. Biol.*, **4**, 404–412.
44. A. M. Lesk, M. Levitt, and C. Chothia (1986). *Protein Eng.*, **1**, 77–78.
45. G. J. Barton and M. J. E. Sternberg (1987). *Protein Eng.*, **1**, 89–94.
46. M. Kanaoka, F. Kishimoto, Y. Ueki, and H. Umeyama (1989). *Protein Eng.*, **2**, 347–351.
47. Z.-Y. Zhu, A. Sali, and T. L. Blundell (1992). *Protein Eng.*, **5**, 43–51.
48. R. F. Smith and T. F. Smith (1992). *Protein Eng.*, **5**, 35–42.

49. J. M. Levin, S. Pascarella, P. Argos, and J. Garnier (1993). *Protein Eng.*, **6**, 849–854.
50. B. Rost and C. Sander (1993). *J. Mol. Biol.*, **232**, 584–599.
51. R. Lüthy, A. D. Mclachlan, and D. Eisenberg (1991). *Proteins*, **10**, 229–239.
52. J. Overington, D. Donnelly, M. S. Johnson, A. Sali, and T. L. Blundell (1992). *Protein Sci.*, **1**, 216–226.
53. D. T. Jones, W. R. Taylor, and J. M. Thornton (1994). *FEBS Lett.*, **339**, 269–275.
54. W. R. Taylor, D. T. Jones, and N. M. Green (1994). *Prot. Struct. Funct. Genet.*, **18**, 281–294.
55. J. U. Bowie, N. D. Clarke, C. O. Pabo, and R. T. Sauer (1990). *Proteins*, **7**, 257–264.
56. J. U. Bowie, R. Lüthy, and D. Eisenberg (1991). *Science*, **253**, 164–170.
57. M. S. Johnson, J. P. Overington, and T. L. Blundell (1993). *J. Mol. Biol.*, **231**, 735–752.
58. W. R. Taylor (1986). *J. Mol. Biol.*, **188**, 233–258.
59. D. Bashford, C. Chothia, and A. M. Lesk (1986). *J. Mol. Biol.*, **196**, 199–216.
60. P. Bork, C. Sander, and A. Valencia (1992). *Proc. Natl. Acad. Sci. USA*, **89**, 7290–7294.
61. A. V. Efimov (1993). *Prog. Biophys. Mol. Biol.*, **60**, 201–239.
62. M. J. Sippl (1990). *J. Mol. Biol.*, **213**, 859–883.
63. G. M. Crippen (1991). *Biochemistry*, **30**, 4232–4237.
64. S. H. Bryant and C. E. Lawrence (1993). *Prot. Struct. Funct. Genet.*, **16**, 92–112.
65. D. T. Jones, W. R. Taylor, and J. M. Thornton (1992). *Nature*, **358**, 86–89.
66. A. Godzik, A. Kolinski, and J. Skolnick (1992). *J. Mol. Biol.*, **227**, 227–238.
67. C. Ouzounis, C. Sander, M. Scharf, and R. Schneider (1993). *J. Mol. Biol.*, **232**, 805–825.
68. M. Wilmanns and D. Eisenberg (1993). *Proc. Natl. Acad. Sci. USA*, **90**, 1379–1383.
69. A. M. Lesk and D. R. Boswell (1992). *Bioessays*, **14**, 407–410.
70. B. W. Matthews and M. G. Rossmann (1985). *Methods in enzymology*, Vol. 115, pp. 397–420, Academic Press, London.
71. C. Sander and G. E. Schulz (1979). *J. Mol. Evol.*, **13**, 245–250.
72. M. G. Rossmann, D. Moras, and K. W. Olsen (1974). *Nature*, **250**, 194–199.
73. W. R. Taylor and C. A. Orengo (1989). *J. Mol. Biol.*, **208**, 1–22.
74. A. Sali and T. L. Blundell (1990). *J. Mol. Biol.*, **212**, 403–428.
75. W. R. Taylor, T. P. Flores, and C. A. Orengo (1995). *Protein Sci.*, **3**, 1858–70.
76. R. Nussinov and H. J. Wolfson (1991). *Proc. Natl. Acad. Sci. USA*, **88**, 10495–10499.
77. G. Vriend and C. Sander (1991). *Proteins*, **11**, 52–58.
78. D. Fischer, H. Wolfson, and R. Nussinov (1993). *J. Biomol. Struct. Dynam.*, **11**, 367.
79. D. Fischer, H. Wolfson, S. L. Lin, and R. Nussinov (1994). *Prot. Sci.*, **3**, 769–778.
80. L. Holm and C. Sander (1993). *J. Mol. Biol.*, **233**, 123–138.
81. M. G. Rossmann and P. Argos (1977). *J. Mol. Biol.*, **109**, 99–129.
82. J. Janin (1979). *Bull. Inst. Pasteur*, **77**, 337–373.
83. O. B. Ptitsyn and A. V. Finkelstein (1980). *Q. Rev. Biophys.*, **13**, 339–386.
84. J. S. Richardson (1981). *Adv. Protein Chem.*, **34**, 167–339.
85. E. G. Hutchinson and J. M. Thornton (1993). *Protein Eng.*, **6**, 233–245.
86. T. L. Blundell and M. S. Johnson (1993). *Protein Sci.*, **2**, 877–883.

87. A. G. Murzin and C Chothia (1992). *Curr. Opin. Struct. Biol.*, **2**, 895–903.
88. A. G. Murzin (1994). *Curr. Opin. Struct. Biol.*, **4**, 441–449.
89. C. A. Orengo, N. P. Brown, and W. R. Taylor (1993). *Prot. Struct. Funct. Genet.*, **14**, 139–167.
90. D. P. Yee and K. A. Dill (1993). *Protein Sci.*, **2**, 884–899.
91. C. A. Orengo, T. P. Flores, W. R. Taylor, and J. M. Thornton (1993). *Protein Eng.*, **6**, 485–500.
92. L. Holm and C. Sander (1993). *Nature*, **361**, 309.
93. C. A. Orengo and W. R. Taylor (1993). *J. Mol. Biol.*, **233**, 488–497.
94. C. A. Orengo, T. P. Flores, D. T. Jones, W. R. Taylor, and J. M. Thornton (1993). *Curr. Biol.*, **3**, 131–139.
95. P. Bork, C. Ouzounis, and C. Sander (1994). *Curr. Opin. Struct. Biol.*, **4**, 393–402.
96. C. Chothia (1988). *Nature*, **333**, 598–599.
97. G. K. Farber and G. Petsko (1990). *Trends Biochem. Sci.*, **15**, 228–234.
98. N. S. Scrutton (1993). *BioEssays*, **16**, 115–122.
99. G. A. Petsko, G. L. Kenyon, J. A. Gerlt, D. Ringe, and J. W. Kozarich (1993). *Trends Biochem. Sci.*, **18**, 372–376.
100. R. L. Dorit, L. Schoenbach, and W. Gilbert (1990). *Science*, **250**, 1377–1382.
101. R. F. Doolittle (1991). *Science*, **253**, 677–679.
102. L. Patthy (1991). *BioEssays*, **13**, 187–191.
103. A. V. Finkelstein, A. M. Gutun, and A. Y. Badretdinov (1993). *FEBS Lett.*, **325**, 23–28.
104. C. A. Orengo (1994). *Curr. Opin. Struct. Biol.*, **4**, 429–440.
105. T. Dandekar and P. Argos (1992). *Protein Eng*, **5**, 637–645.
106. D. T. Jones (1994). *Protein Sci.*, **3**, 567–574.
107. T. Dandekar and P. Argos (1994). *J. Mol. Biol.*, **5**, 637–645.
108. J. A. Shapiro, S. L. Adhya, and A. I. Bukhari (1997). In *DNA insertion elements, plasmids and episomes* (ed. A. I. Bukhari, J. A. Shapiro, and S. L. Adhya), pp. 3–11. Cold Spring Harbor Lab.
109. A. D. McLachlan (1979). *J. Mol. Biol.*, **128**, 49–79.
110. G. E. Schulz (1980). *J. Mol. Biol.*, **138**, 335–347.
111. H. Nojima (1987). *FEBS Lett.*, **217**, 187–190.
112. A. D. McLachlan (1983). *J. Mol. Biol.*, **169**, 15–30.
113. D. G. Higgins, S. Labeit, M. Gautel, and T. J. Gibson (1990). *J. Mol. Evol.*, **38**, 395–404.
114. J. G. Piatigorsky and G. Wistow (1991). *Science*, **252**, 1078–1079.
115. G. Dover (1982). *Nature*, **299**, 111–117.
116. T. L. Blundell, J. A. Jenkins, B. T. Sewell, L. H. Pearl, J. B. Cooper, I. J. Tickle, *et al.* (1990). *J. Mol. Biol.*, **211**, 919–941.
117. J. Tang, M. N. G. James, I. N. Hsu, J. A. Jenkins, and T. L. Blundell (1978). *Nature*, **271**, 619–621.
118. H. Toh, M. Ono, K. Saigo, and T. Miyata (1985). *Nature*, **315**, 691.
119. A. Wlodawer, M. Miller, M. Jaskolski, B. K. Sathyanarayana, E. Baldwin, I. T. Weber, *et al.* (1989). *Science*, **245**, 616–621.
120. J.M. Thornton and B.L. Sibanda (1983). *J. Mol. Biol.*, **167**, 443–460.
121. J. H. Wilson (1985). *Genetic recombination.* Benjamin-Cummings, Menlo Park, CA.
122. W. Gilbert (1978). *Nature*, **271**, 501.

123. C. C. F. Blake (1978). *Nature*, **273**, 267.
124. M. Gõ and M. Nosaka (1978). *Cold Spring Harbor Symp. Quant. Biol.*, **LII**, 915–924.
125. A. Newman (1994). *Curr. Opin. Cell Biol.*, **6**, 360–367.
126. J. A. Steitz (1988). *Sci. Am.*, **258**, 36–41.
127. W. Gilbert (1978). *Cold Spring Harbor Symp. Quant. Biol.*, **LII**, 901–905.
128. P. J. Artymiuk, C. C. F. Blake, and A. E. Sippel (1981). *Nature*, **290**, 287–288.
129. W. F. Doolittle (1987). *Cold Spring Harbor Symp. Quant. Biol.*, **LII**, 907–913.
130. R. F. Doolittle (1985). *Trends Biochem. Sci.*, **2**, 233–237.
131. L. Patthy (1985). *Cell*, **41**, 657–663.
132. C. C. F. Blake, K. Harlos, and S. K. Holland (1978). *Cold Spring Harbor Symp. Quant. Biol.*, **LII**, 925–929.
133. R. F. Doolittle, D.-F. Feng, M. S. Johnson, and M. A. McClure (1986). *Cold Spring Harbor Symp. Quant. Biol.*, **LI**, 447–455.
134. M. Belfort (1993). *Science*, **262**, 1009–11010.
135. J. M. (Jr.) Logsdon and J. D. Palmer (1994). *Nature*, **369**, 526–527.
136. N. J. Dibb and A. J. Newman (1989). *EMBO J.*, **8**, 2015–2021.
137. J. D. Palmer and J. M. Logson (1991). *Curr. Opin. Gen. Dev.*, **1**, 470–477.
138. L. Patthy (1994). *Curr. Opin. Struct. Biol.*, **4**, 383–392.
139. T. Cavalier-Smith (1989). *Trends Genet.*, **7**, 145–148.
140. D. L. Ollis, E. Cheah, M. Cygler, B. Dijkstra, F. Frolow, S. M. Franken, *et al.* (1992). *Protein Eng.*, **5**, 197–211.
141. A. G. Murzin (1993). *Trends Biochem. Sci.*, **18**, 403–405.
142. S. A. Benner and A. D. Ellington (1990). *Bioorg. Chem. Front.*, **1**, 1–70.
143. S. A. Benner, M. A. Cohen, G. H. Gonnet, D. B. Berkowitz, and K. P. Johnsson (1993). In *The RNA world: the nature of modern RNA suggests a prebiotic RNA world* (ed. R. F. Gesteland and J. A. Atkins), pp. 27–70. Cold Spring Harbor Lab.
144. G. Wistow and J. G. Piatigorsky (1987). *Science*, **236**, 1554–1556.
145. R. Unger and J. Moult (1993). *J. Mol. Biol.*, **231**, 75–81.
146. D. Baker, H. S. Chan, and K. A. Dill (1993). *J. Chem. Phys.*, **98**, 9951–9962.
147. A. Godzik, A. Kolinski, and J. Skolnick (1993). *J. Comput. Chem.*, **14**, 1194–1202.
148. W. R. Taylor (1991). *Protein Eng.*, **4**, 853–870.
149. A. Aszódi and W. R. Taylor (1994). *Protein Eng.*, **7**, 633–644.
150. O. B. Ptitsyn (1983). In *Conformation in biology* (ed. R. Srinivasan and R. M. Sarma), pp. 49–58. Academic Press, New York.
151. A. V. Finkelstein (1991). *Prot. Struct. Funct. Genet.*, **9**, 23–27.
152. A. V. Finkelstein (1994). *Curr. Opin. Struct. Biol.*, **4**, 422–428.
153. A. R. Davidson and R. T. Sauer (1994). *Proc. Natl. Acad. Sci. USA*, **91**, 2146–2150.
154. J. M. Thornton (1992). *Curr. Opin. Struct. Biol.*, **2**, 888–894.

A1

List of suppliers

Biosoft
   BioSoft, 49 Bateman Street, Cambridge CB2 1LR, UK.
   BioSoft, PO Box 10938, Ferguson, MO 63135, USA.
Christian Marck, Service de Biochimie de Genetique Moleculaire, Bâtiment 142, Centre d'Etudes de Saclay, 91191 Gif-Sur-Yvette Cedex, France.
DNAStar
   DNAStar Ltd, Abacus House, Manor Road, West Ealing, London W13 0AS, UK.
   DNAStar Inc., 1228 South Park Street, Madison, WI 53715, USA.
Eastman Kodak Co., 343 State Street, Building 642, Rochester, NY 14652-3512, USA.
GBF, Mascheroder Weg 1, 38124 Braunschweig, Germany.
Genes Code Corporation, 2901 Hubbard Street, Ann Arbor, MI 48105, USA.
IntelliGenetics Inc., 700 East El Camino Real, Mountain View, CA 94040, USA.
MedProbe A.S., Postboks 2640, St Hanshaugen, N-0131 Oslo, Norway.
MRC-LMB, Hills Road, Cambridge CB2 2QH, UK.
National Biosciences Inc., 3650 Annapolis Lane North, #140 Plymouth, MN 55447-5434, USA.
Oxford Molecular Ltd, The Magdelen Centre, Oxford Sciences Park, Oxford, OX4 4GA, UK.
Scientific Imaging Systems, 36 Clifton Road, Cambridge, CB1 4ZR, UK.
Textco Inc., 27 Gilson Road, West Lebanon, NH 03784, USA.

Glossary

backtranslation
Translation from a protein sequence to a partially degenerate nucleotide sequence according to the genetic code.

Backus-Naur-Form
A notation in which rules of a language can be written. It was first used to define the rules of the algol computer language.

bandwidth
Capacity of a communication channel, usually measured in bits per second.

BioGopher
Biological Gopher.

bioinformatics
Informatics applied to biology, especially biological information encoded in macromolecular sequences of DNA, RNA, and proteins.

centromere
Point of attachment of sister chromatids prior to mitosis, and to the spindle during mitosis. The region contains characteristic repetitive sequences.

check-box
An item in a form which can be set in one of two states, often by clicking on it with the mouse pointer.

cladistics
A method of classification of organisms based on a hypothesis of their phylogenetic relationships.

client-server
A model of computing where resources are distributed between a client computer which makes requests and a server computer which responds to them over a network.

contig
A set of overlapping DNA fragments which represent a contiguous region of DNA.

copy-protection
Process by which computer data is made unreadable, except with the necessary authorization.

dendrogram
A tree diagram (in the graph theoretical sense).

dicodon
An adjacent pair of codons formed by six nucleotides coding for two amino acids.

dinucleotide
Two adjacent nucleotides.

dipeptide
Two adjacent amino acids.

diskette
A small removable computer storage device which contains a flexible plastic disk with two magnetizable surfaces.

dongle
A security device implementing copy-protection.

dotplot
A method of displaying pairwise sequence similarity as a rectangular diagram with one sequence drawn on the horizontal axis and the other on the vertical axis of a two-dimensional matrix. A mark is made in every cell where the sequence symbols on the axes are identical. Regions of similarity appear as lines parallel to the diagonal.

drag and drop
A paradigm of computing using icons. A data item is dragged on to a program icon using the mouse pointer. The program then processes the data. A familiar example is dragging a file icon on to a wastebasket icon in order to delete the file, an action which is performed when you empty waste.

epitope
A peptide subsequence of a protein capable of inducing an immune response.

ethernet
A method by which electronic signals can be placed on a local area network. It uses a bus configuration of a single strand of cable to which each node connects. Only one signal can travel on the cable at one time and the bandwidth is 10 Mbps for standard ethernet and 100 Mbps for fast ethernet.

exon
The DNA coding for proteins in archaebacteria and eukaryotes is not necessarily continuous as it is in prokaryotes. The protein coding sequences are called exons and the intervening sequences are called introns.

extension (as in 'file name extension')
A file name extension is designed to give some information about what to expect in a file in terms of its content or format. For example, <name>.txt for a text file and <name>.xls for an Excel spreadsheet file.

fractal

A geometrical shape which is complex and detailed in structure at any level of magnification. A fractal cannot be treated as existing in an integer number of dimensions but rather handled mathematically as though it existed in a fractional dimension (fractal).

frameshift

RNA is translated into protein in units of three nucleotides each of which codes for an amino acid. If a number of nucleotides not divisible by three are inserted or deleted the reading frame for translation is changed and the amino acids encoded are changed resulting in a frameshift.

freeware

Software which is freely available, usually for no charge.

G-banding

The pattern of banding in chromosomes stained with Giemsa after trypsin digestion. Reveals AT-rich regions.

Gopher

A service to deliver documents worldwide from a multitude of servers at centres which cooperate in providing this service.

hnRNA

Heterogenous nuclear RNAs (hnRNAs) are the transcripts produced in the nucleus of eukaryotes by the enzyme RNA polymerase II and first characterized by the variety of their sizes.

homopolymeric

A polymer with the same repeating unit.

hotlist

An organized list of addresses on the Internet (often WWW URLs) which are considered to be of interest by the compiler of the list.

hotspot

A region of DNA which is more mutable than the surrounding areas.

hydropathicity

Measure of propensity of an amino acid residue in a protein to occupy the core rather than the surface of the structure. Core residues are hydrophobic and surface residues hydrophilic.

indel

Insertion or deletion in a molecular (or other) sequence.

informatics

Study of information processing including computing.

intron
The DNA coding for proteins in archaebacteria and eukaryotes is not necessarily continuous as it is in prokaryotes. The protein coding sequences are called exons and the intervening sequences are called introns.

LISTSERV
A program on an IBM mainframe which manages a mail distribution list. Subscription is automatic and is done by sending a message to LISTSERV. Mail is distributed to individual mailboxes.

logical operator
An operation permitted in binary or Boolean algebra where the values are TRUE or FALSE. The operators are AND, OR, and NOT.

midisatellite
DNA sequence with a middle-sized oligonucleotide repeat element.

minisatellite
DNA sequence with a small oligonucleotide repeat element.

Netscape
The registered trade mark of software products from Netscape Communications Corporation. Often used to mean the WWW client product Netscape Navigator.

oligonucleotide
A polymer consisting of a few nucleotide residues.

oligopeptide
A polymer consisting of a few amino acid residues.

PICT
An object-oriented graphics format than can be manipulated by a variety of graphics programs.

prion
Class of infectious protein particles, probably an altered form of a normal protein.

R-banding
Essentially the reverse of G-banding. Before staining with Giemsa the chromosomes are heat-treated. Reveals GC-rich regions.

ResEdit
A software utility from Apple Computer for Resource Editing on the Macintosh.

SEQNET
An online molecular biology computing resource at Daresbury Laboratory in the UK. URL: `http://www.dl.ac.uk/SEQNET`

shareware
Software which is freely available, usually for a small charge.

spliceosome
RNA and protein complex involved in the post-transcriptional processing of RNA to remove introns.

spread-sheet
A computer program designed to deal with numerical data in tabular form, rows, and columns.

Telnet
A program which enables you to log in from a terminal session to a remote host on the Internet.

telomere
The structure at the end of a eukaryotic chromosome which maintains the integrity of the chromosome by preventing its ends from fraying. The region contains characteristic repetitive sequences.

traceback
In the context of sequence alignment, this is the pathway through the score matrix which will be traversed to determine the optimal alignment.

transposon
A DNA sequence which can be excised from its location and reinserted in a different part of the genome.

transversion
Substitution of a purine by a pyrimidine nucleotide (or vice versa) in a DNA sequence.

Usenet
Network news, a bulletin board system on the Internet.

uuencode and uudecode
Programs which will convert binary files into ASCII and vice versa so that they can be e-mailed to systems which would corrupt binary mail.

VMS
Operating system produced by Digital Equipment Corporation and currently called OpenVMS.

Index